The New Science of Technical Analysis

Using the Statistical Techniques of Neuroscience to Uncover Order and Chaos in the Markets

Clifford J. Sherry, Ph. D.

AN AUTHORS GUILD BACKINPRINT.COM EDITION

The New Science of Technical Analysis
Using the Statistical Techniques of Neuroscience to Uncover Order and Chaos in the Markets
All Rights Reserved © 1994, 2004 by Clifford J. Sherry

AN AUTHORS GUILD BACKINPRINT.COM EDITION
Published by iUniverse, Inc.

For information address:
iUniverse, Inc.
2021 Pine Lake Road, Suite 100
Lincoln, NE 68512
www.iuniverse.com

Originally published by McGraw Hill

ISBN: 0-595-31438-4

Printed in the United States of America

With Love and Respect
This Book Is Dedicated to My Children
Lori Beth, Jason William, and Christopher Joseph
Without Them My Work Would Be Meaningless

Table of Contents

Acknowledgments

I would like to thank my wife Nancy, who helped type, edit, and proof the manuscript and render the illustrations. Without her love and encouragement, I could accomplish nothing.

This book would not have been possible without the interactions I have had over the last 25 years with many friends and colleagues. They include Drs. Heinz Von Foerster, W.R. Klemm, D.L. Barrow, T.J. Marczynski, G. Karmos, D.J.Wolf, and Mr. S. Brudno. Each one, in his own special way, provided advice, counsel, and constructive criticism during the initial development of the ideas and techniques described in this book, when these techniques were originally being developed to study information processing in the nervous system.

Mr. John Sweeney and Jack K. Hutson, the editor and publisher, respectively, of *Technical Analysis of Stocks and Commodities* helped me transition these techniques from neurobiology applications to the study of economically important time series.

The majority of the tables and graphs in this book were created in StatView (Abacus Concepts, 1984 Bonita Ave., Berkeley, CA 97404). I would like to thank Daniel S. Feldman, Jr. and Will Scoggin, the president and vice president-marketing, respectively, for providing me with a copy of StatView.

The remaining graphs were created with BEHOLD! (Investors' Technical Services, Inc., P.O. Box 164075, Austin, TX 78716). I would also like to thank Dr. J.S. Payne, the president of Investors' Technical Services for providing me with a copy of this program and its utility programs.

I would like to thank Mr. Dean S. Barr of LBS Capital Management for providing me with much of the data used in the illustrations.

Last, but certainly not least, I would like to thank all of the people at Probus Publishing Company, especially my editors, Kevin Commins and Kevin Thornton, for being patient and supportive.

Introduction

As the title of this book implies (although my Ph.D. is in psychology), I consider myself to be a neuroscientist, more specifically a neurophysiologist. That means that I study the electrical activity of the nervous system. My own interests center on how the electrical activity of the nervous system and behavior are related. In the early 1970s, my colleagues and I were studying an electroencephalographic (EEG or *brain wave*) pattern that was correlated with behavior and decided to look at this pattern on a more basic level, the output of individual neurons.

In the early 1970s when I started to think about these issues, I discovered that many neurobiologists believed that the output of individual neurons, the basic building block of the nervous system, was a *random walk*. That is, these neurobiologists believed that the output of individual neurons was *random* and *independent*(see below).

These two ideas: (1) that action potentials are the main mechanism for communication between neurons and this communication is the underlying basis of behavior and (2) that action potential intervals demonstrated a *random walk* seemed incompatible. Therefore, I decided to try to develop a method for testing these ideas empirically using statistics and probability theory.

It is important that we agree on a nonmathematical definition of the terms *independent* and *random*. The mathematical definitions are a bit complicated and require a strong background in mathematics and probability theory and are not really required for the basic understanding of the phenomena.

Independent means that the outcome of an event is not impacted by the outcome of any previous event and does not have any impact on any succeeding event. Flipping a fair coin is a classic example of statistical *independence*. Each time you flip the coin, the probability of a head or a tail is 50:50. If you flip a fair coin 10 times and it comes up heads each time, what is the probability that the outcome of the eleventh flip will be a head or a tail? Is there some *process* (fate, Mother Nature, . . .) that *adjusts* the relative probabilities to favor the occurrence of a tail? Enough people believe that there is, that the phenomenon has been given a name, the gambler's paradox. Unfortunately, for gamblers and everyone else, this process does not exist. The probability of a head or a tail on the eleventh toss is still 50:50.[1]

Random means that the outcome of an event is determined by chance. In the case of our fair coin, the probability of a head or a tail is 50:50. It is important to note that this does not mean that if you flip a coin twice, you will get one head and one tail or if you flip a coin 50 times, you will get 25 heads and 25 tails. On the other hand, if you flip a coin 10,000 or 100,000 times, you will approach 50:50 more and more closely. Probability theorists call this phenomenon the law of large numbers.

There is another important characteristic of a time series, *stationarity*. Stationarity, again in nonmathematical terms, means that the underlying *rules* do not change over time. If you flip a coin today and one of your ancestors flips a coin 100 years or 1,000 years from now (assuming our species manages to survive that long!), the *rules* would be the same. That is, the probability of occurrence of a head or a tail in the long run would still be 50:50 and the process would still be independent, etc.

From the early 1970s through the mid-1980s, with input from a number of different colleagues, I developed a number of simple, but statistically rigorous tests for determining if a time series made up of action potential intervals was:

---

1. This seems contrary to common sense. But, if you think about it, if a universe existed where this were not true, it would have a profound effect on virtually everything that goes on, and I suspect that life itself would not be possible in such a universe.

1. Independent

Sherry, C. J., and Marczynski, T. J. "A new analysis of neuronal inter-spike intervals based on inequality tests." *Int J Neurosci* 3: 259–270, 1972.

Sherry, C. J., and Marczynski, T. J. "The differential spectrum of neuronal interspike intervals: a one dimensional analysis of sequential relationships." *Brain Res* 40: 513–515, 1972.

Sherry, C. J., Barrow, D. L., and Klemm, W. R. "Serial dependencies and Markov properties of neuronal interspike intervals from rat cerebellum." *Brain Res Bull* 8: 169–182, 1982.

2. Random/Stationary

Sherry, C. J., and Klemm, W. R. "Stationarity, randomness, and serial dependence in neuronal spike trains." *J. Electrophysiol Meth* 10: 59–83, 1983.

Sherry, C. J. *How random is random?* Correlations Temporelles et Codage de l'Information dans le cerveau. High Level Course and Workshop Extended Abstracts, 11–13, Paris, 1990.

In addition, my colleagues and I used Information Theory[2] as a method to demonstrate the neurons are *byte* processors, rather than *bit* processors:

3. Information Theory

Sherry, C. J., and Klemm, W. R. "Entropy correlations with ethanol-induced changes in specified patterns of nerve impulses: evidence for *Byte* processing in the nervous system." *Prog Neuropsychopharmacol* 4: 261–267, 1980.

Klemm, W. R., and Sherry, C. J. "Entropy measures of signal in the presence of noise: evidence for *Byte* vs *Bit* processing in the nervous system." *Experientia* 37: 55–58, 1981.

---

2. An excellent relatively nonmathematical introduction to information theory can be found in J.R. Pierce's book, *Symbold, Signals, and Noise: The Nature And Process Of Communication.* New York: Harper Torchbooks, 1961.

It is important to distinguish between the *signal* (that is, in the case of the nervous system, the patterns of action potential intervals in time) and the *message* that the signal is carrying. Most of my work and the work of most other neurophysiologists that work in this area have focused on the *signal*. My colleagues and I have suggested that the more ways you look at the *signal*, the better you can describe it and the more likely you are to be able to ultimately detect and decode the *message*. Our ideas are highlighted in:

4. Reviews

Klemm, W. R., and Sherry, C. J. "Serial ordering in spike trains: What's it 'trying to tell us'?" *Int J Neurosci* 14: 15–33, 1981.

Sherry, C. J., and Klemm, W. R. "What is the meaningful measure of neuronal spike train activity?" *J Neurosc Meth* 10: 205–213, 1984.

One day while I was browsing at the library, I came across a book dealing with technical analysis. I discovered that people who were interested in analyzing economically important time series, were interested in many of the same questions that I was dealing with in the time series that I was studying (i.e., action potential intervals). I soon located Paul Cootner's book, *The Random Character Of Stock Prices* (Cambridge: M.I.T. Press, 1964). I was then introduced to *Technical Analysis Of Stocks And Commodities* and found Jack Hudson and particularly John Sweeney to be excellent mentors to help me convey my ideas to this new audience. I published a series of articles in *Technical Analysis*:

"Gambler's Paradox," October 1985, 30–1, 44

"Detecting A Dependent Process," April 1986, 21–3, 29

"Detecting Dependence," June 1986, 16–9

"How Random Is Random," August 1990, 50–3.

Questions about these articles and others that I published in *Technical Analysis* ultimately led me to write my first book:

Sherry, C. J. *Mathematics of Technical Analysis: Applying Statistics to Trading Stocks, Options, and Futures.* Chicago, IL: Probus Publishing Co., 1992.

In the succeeding chapters, I will describe these techniques and apply them, in some detail, to economically important time series.

Methods: Stationarity, Independence and Randomness

1

The statistical tests used to determine if an economically important time series is stationary, random, and/or independent have been described in detail in my book, *Mathematics of Technical Analysis: Applying Statistics to Trading Stocks, Options, and Futures* (Chicago, IL: Probus Publishing Co., 1992). The methods will be briefly described here and then applied in the succeeding chapters to individual economically important time series.

I will use artificial data to illustrate the use of these techniques. These data were drawn from two tables from an excellent book by David Salsburg, *Understanding Randomness: Exercises For Statisticians*[1] (New York: Marcel Dekker, 1983). They are Tables 2.1, Symmetric and 2.4, Extremely Asymmetric (26.9 was added to each value in this table to eliminate negative numbers). These data are shown in Table 1.1. The summary statistics for these data are shown in Table 1.2 a-b. I will present summary statistics for each time series used. There are two reasons for doing this. First, if you choose to work with the same time series as described in this and succeeding chapters, it provides a simple method for determining if you have the same data set that I used. Second, it provides a simple and quick method to compare two or more time series or portions of time series and confirm the finding presented. For example, if we do an analysis and find that two prob-

---

1. Dr. Salsburg's book is an excellent introduction to the concept of randomness and he provides a series of tables that contain distributions of numbers that have specific statistical characteristics. These sample distributions potentially can be used to determine the impact of the underlying distribution on technical tools.

Table 1.1

	Symmetric	Asymmetric		Symmetric	Asymmetric
1	52.0	70.3	38	52.9	47.5
2	40.5	156.2	39	65.0	54.0
3	51.2	87.5	40	56.7	124.9
4	56.7	57.1	41	33.3	150.9
5	37.8	40.5	42	49.3	65.0
6	45.9	18.0	43	63.5	40.9
7	48.0	68.1	44	61.0	37.0
8	54.2	77.1	45	52.5	144.1
9	62.0	116.2	46	45.4	93.7
10	57.2	72.6	47	45.3	127.4
11	54.0	92.6	48	42.5	55.2
12	56.7	28.0	49	33.8	44.0
13	14.0	189.7	50	42.4	34.7
14	48.4	88.4	51	30.0	78.9
15	42.1	50.2	52	39.8	85.1
16	51.7	84.0	53	50.4	41.6
17	43.6	74.9	54	51.0	76.5
18	61.3	82.1	55	36.2	39.1
19	37.3	37.8	56	53.5	126.2
20	50.9	47.6	57	46.5	46.2
21	54.3	5.5	58	66.9	107.9
22	68.3	50.8	59	35.8	72.5
23	49.8	127.7	60	46.5	72.2
24	43.1	163.8	61	51.9	51.3
25	26.2	98.1	62	46.0	66.9
26	49.1	63.7	63	44.4	38.3
27	69.1	104.4	64	52.7	59.6
28	37.6	29.1	65	46.5	76.1
29	35.4	115.5	66	54.6	37.5
30	58.9	101.5	67	57.0	148.3
31	74.8	82.4	68	49.2	67.3
32	37.8	76.8	69	55.8	86.6
33	45.1	51.5	70	64.1	58.2
34	47.5	126.7	71	50.2	151.2
35	51.4	123.4	72	59.7	121.8
36	60.8	82.5	73	43.7	80.8
37	63.6	85.5	74	53.8	76.2

Table continues

Table 1.1   Continued

	Symmetric	Asymmetric		Symmetric	Asymmetric
75	56.0	64.7	112	48.2	48.2
76	41.4	44.2	113	62.6	107.5
77	52.4	150.7	114	56.1	105.2
78	51.2	89.8	115	41.6	35.2
79	39.9	72.0	116	43.3	86.7
80	35.0	79.1	117	64.7	105.6
81	44.2	65.5	118	37.1	66.2
82	72.5	74.1	119	61.2	53.2
83	37.8	169.1	120	63.5	88.0
84	68.5	68.6	121	49.6	110.8
85	45.7	124.2	122	36.1	78.2
86	70.6	30.8	123	62.9	58.6
87	59.7	108.3	124	58.2	94.4
88	49.8	50.7	125	27.0	92.0
89	54.3	66.7	126	42.8	155.3
90	52.6	56.0	127	45.7	159.9
91	45.2	86.4	128	38.2	87.3
92	55.1	49.2	129	54.1	76.4
93	68.9	93.5	130	63.3	123.3
94	48.4	83.8	131	39.4	67.1
95	54.7	52.5	132	49.5	23.9
96	57.0	87.5	133	48.8	44.4
97	38.2	58.8	134	51.2	13.6
98	68.1	55.7	135	38.8	16.6
99	57.3	73.2	136	31.0	54.0
100	53.0	86.4	137	56.3	89.6
101	49.1	62.6	138	37.5	115.1
102	48.9	90.2	139	52.6	60.1
103	65.8	15.6	140	64.3	48.0
104	54.9	64.0	141	47.8	111.1
105	58.4	48.9	142	41.3	42.7
106	44.4	28.6	143	68.7	47.6
107	46.5	45.5	144	38.1	103.5
108	45.2	112.0	145	67.3	83.3
109	24.1	123.5	146	38.2	139.7
110	42.2	74.2	147	51.9	85.4
111	53.6	110.0	148	56.5	86.5

Table continues

Table 1.1   Continued

	Symmetric	Asymmetric		Symmetric	Asymmetric
149	59.7	49.1	186	45.9	68.8
150	47.6	35.1	187	50.0	131.1
151	39.2	110.9	188	40.9	32.0
152	33.6	102.3	189	40.1	90.5
153	59.2	74.4	190	44.1	67.9
154	47.6	109.3	191	72.5	68.9
155	52.8	64.6	192	43.4	95.9
156	38.4	83.8	193	56.0	67.7
157	29.4	67.5	194	51.9	76.5
158	71.3	12.1	195	39.5	97.9
159	57.7	80.6	196	49.9	26.9
160	57.9	66.1	197	49.5	72.5
161	47.7	60.2	198	38.8	52.3
162	43.5	68.7	199	60.3	166.7
163	33.7	73.9	200	37.4	89.2
164	26.7	101.8	201	72.8	24.0
165	60.1	56.2	202	47.0	72.9
166	50.8	89.8	203	47.7	60.1
167	18.7	86.6	204	46.9	58.2
168	36.8	82.3	205	47.0	54.7
169	57.6	74.9	206	53.1	83.4
170	49.0	71.5	207	45.0	30.9
171	42.2	79.8	208	69.7	32.5
172	61.3	76.0	209	57.4	25.5
173	48.6	128.2	210	42.5	45.4
174	46.5	127.1	211	50.2	109.7
175	57.9	93.2	212	53.8	82.2
176	35.9	45.1	213	40.2	61.0
177	63.7	66.1	214	49.3	33.0
178	49.9	73.4	215	67.5	84.2
179	46.5	109.3	216	36.3	113.7
180	37.0	74.8	217	60.4	53.3
181	61.1	10.6	218	54.2	60.6
182	55.4	148.7	219	61.3	117.0
183	43.4	104.0	220	56.8	105.5
184	46.2	1.0	221	55.4	84.7
185	46.5	84.7	222	36.7	20.9

Table continues

Table 1.1   Continued

	Symmetric	Asymmetric		Symmetric	Asymmetric
223	50.1	66.6	237	43.5	114.1
224	50.7	135.6	238	59.9	114.1
225	47.8	34.1	239	52.1	63.4
226	53.1	54.7	240	42.5	81.1
227	39.7	79.6	241	36.2	57.9
228	44.7	75.7	242	27.4	81.7
229	54.0	69.9	243	48.1	47.4
230	33.5	87.6	244	50.7	14.3
231	26.7	97.2	245	46.1	52.9
232	43.7	81.0	246	50.1	75.5
233	44.1	36.6	247	59.1	46.0
234	43.9	74.4	248	46.7	24.7
235	60.5	49.9	249	54.6	43.5
236	46.4	135.5	250	47.7	88.4

ability density functions (see below) are similar or identical and the summary statistics show that the mean and standard deviation or the medians are very different, then something is wrong and the work should be checked.

A plot of these prices is shown in Figures 1.1 and 1.2 for the symmetric and asymmetric data respectively. One way to summarize price data is to collect it into a frequency histogram. The first step in this process is to lay down a horizontal axis and label it Price and divide it into a series of bins. The size of the bins can vary. In this case, the bin width is $1.00. Therefore, all of the prices from $0.01 to $0.99 go in the first bin, labeled Bin-$0.00. All of the prices from $1.00 to $1.99 would go in the second bin labeled Bin-$1.00. The third bin, labeled Bin-$2.00, would contain all of the prices from $2.00 to $2.99, etc. The vertical axis is labeled Count. The first price in our time series is $52.00, so we move along the Price axis until we reach Bin-$52.00 and increase the count in this bin from 0 to 1, as indicated by the small 1 in this bin in Figure 1.3, top. The small numbers in this figure do not represent the counts contained in these bins, but rather the number of the price in the time series that causes the count in the histogram. So, the next price in our time series is $40.50 and we move along the Price axis on the frequency histogram until we reach Bin-

Table 1.2

X_1: *Symmetric*

Mean	Std. Dev.	Std. Error	Variance	Coef. Var.	Count
49.401	10.5	.664	110.241	21.254	250

Minimum	Maximum	Range	Sum	Sum of Sqr.	# Missing
14	74.8	60.8	12350.2	637559.68	226

# < 10th %	10th %	25th %	50th %	75th %	90th %
25	36.75	42.8	49.3	56.5	63.4

# > 90th %	Mode	Geo. Mean	Har. Mean	Kurtosis	Skewness
25	46.5	48.173	46.735	.177	–.129

X_2: *Asymmetric*

Mean	Std. Dev.	Std. Error	Variance	Coef. Var.	Count
76.558	34.254	2.166	1173.341	44.743	250

Minimum	Maximum	Range	Sum	Sum of Sqr.	# Missing
1	189.7	188.7	19139.53	1757448.309	226

# < 10th %	10th %	25th %	50th %	75th %	90th %
25	34.9	52.5	74.4	93.5	123.85

# > 90th %	Mode	Geo. Mean	Har. Mean	Kurtosis	Skewness
25	•	67.239	46.659	.241	.505

$40.00 and we increase the count contained in it from 0 to 1, as shown by the small 2 in Bin-$40.00 which represents the second price in our time series. The third price is $51.20, so we increase the count in Bin-$51.00 from 0 to 1 and enter a small 3 in that bin. The completed frequency histogram for the symmetric data is shown in Figure 1.3

Figure 1.1

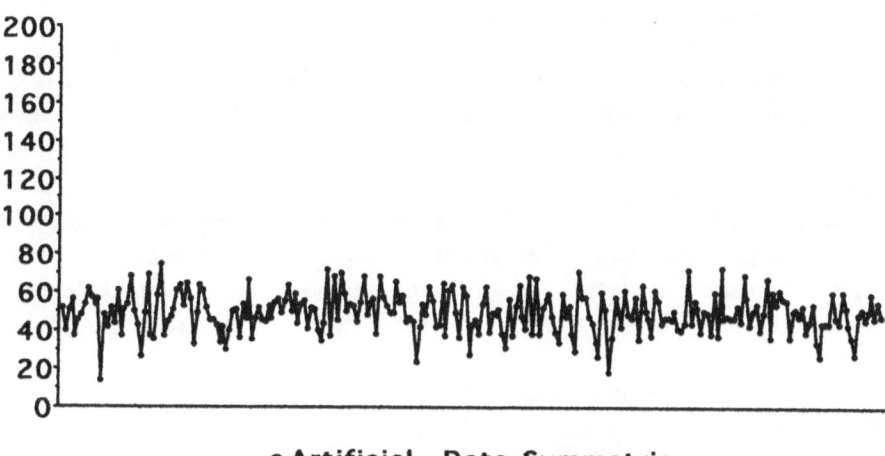

○ Artificial   Data-Symmetric

Figure 1.2

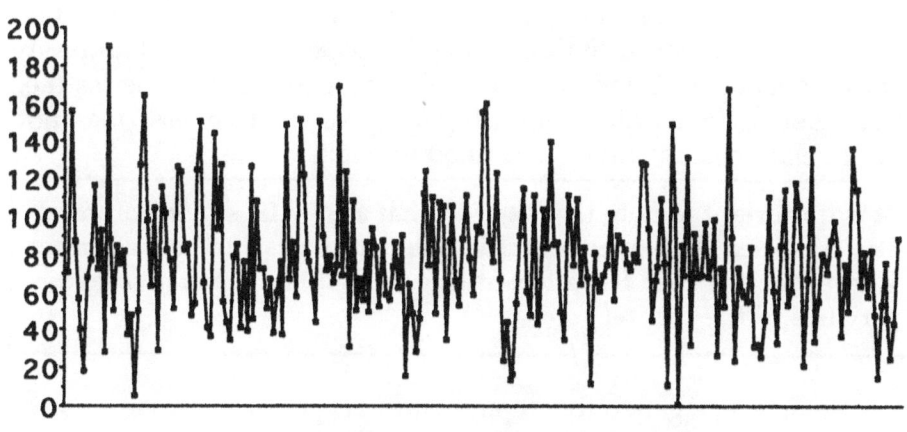

□ Artificial   Data-Assymmetric

bottom, while the completed frequency histogram for the asymmetric data is shown in Figure 1.4.

If you choose to use the techniques described below and in the remaining chapters in this book, it is generally appropriate to start by determining if your time series is stationary or not stationary. If you find that your time series is not stationary, then it is a good idea to reexamine your investment strategy because nonstationarity implies that the *rules* that generate the time series change over time. These changes often occur without any outward sign or indication. It is unlikely that the tools that technicians use would work with a nonstationary time series. On the other hand, if you find that your time series is stationary, then it is appropriate to test your data to determine if it contains serial dependencies (that is, that it is not independent).

If you find that your time series is independent, then it is best to reexamine your investment strategy. This finding implies that the current price was not impacted by any previous price and will not impact on any succeeding price. This also means that the standard tools that technicians use will probably not work very well or at all, since they are based on the assumption that prices do contain serial dependencies. In parallel, you can determine if your time series is random.

Stationarity, independence, and randomness are independent parameters of a time series, so a time series can potentially be stationary, independent, and nonrandom; stationary, dependent, and random; as well as all of the other possible combinations. Stationarity and independence refer to sequential aspects of the time series (i.e., they refer to temporal relationships), while randomness does not. These relationships will be described in detail below.

> It is important to note that the fact that a specific stock, commodity, or index displayed specific characteristics in the past does not guarantee that it will continue to display the same characteristics in the future!

Statisticians generally recommend that your sample be as large as possible. The same rule of thumb applies here. It is best to use as many data points (prices or price changes) in your analysis as possible. For example, if you have 10–20 years of daily data, use it all. Then, repeat the analysis using the most recent five years and then

Figure 1.3

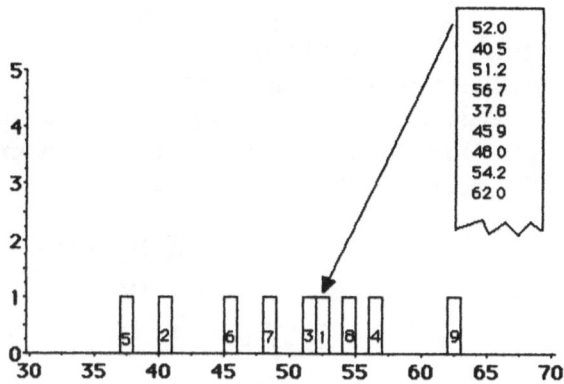

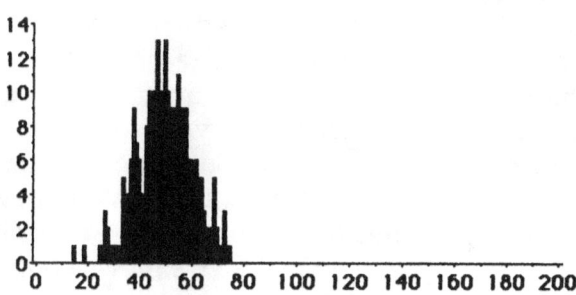

Figure 1.4

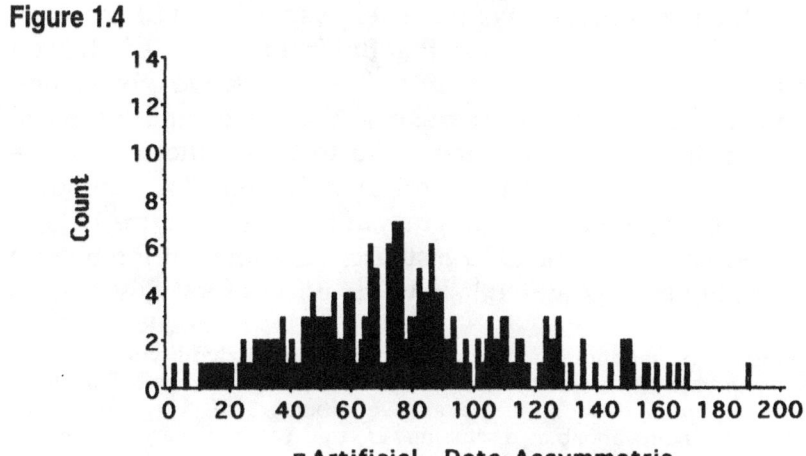

the most recent year. It is also important to keep updating your data set and redoing your analysis. It is also a good idea to analyze your data in the same time frame as you plan to trade. For example, if you are an intraday trader, you would want to do your analysis of tick data. On the other hand, if you are an interday trader, you would do your analysis on daily prices. It is also important to analyze your data on longer time lines. For example, you might want to analyze weekly or monthly data in addition to daily data.

Random/Nonrandom

As indicated above, it is best to start an analysis program by testing for stationarity, but it is easier to explain the principles using the test for randomness. The first step in this analysis is to set up two frequency histograms. The horizontal axis is labeled Price and is divided into a series of bins that are labeled 'bin-1', 'bin-2', 'bin-3', etc. As indicated above, you can vary the size of the bins in a specific histogram. The vertical axis is labeled Count. Next, we will divide our time series into parts by some unvarying rule.[2] For example, we can divide it in half and collect the Odd prices in one frequency histogram and the Even prices into another histogram as shown in Figure 1.5. The bin size in this histogram will be $1.00. The first price in our symmetric time series is $52.00, so we move along the horizontal axis of our Odd frequency histogram until we reach the bin labeled Bin-$52.00 and increase the count in that bin from 0 to 1, as indicated by the small 1 in Bin-$52.00. The next price is $40.5, so we move along the horizontal axis of our Even histogram until we reach $40.00 and we increase the count in that Bin from 0 to 1, as indicated by the small 2 in that bin. The next price is $51.20, so we move along the horizontal axis of the Odd frequency histogram until we reach bin-$51.00 and increase the count from 0 to 1 as indicated by the small 3 (the third price in our time series) in that bin. An 'exploded' version of the Odd frequency histogram at the top of Figure 1.6 and the completed version of the Odd histogram is shown at the bottom of Figure 1.6. Figures 1.7 and 1.8 show the effect of varying the bin

---

2. This definition of randomness follows the work of Richard von Mises, a well-known probability theorist. He explains his views in his book, *Probability, Statistics, And Truth*, New York: Dover Publications, 1957. He also has an excellent and understandable discussion of the Law of Large Numbers mentioned above.

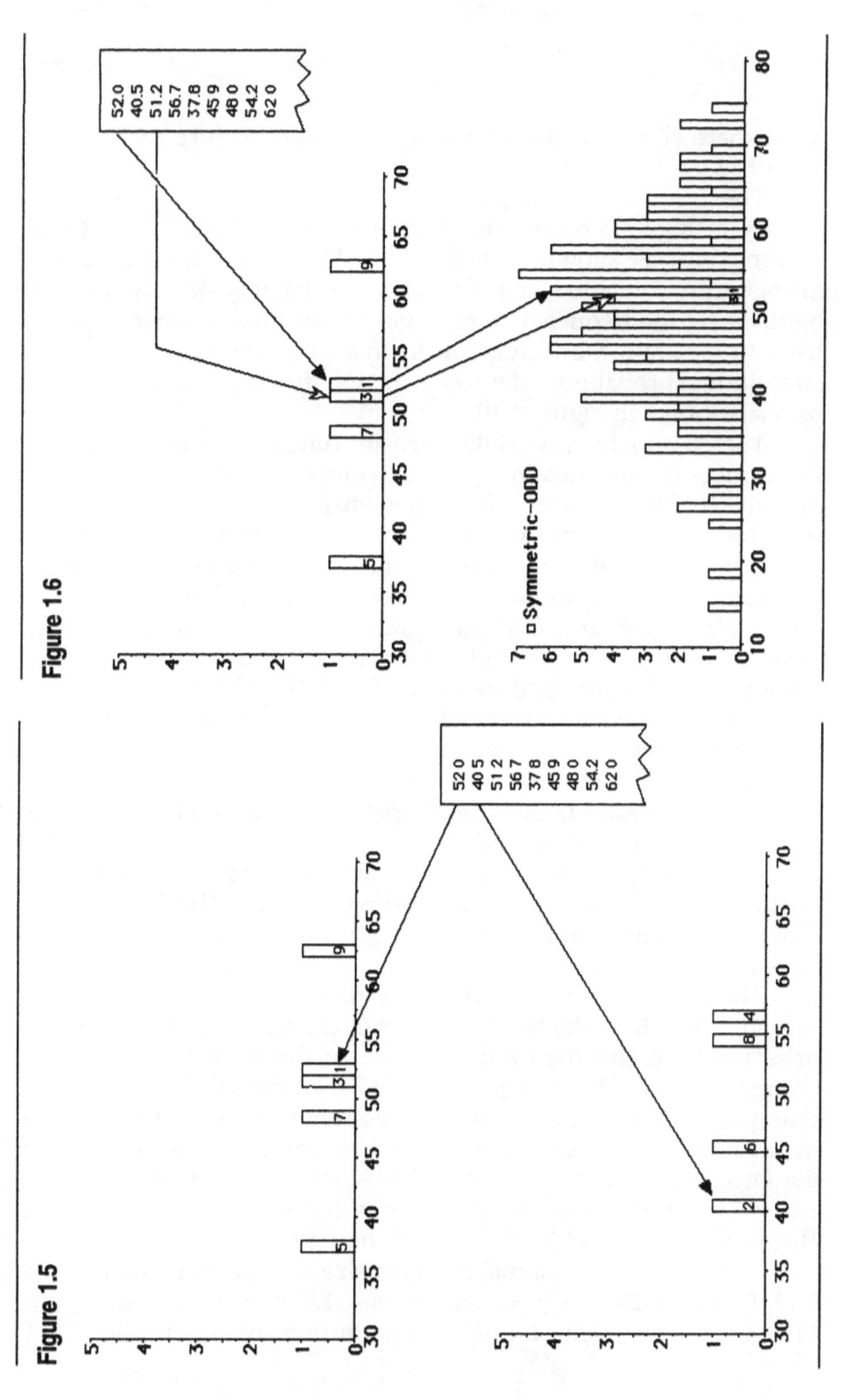

Figure 1.5

Figure 1.6

size, where Figure 1.7 has a bin size of $0.25 and Figure 1.8 has a bin size of $2.00.

The frequency histograms are converted to cumulative frequency histograms by summing the number of counts in each succeeding bin, as shown in Figure 1.9. The cumulative frequency histogram is converted to a cumulative probability density function by dividing the frequency in each bin by the total number of prices used to generate the histogram. In this case, the total is 125. The cumulative probability density function for the symmetric Odd prices is shown in Figure 1.10.

The cumulative probability density function for the symmetric Even prices is generated in the same manner and the two cumulative probability density functions are shown in Figure 1.11. These two cumulative probability density functions are essentially identical. Therefore, we would conclude that the underlying price time series was random. The cumulative probability density functions for the asymmetric Even and Odd histograms are shown in Figure 1.12. Therefore, we would conclude that the asymmetric prices are also random. This is confirmed by examining Table 1.3a-d, which shows the summary statistics for the Odd and Even symmetric and asymmetric time series.

It is possible that the prices in an economic time series might contain a significant trend. One simple way to detrend price data is to use price changes rather than prices as the basis of your calculations. For example, the Delta-1 price changes are generated by subtracting price1 from price2, price2 from price3, etc. The Delta-5 price changes are generated by subtracting price1 from price5, price2 from price6, etc.

The first price in our symmetric time series is $52.00 and the second is $40.50, so the first Delta-1 price change is −11.5. The second price is $40.50 and the third is $51.20, so the second Delta-1 price change is +10.7. The next price is $56.70, so the third Delta-1 price change is +5.5. We continue this process until we reach the last price in our time series. We have created a new time series that consists of the Delta-1 price changes. We will now use this new time series to create two frequency histograms, one for the Odd Delta-1 price changes and a second for the Even Delta-1 price changes.

In this case, we will make the bin size $1.00, as shown in Figure 1.13. Our first Delta-1 price change is −11.5, so we move along the Price axis of the Delta-1 Odd histogram until we get to Bin-$11.00

Figure 1.7

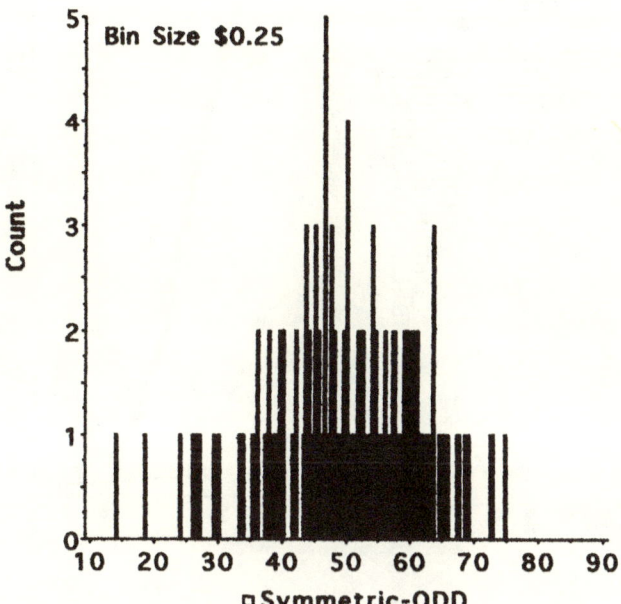

Figure 1.8

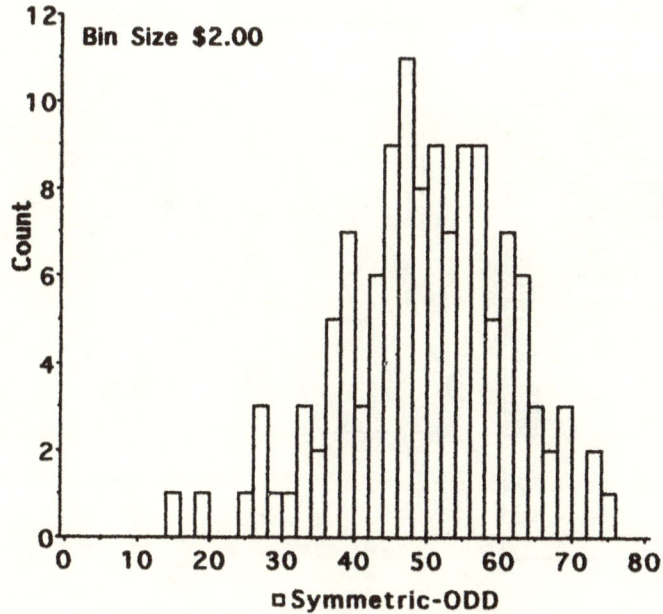

Figure 1.9

□ Symmetric-ODD

Figure 1.10

□ Symmetric-ODD

□ Symmetric-ODD

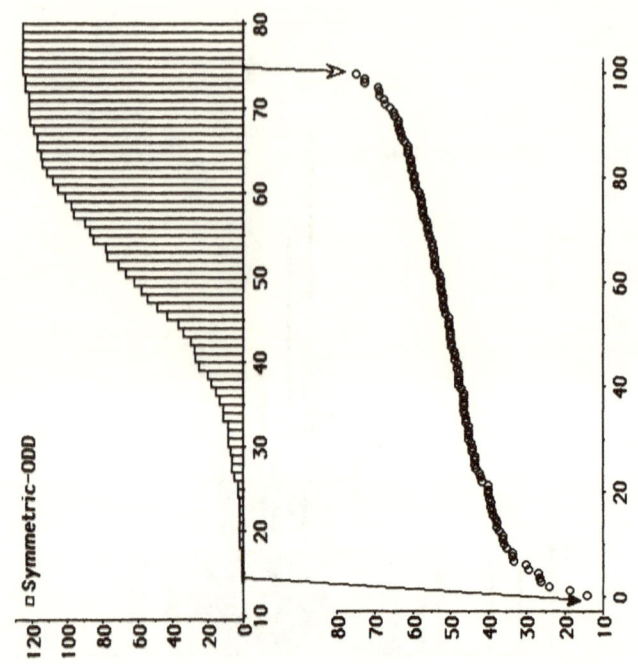

Figure 1.11

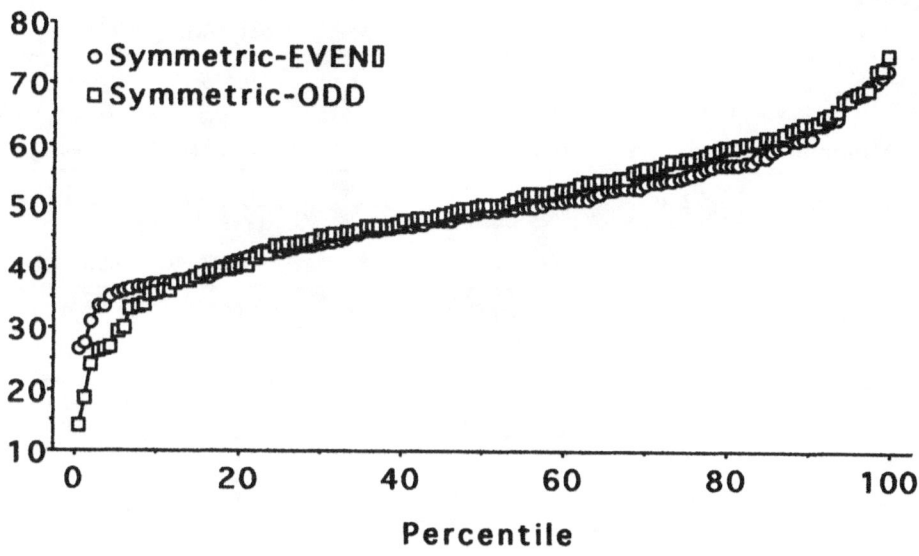

Figure 1.12

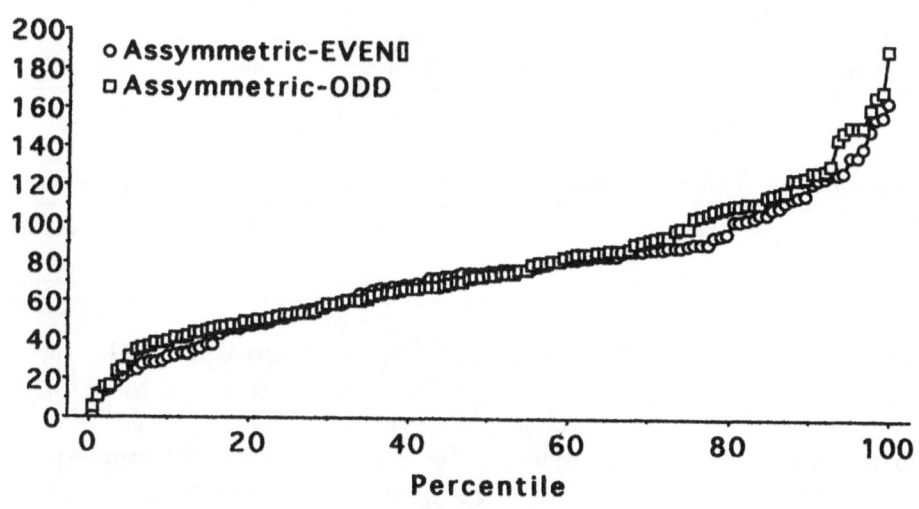

Table 1.3a-b

X_1: Symmetric–Even

Mean	Std. Dev.	Std. Error	Variance	Coef. Var.	Count
49.142	9.546	.854	91.133	19.426	125

Minimum	Maximum	Range	Sum	Sum of Sqr.	# Missing
26.7	72.5	45.8	6142.8	313172.44	225

# < 10th %	10th %	25th %	50th %	75th %	90th %
12	37.1	42.5	49	54.95	61.3

# > 90th %	Mode	Geo. Mean	Har. Mean	Kurtosis	Skewness
12	•	48.212	47.264	–.171	.255

X_2: Symmetric–Odd

Mean	Std. Dev.	Std. Error	Variance	Coef. Var.	Count
49.659	11.406	1.02	130.103	22.969	125

Minimum	Maximum	Range	Sum	Sum of Sqr.	# Missing
14	74.8	60.8	6207.4	324387.24	225

# < 10th %	10th %	25th %	50th %	75th %	90th %
12	35.8	43.475	50	57.625	63.6

# > 90th %	Mode	Geo. Mean	Har. Mean	Kurtosis	Skewness
12	46.5	48.133	46.218	.253	–.372

and increase the count contained in this bin from 0 to 1. The next Delta-1 price change is 10.7 and we increase the count in Bin-$10.00 of the Even histogram from 0 to 1 (Figure 1.14). Our next Odd Delta-1 price change is 5.5, so we increase the count in Bin-+$5.00 from 0 to 1 as indicated in Figure 1.13. The completed frequency histogram for the Odd Delta-1 price changes is shown in Figure 1.15, while the

Table 1.3c-d

X_3: *Asymmetric–Even*

Mean	Std. Dev.	Std. Error	Variance	Coef. Var.	Count
74.521	32.849	2.938	1079.053	44.08	125

Minimum	Maximum	Range	Sum	Sum of Sqr.	# Missing
1	163.8	162.8	9315.1	827971.27	225

# < 10th %	10th %	25th %	50th %	75th %	90th %
12	30.8	51.925	75.5	89.35	121.8

# > 90th %	Mode	Geo. Mean	Har. Mean	Kurtosis	Skewness
12	•	64.879	39.455	.03	.301

X_4: *Asymmetric–Odd*

Mean	Std. Dev.	Std. Error	Variance	Coef. Var.	Count
78.595	35.619	3.186	1268.723	45.32	125

Minimum	Maximum	Range	Sum	Sum of Sqr.	# Missing
5.5	189.7	184.2	9824.43	929477.039	225

# < 10th %	10th %	25th %	50th %	75th %	90th %
12	39.1	52.8	73.2	99.575	127.4

# > 90th %	Mode	Geo. Mean	Har. Mean	Kurtosis	Skewness
12	•	69.684	57.083	.267	.641

Even Delta-1 price change frequency histogram is shown in Figure 1.16. The cumulative probability density functions for the Symmetric Delta-1 Odd and Even price changes is shown in Figure 1.17. Figure 1.18 shows the cumulative probability density functions for the Asymmetric Delta-1 Odd and Even price changes. The respective cumulative probability density functions are virtually identical. This

Figure 1.13

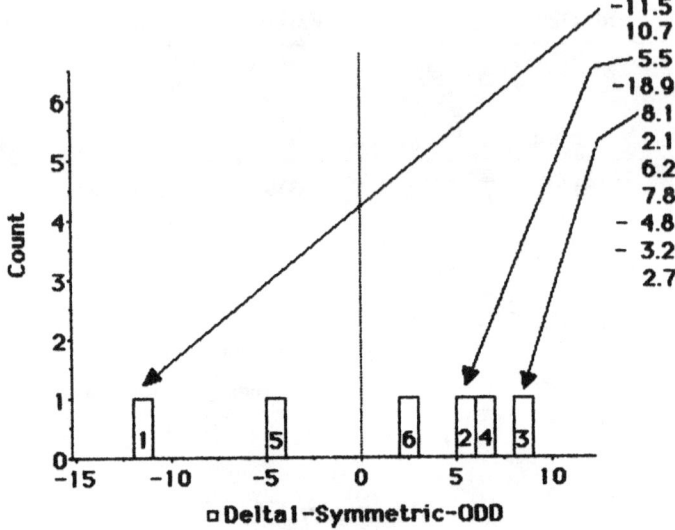

☐ Delta-Symmetric-ODD

Figure 1.14

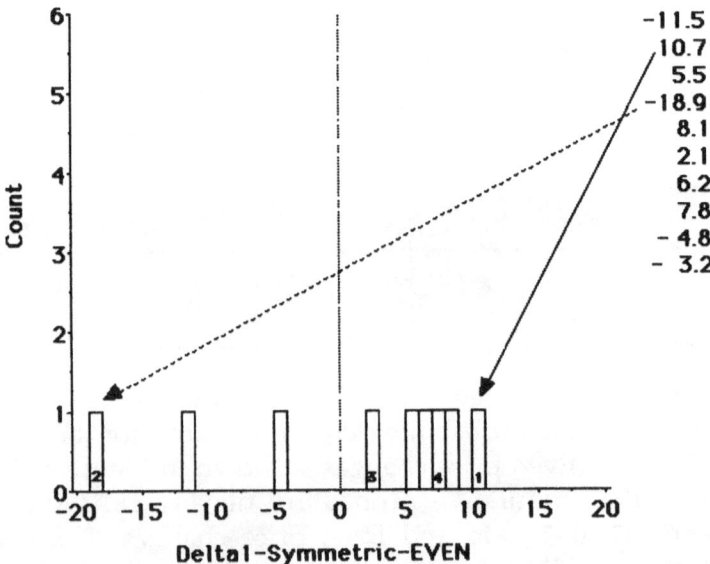

Delta-Symmetric-EVEN

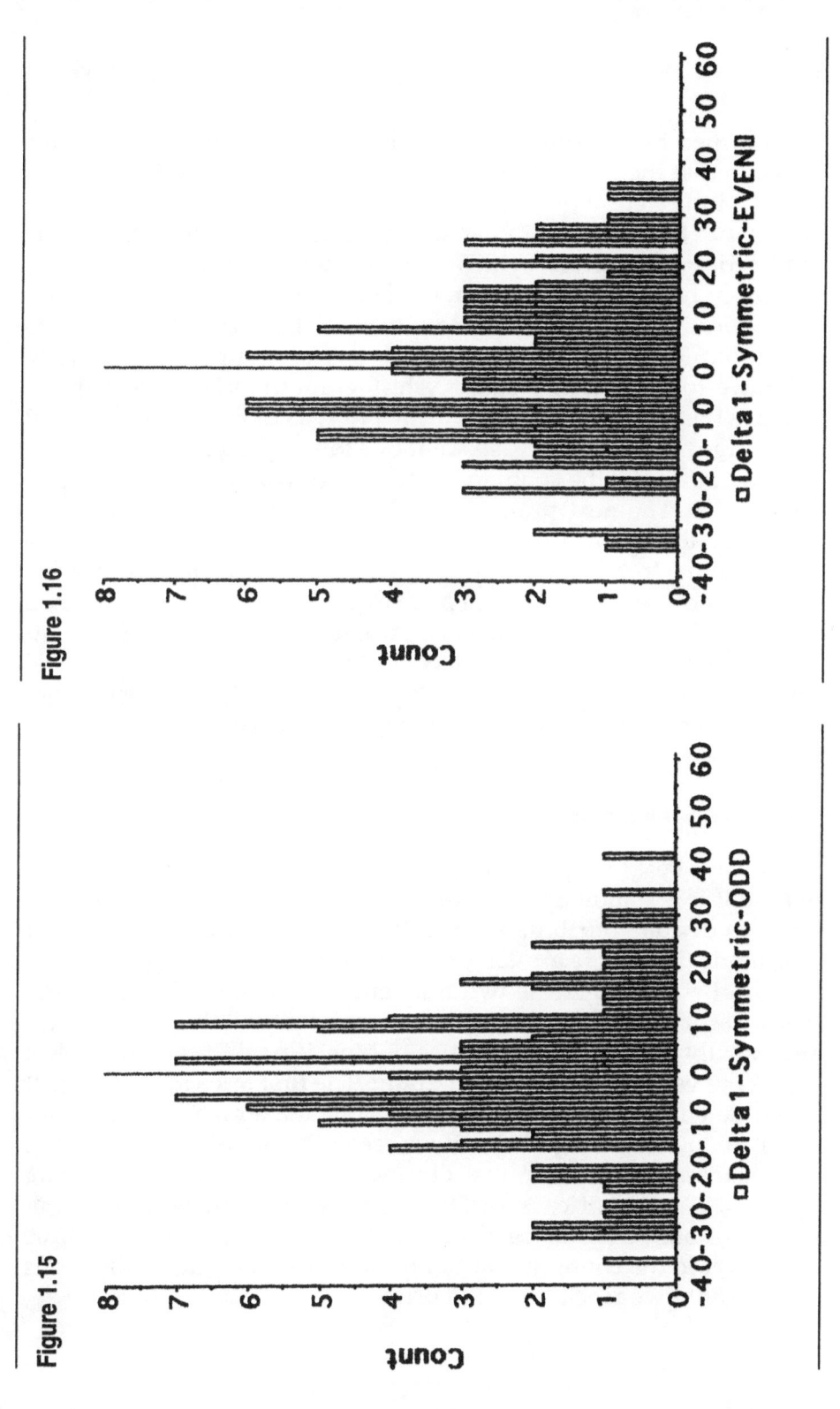

Figure 1.15

Figure 1.16

is confirmed by the summary statistics shown in Table 1.4 a-d. This suggests that the underlying Delta-1 time series are also random.

As indicated above, we can use some other unvarying rule to divide our time series into parts. For example, we could divide the time series into thirds. In this case, we will need to set up three frequency histograms, as shown in Figure 1.19. The first price in our time series is $52.00, so we increase the count in Bin-$52.00 from 0 to 1, as indicated by the small 1 in that bin. The next price is $40.50, so we move to the second frequency histogram (middle of figure) and increase the count in Bin-$40.00 from 0 to 1, as indicated by the small 2. The third price is $51.20, so we move to the third frequency histogram and increase the count in Bin-$51.00 from 0 to 1, as indicated by the small 3. The next price is $56.70, so we move back to the first frequency histogram (top of figure) and increase the count in Bin-$56.00 from 0 to 1, as indicated by the small 4 (which means that this count was caused by the fourth price in our time series). We continue this process until we reach the last price in our time series. The summary statistics for the three frequency histograms are shown in Table 1.5 a-c and the cumulative probability density functions derived from the frequency histograms are shown in Figure 1.20.

Stationary/Nonstationary

It is always best to begin your analysis of a new time series by determining if it is stationary or nonstationary. The first step in this process is to divide your time series into two halves. In the case of our symmetric data, there are 250 prices. So, prices 1–125 and prices 126–250 will be used to create two frequency histograms, for Half1 and Half2, respectively. The horizontal axis of each will be labeled Price and it will be divided into a series of bins. We will use a bin size of $1.00. The vertical axis is labeled Count. The first price in the first half of our data is $52.00, so we will move along the horizontal axis of our Half1 frequency histogram until we reach the bin labeled Bin-$52.00 and increase the count in that bin from 0 to 1, as shown in Figure 1.21, top. The next price is $40.50, so we will move along the horizontal axis of our Half1 histogram until we reach the bin labeled $40.00 and increase the count in that bin from 0 to 1. We will continue this process until we reach the 125th price.

Figure 1.17

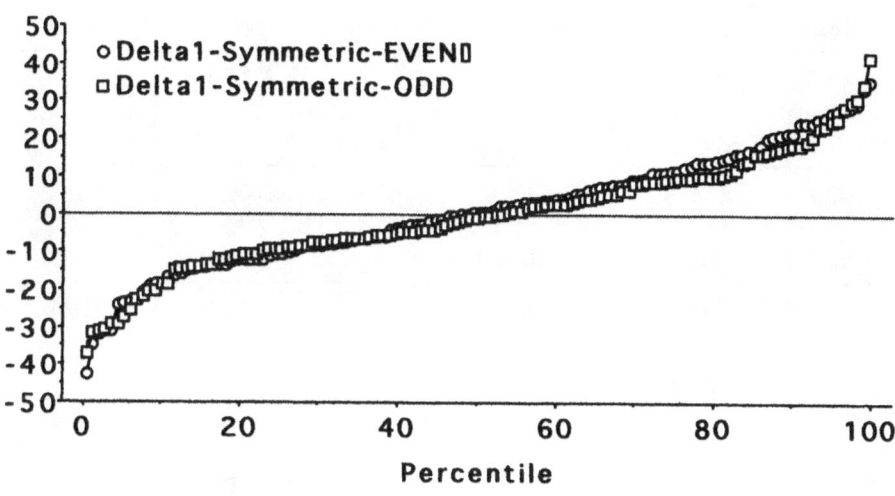

Figure 1.18

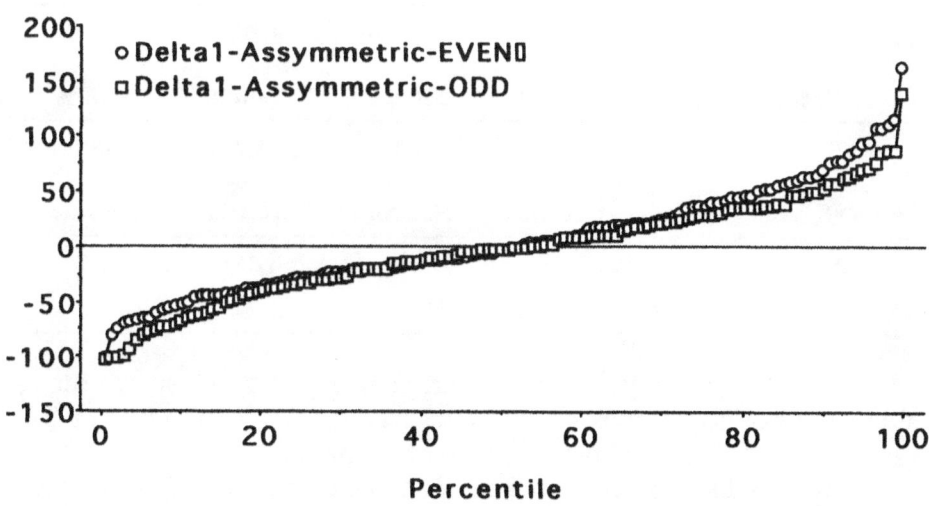

Table 1.4a-b

X_1: Delta-1–Symmetric–Even

Mean	Std. Dev.	Std. Error	Variance	Coef. Var.	Count
.486	15.625	1.403	244.134	3213.056	124

Minimum	Maximum	Range	Sum	Sum of Sqr.	# Missing
–42.7	35.4	78.1	60.3	30057.77	226

# < 10th %	10th %	25th %	50th %	75th %	90th %
12	–18.53	–10.85	.2	11.4	21.76

# > 90th %	Mode	Geo. Mean	Har. Mean	Kurtosis	Skewness
12	–12.4	•	•	–.285	–.057

X_2: Delta-1–Symmetric–Odd

Mean	Std. Dev.	Std. Error	Variance	Coef. Var.	Count
–.517	14.724	1.317	216.791	–2849.036	125

Minimum	Maximum	Range	Sum	Sum of Sqr.	# Missing
–37	41.9	78.9	–64.6	26915.44	225

# < 10th %	10th %	25th %	50th %	75th %	90th %
12	–18.8	–9.35	–1.2	9.25	18.1

# > 90th %	Mode	Geo. Mean	Har. Mean	Kurtosis	Skewness
12	•	•	•	.141	.11

The 126th price in our symmetric prices is $42.80, so we move along the horizontal axis of the Half2 frequency histogram until we reach Bin-$42.00 and increase the count in that histogram from 0 to 1, as shown in Figure 1.21, bottom. The next price is $45.70, so we increase the count in bin-$45.00 from 0 to 1. We continue this process until we reach the last price.

Table 1.4c-d

X₃: Delta-1–Asymmetric–Even

Mean	Std. Dev.	Std. Error	Variance	Coef. Var.	Count
4.253	48.013	4.312	2305.213	11.28.788	124

Minimum	Maximum	Range	Sum	Sum of Sqr.	# Missing
−101.5	161.7	263.2	527.43	285784.667	226

# < 10th %	10th %	25th %	50th %	75th %	90th %
12	−52.68	−28.65	−3.4	35.85	69.23

# > 90th %	Mode	Geo. Mean	Har. Mean	Kurtosis	Skewness
12	−37.4	•	•	.117	.555

X₄: Delta-1–Asymmetric–Odd

Mean	Std. Dev.	Std. Error	Variance	Coef. Var.	Count
−4.075	45.688	4.086	2087.407	−1121.281	125

Minimum	Maximum	Range	Sum	Sum of Sqr.	# Missing
−103	138.1	241.1	−509.33	260913.795	225

# < 10th %	10th %	25th %	50th %	75th %	90th %
12	−68.4	−33.425	−3.4	27.675	51.5

# > 90th %	Mode	Geo. Mean	Har. Mean	Kurtosis	Skewness
12	•	•	•	−.013	.026

The completed frequency histogram for our symmetric data Half1 is shown in Figure 1.22, while the completed frequency histogram for Half2 is shown in Figure 1.23. We will convert these two frequency histograms into probability density functions and cumulative probability density functions using the methods described above. The two cumulative probability density functions are shown

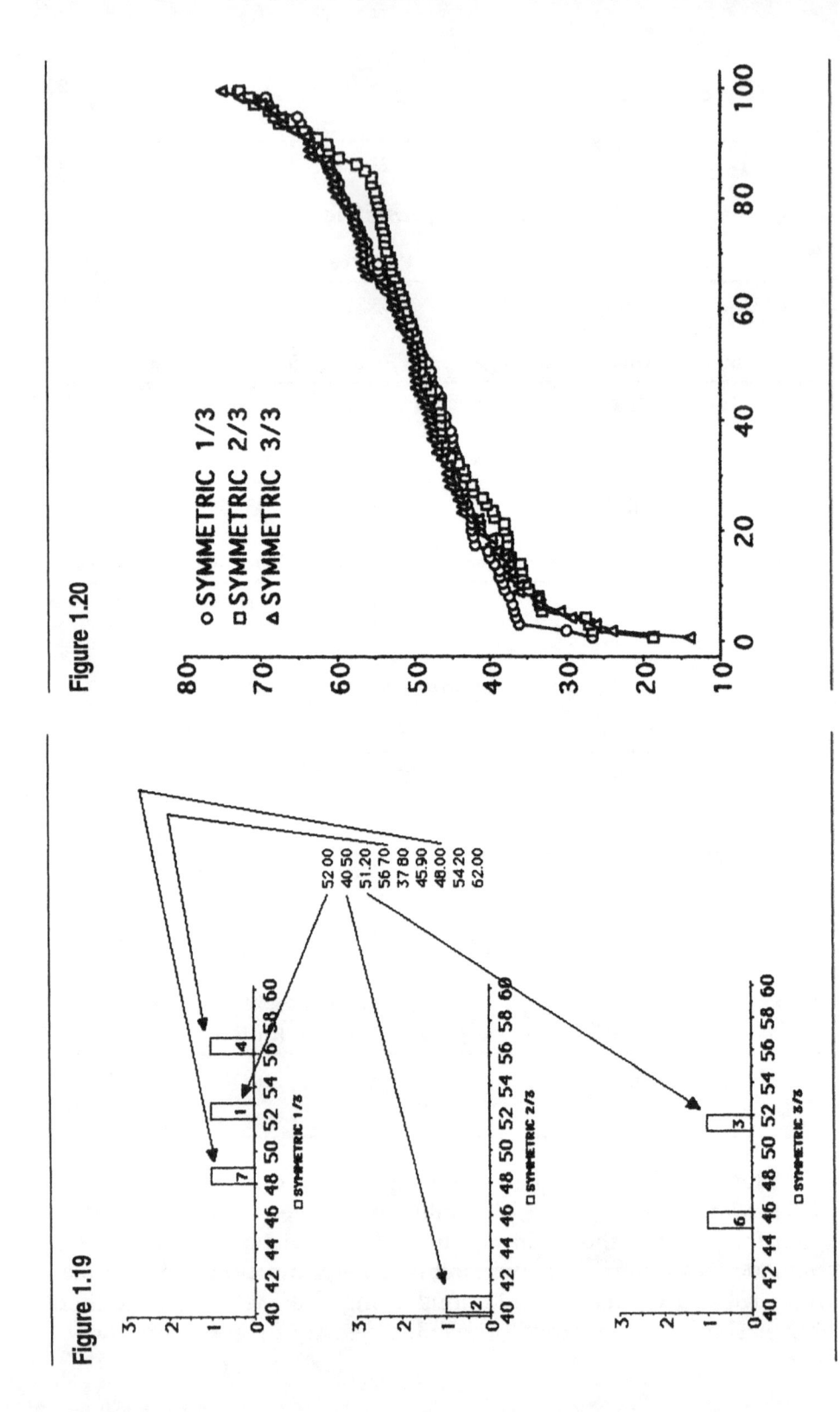

Figure 1.19

Figure 1.20

Table 1.5a-b

X_1: Symmetric 1/3

Mean	Std. Dev.	Std. Error	Variance	Coef. Var.	Count
49.939	9.583	1.052	91.841	19.19	83

Minimum	Maximum	Range	Sum	Sum of Sqr.	# Missing
26.7	72.8	46.1	4144.9	214521.29	225

# < 10th %	10th %	25th %	50th %	75th %	90th %
8	38.34	43.175	48.1	57.225	63.02

# > 90th %	Mode	Geo. Mean			
8	•	49.017			

X_2: Symmetric 2/3

Mean	Std. Dev.	Std. Error	Variance	Coef. Var.	Count
48.202	10.625	1.166	112.894	22.043	83

Minimum	Maximum	Range	Sum	Sum of Sqr.	# Missing
18.7	72.5	53.8	4000.8	202105.54	225

# < 10th %	10th %	25th %	50th %	75th %	90th %
8	35.32	40.6	49.1	54.15	61.48

# > 90th %	Mode	Geo. Mean			
8	46.5	46.945			

in Figure 1.24. These two cumulative probability density functions are virtually identical, which indicates that the underlying time series is stationary.

The frequency histograms for Half1 and Half2 for the asymmetric time series are shown in Figures 1.25 and 1.26, respectively. The two cumulative probability density functions are shown in Figure

Table 1.5c

X_3: *Symmetric 3/3*

Mean	Std. Dev.	Std. Error	Variance	Coef. Var.	Count
50.054	11.238	1.226	126.296	22.452	84

Minimum	Maximum	Range	Sum	Sum of Sqr.	# Missing
14	74.8	60.8	4204.5	220932.85	224

# < 10th %	10th %	25th %	50th %	75th %	90th %
7	36.2	43.9	49.95	57.75	63.65

# > 90th %	Mode	Geo. Mean
8	•	48.576

1.27 and are virtually identical, so the underlying asymmetric time series is also stationary. The summary statistics for symmetric Half1 and Half2 are shown in Table 1.6a-b, while the summary statistics for asymmetric Half1 and Half2 are shown in Table 1.6c-d.

Independence/Dependence

If we find that our time series is stationary, then the next step is to determine if it contains serial dependencies. There are a number of different methods for detecting serial dependencies. Each method makes specific assumptions about the underlying time series. We can use one method or a series of different methods. If we use one method and find that our time series is independent, that does not mean that if we chose another method that makes a different set of assumptions, that we would necessarily find that the second method will also find that our time series is independent. If we use one or more methods and find that our time series is independent (that is, it does not appear to contain serial dependencies), then it is probably best to select another time series. If a time series is independent, it means that the tools technicians use to detect patterns that they use to develop a trading strategy will probably not *work*. If, on the other

Figure 1.21

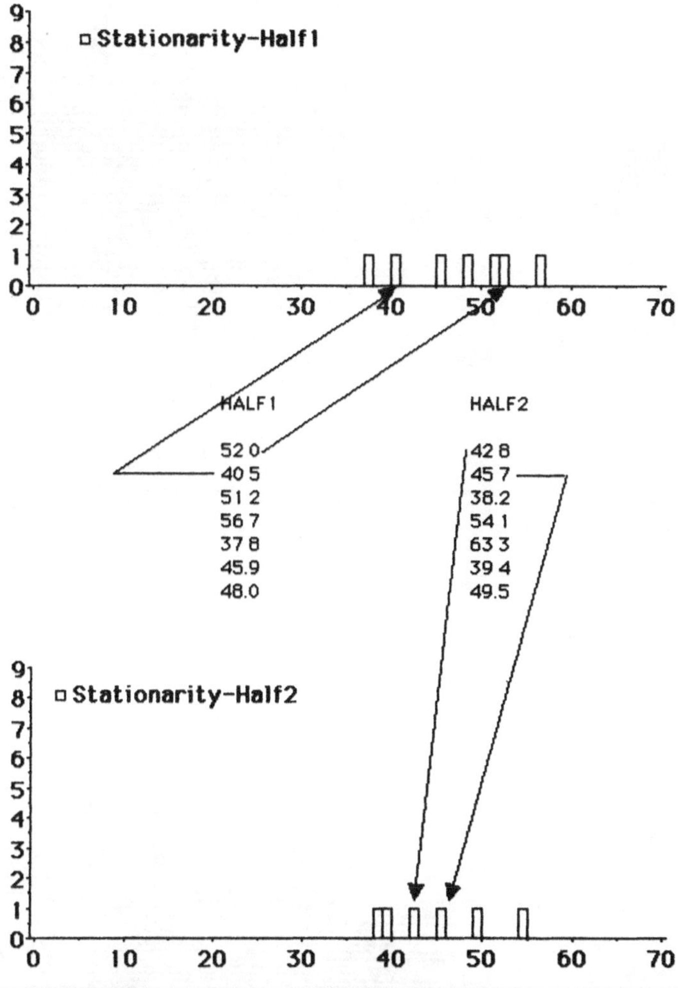

HALF 1	HALF 2
52 0	42 8
40 5	45 7
51 2	38.2
56 7	54 1
37 8	63 3
45.9	39 4
48.0	49.5

hand, we use one or more of these methods and discover that our time series does contain serial dependencies, then it should be possible to develop a workable and potentially profitable trading strategy. The three methods described below make the fewest assumptions about the underlying time series: Other methods, such as runs tests/persistence and serial- and autocorrelation are described in detail in *The Mathematics of Technical Analysis* and will not be described or used here.

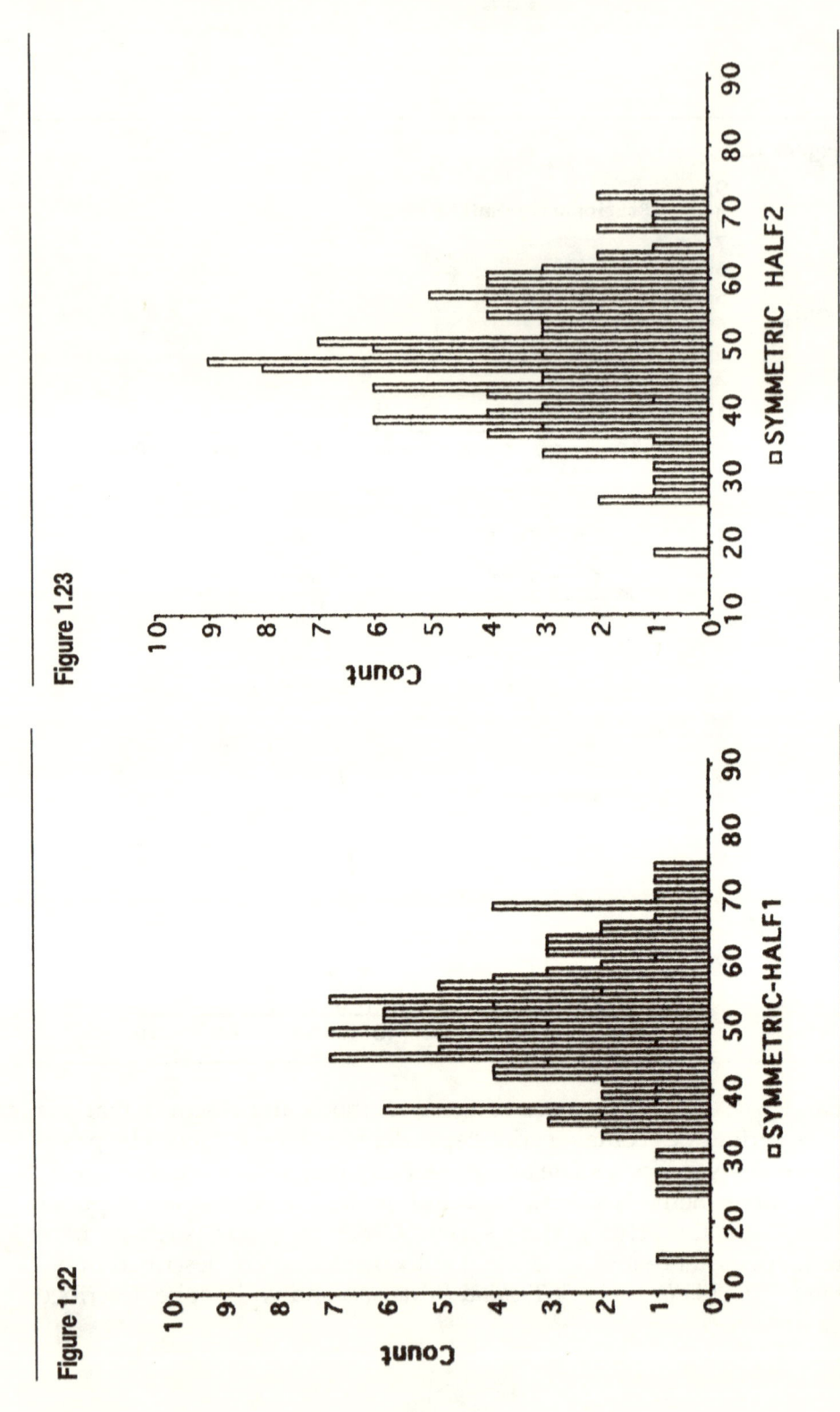

Figure 1.22

Figure 1.23

Figure 1.24

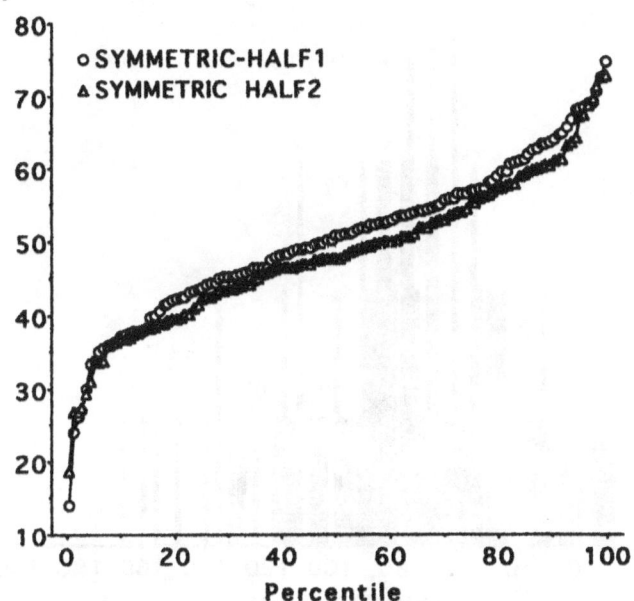

Figure 1.25

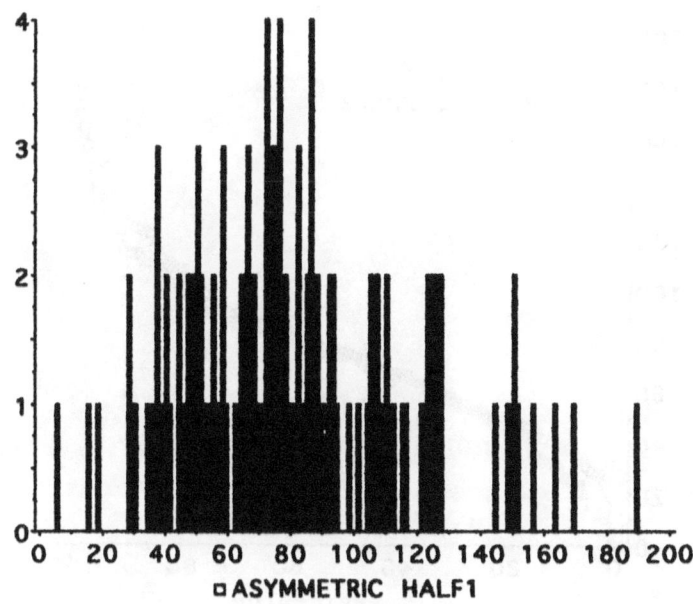

Figure 1.26

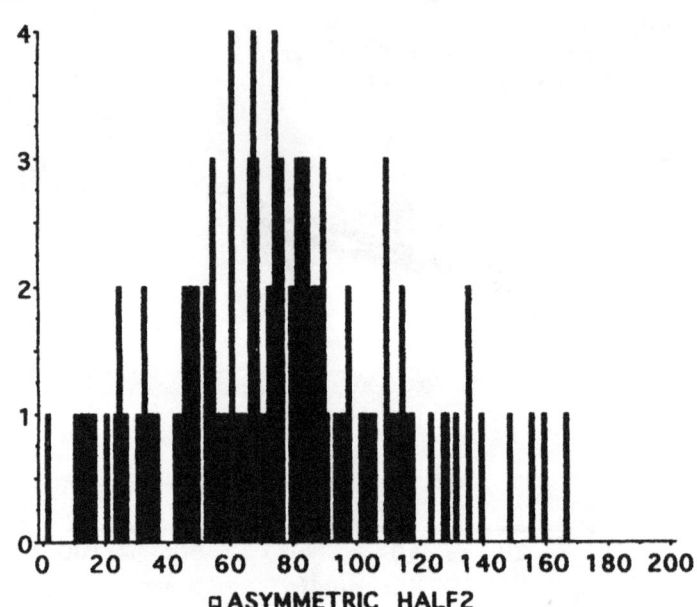

Figure 1.27

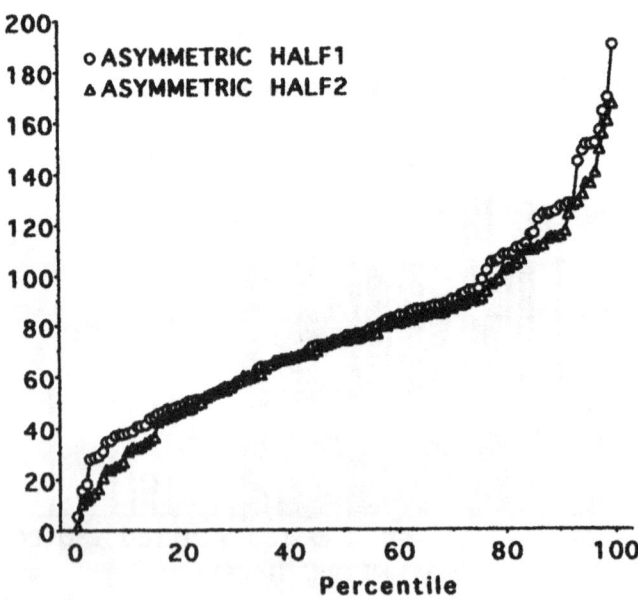

Table 1.6a-b

X_1: Symmetric–Half1

Mean	Std. Dev.	Std. Error	Variance	Coef. Var.	Count
50.326	10.752	.962	115.614	21.365	125

Minimum	Maximum	Range	Sum	Sum of Sqr.	# Missing
14	74.8	60.8	6290.8	330929.46	0

# < 10th %	10th %	25th %	50th %	75th %	90th %
12	37.1	43.675	50.9	57	64.1

# > 90th %	Mode	Geo. Mean	Har. Mean	Kurtosis	Skewness
12	46.5	49.019	47.397	.354	−.294

X_2: Symmetric–Half2

Mean	Std. Dev.	Std. Error	Variance	Coef. Var.	Count
48.475	10.199	.912	104.029	21.041	125

Minimum	Maximum	Range	Sum	Sum of Sqr.	# Missing
18.7	72.8	54.1	6059.4	306630.22	0

# < 10th %	10th %	25th %	50th %	75th %	90th %
12	36.7	41.975	47.8	55.4	61.1

# > 90th %	Mode	Geo. Mean	Har. Mean	Kurtosis	Skewness
12	•	47.341	46.092	.081	.03

Differential Spectrum

The differential spectrum is a method that allows one to make one sweep through our time series and determine if the entire time series is independent or not. We collect our time series into a Delta-1 price change frequency histogram in the same manner as shown in Figure

Table 1.6c-d

X_3: Asymmetric–Half1

Mean	Std. Dev.	Std. Error	Variance	Coef. Var.	Count
78.879	34.977	3.128	1223.408	44.343	125

Minimum	Maximum	Range	Sum	Sum of Sqr.	# Missing
5.5	189.7	184.2	9859.83	929432.579	0

# < 10th %	10th %	25th %	50th %	75th %	90th %
12	38.3	52.25	74.9	95.325	126.2

# > 90th %	Mode	Geo. Mean	Har. Mean	Kurtosis	Skewness
12	•	70.581	59.213	.297	.672

X_4: Asymmetric–Half2

Mean	Std. Dev.	Std. Error	Variance	Coef. Var.	Count
74.238	33.494	2.996	1121.88	45.118	125

Minimum	Maximum	Range	Sum	Sum of Sqr.	# Missing
1	166.7	165.7	9279.7	828015.73	0

# < 10th %	10th %	25th %	50th %	75th %	90th %
12	30.9	52.75	74.4	89.975	115.1

# > 90th %	Mode	Geo. Mean	Har. Mean	Kurtosis	Skewness
12	•	64.055	38.497	.037	.302

1.14. Basic probability theory suggests that if a time series is independent, then the distribution of positive and negative price changes should be symmetrical. That is, the distribution of counts in the Delta-1 price change histogram should be symmetrical around '0'. If they are, then the underlying time series may not contain any serial dependencies. However, if they are not symmetrical, then this suggests that the underlying time series is not independent and thus may contain serial dependencies. It is important to note that we can do the differential spectrum on other levels of price change data. For example, we can do a Delta-5 price change differential histogram or a Delta-10.

The choice of the bin width for the differential spectrum can be varied and will determine the sensitivity of the test and the computational effort involved. As bin size increases, the sensitivity of the differential spectrum typically decreases. This means that it is less likely to detect serial dependencies, if they exist. The bin width should not be smaller than the minimum price change that can occur. For example, if the minimum price change is $0.10 (that is, the price can jump from $1.00 to $1.10, but not from $1.00 to $1.06 or $1.08), then the minimum bin size should not be less than $0.10. However, a bin size this small might mean that your differential spectrum histogram would contain many bins and thus would require a good deal of computation effort. As a general rule of thumb, it is best to begin with a relatively large bin size, such as $0.50 or $1.00, and collect this differential spectrum histogram. Then repeat with a smaller bin size.

The number of price changes in our time series is also a critical factor. As the number of price changes included in the histogram increases, the importance of lack of symmetry also increases. For example, if we have a differential spectrum with 50 prices, we would expect it to look somewhat asymmetrical. However, if we have a differential spectrum with 1,000 price changes and it is asymmetrical, this would strongly suggest that the underlying time series is not independent and does contain serial dependencies. Therefore, it is best to use as many price changes as are available to construct our differential spectrum.

The method used to collect the Delta-1 price change differential spectrum is similar to that illustrated in Figure 1.14, except that we collect all of the available price changes in a single frequency histogram. The Delta-1 price change differential spectrum using a bin size of $1.00 is shown in Figure 1.28. The differential spectrum with a bin size of $0.50 is shown in Figure 1.29, while the differential spectrum with a bin size of $0.25 and $0.10 is shown in Figures 1.30 and 1.31, respectively. Note that as bin size decreases, the number of bins with a count of 1 increases.

There are a number of different methods that we can use to evaluate our differential spectrums. We will use the differential spec-

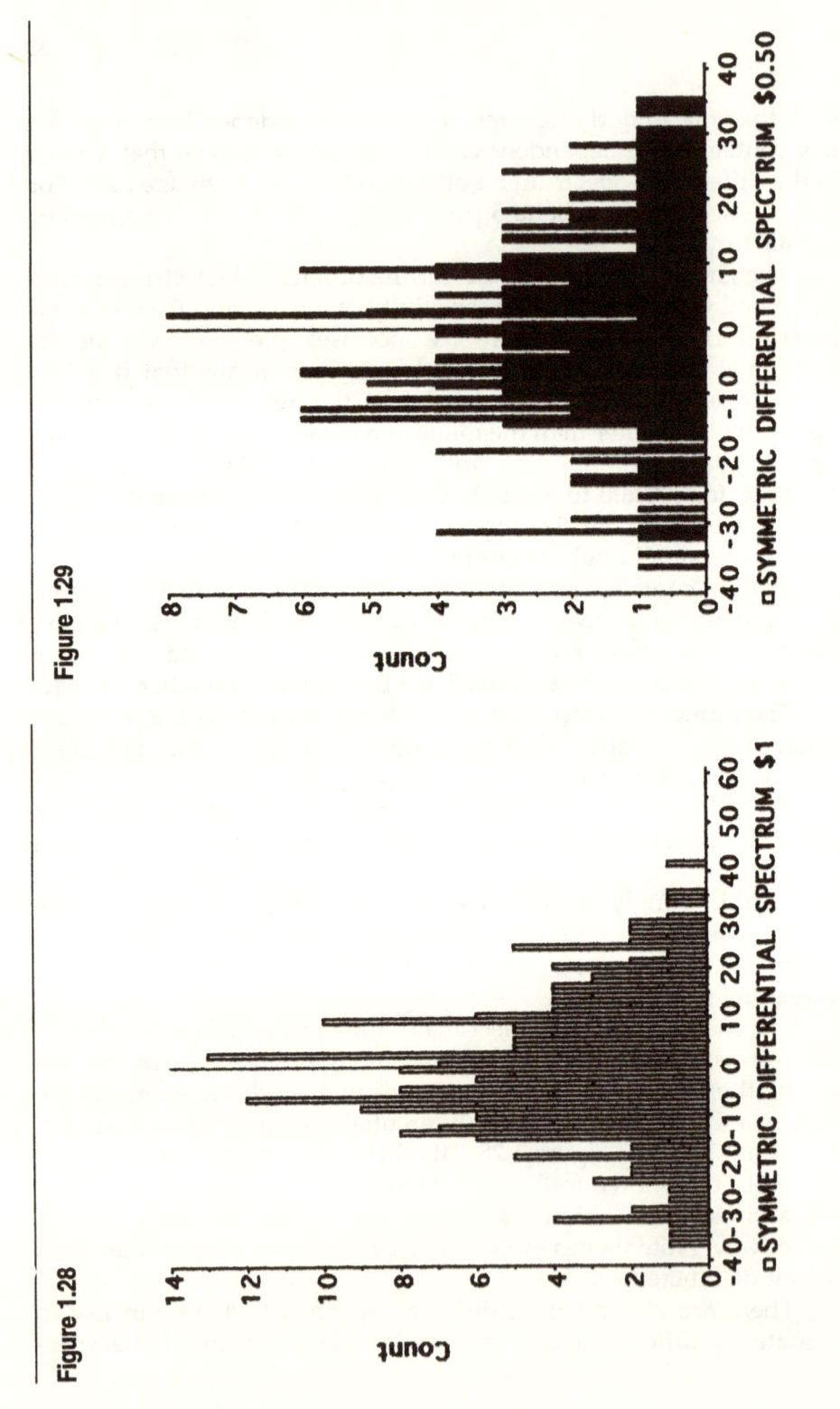

Figure 1.28

Figure 1.29

Figure 1.30

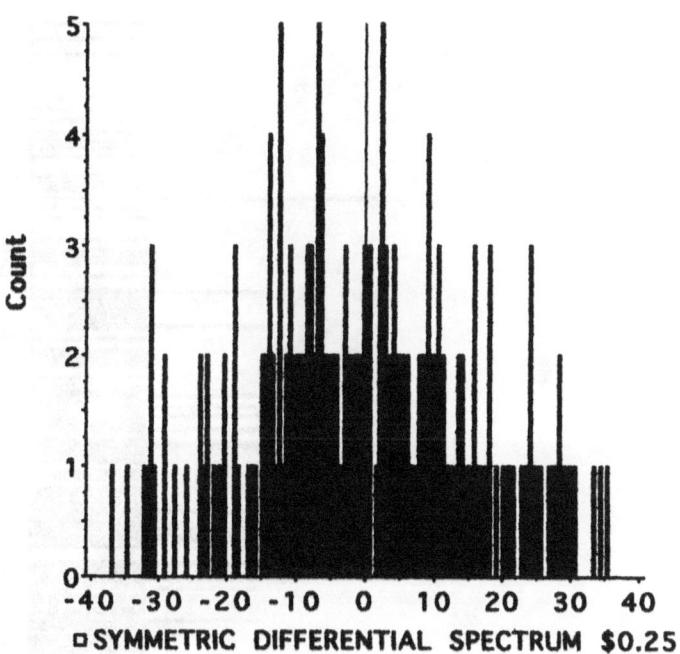

□ SYMMETRIC DIFFERENTIAL SPECTRUM $0.25

trum with a bin size of $1.00 for our evaluations (see Figure 1.28). First, we can examine the histogram. In this case, the tails of the histogram look somewhat symmetrical, but the bins near the 0 appear less symmetrical. For example, the Bin-+$2.00 contains 13 counts, while the Bin—$2.00 contains only five counts.

Or we can examine the summary statistics for the positive and negative price changes as shown in Table 1.7a-b. In this case, the means and standard deviations are very similar, as are the ranges. This would suggest that the positive and negative price changes are symmetrical around the 0 and thus the underlying time series is probably independent and may not contain any serial dependencies.

We can replot the histograms as cumulative probability density functions, as described above. In order to plot both the positive and negative Delta-1 price changes on the same axis, we will need to convert the negative Delta-1 price changes to their absolute values. The cumulative probability density function for the symmetric time series is shown in Figure 1.32. The probability density functions for

Figure 1.31

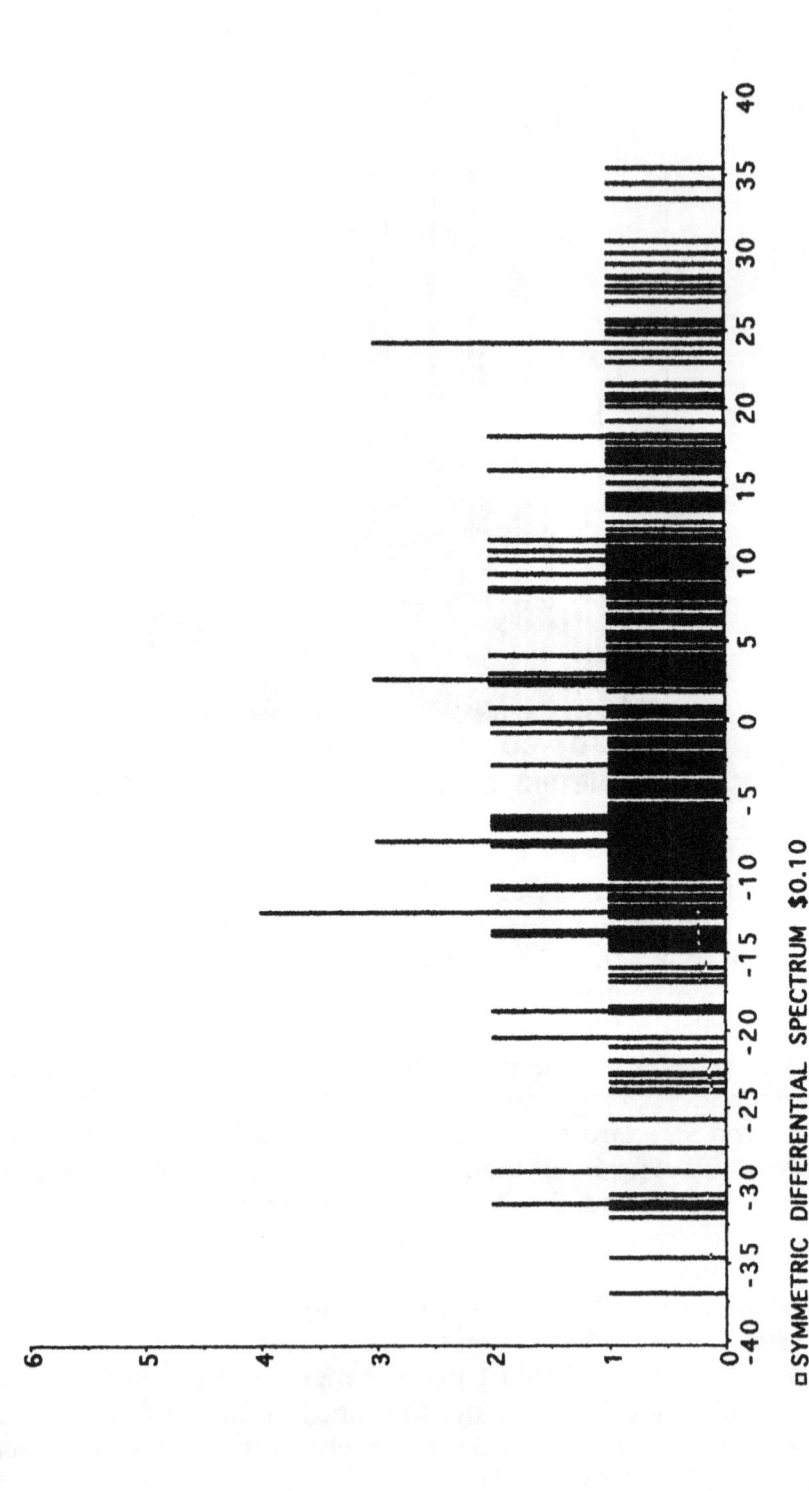

□ SYMMETRIC DIFFERENTIAL SPECTRUM $0.10

Table 1.7a-b

X_1: *Symmetric Differential Spectrum +*

Mean	Std. Dev.	Std. Error	Variance	Coef. Var.	Count
12.412	9.284	.844	86.196	74.803	121

Minimum	Maximum	Range	Sum	Sum of Sqr.	# Missing
.1	41.9	41.8	1501.8	28983.2	106

# < 10th %	10th %	25th %	50th %	75th %	90th %
11	2.3	4.575	10.1	18.1	26.08

# > 90th %	Mode	Geo. Mean	Har. Mean	Kurtosis	Skewness
12	2.4	8.21	2.91	−.039	.798

X_2: *Symmetric Differential Spectrum −*

Mean	Std. Dev.	Std. Error	Variance	Coef. Var.	Count
11.766	8.992	.795	80.855	76.421	128

Minimum	Maximum	Range	Sum	Sum of Sqr.	# Missing
.1	42.7	42.6	1506.1	27990.01	99

# < 10th %	10th %	25th %	50th %	75th %	90th %
13	1.82	5.65	9.55	14.7	25.26

# > 90th %	Mode	Geo. Mean	Har. Mean	Kurtosis	Skewness
13	12.4	7.892	2.82	.854	1.119

the positive and negative (absolute value) Delta-1 price changes are very similar.

Or we can be a bit more rigorous. For example, we can determine the median (50th percentile) positive and negative price change. The median positive price change for the positive price changes is $10.10, while the median price change for the negative

Figure 1.32

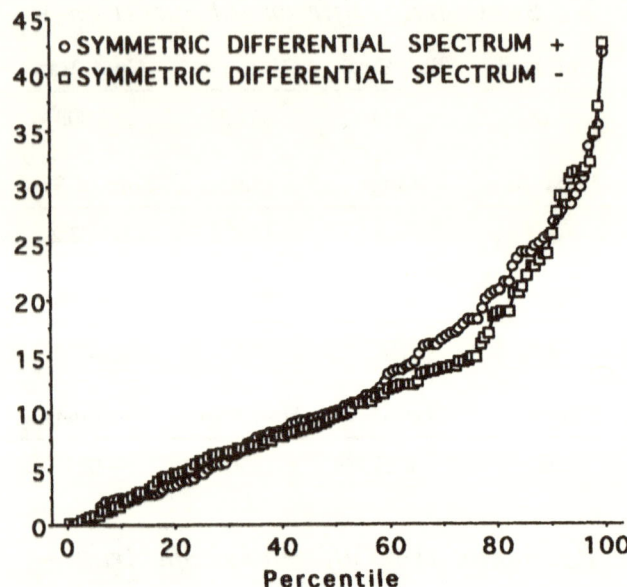

Figure 1.33

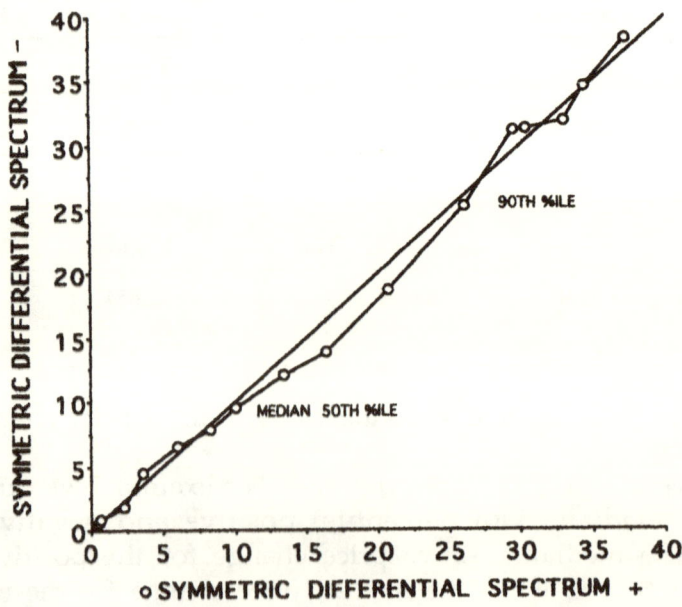

price changes is $9.55. We can plot this information on a scatter diagram, such as the one shown in Figure 1.33. In this case, we move along the horizontal axis (representing the + price changes) until we reach Cell-$10.00 and we move up this column of cells until we reach the Cell-$9.00 on the vertical axis (representing the − price changes). We then increase the count in this cell (10,9) from 0 to 1, as indicated by the point labeled median in this scatter diagram. We can also determine the price changes associated with other deciles, such as the 10th percentile, the 20th, . . . the 90th and plot them on the same scatter diagram. If the positive and negative price changes were perfectly symmetrical, all of the points plotted in this scatter diagram would fall on the diagonal line plotted in the scatter diagram. Clearly, the points do not all fall on the diagonal line. This also suggests that the distributions of positive and negative price changes are fairly similar.

We can also plot the positive and negative price changes as a box and whiskers plot, as shown in Figure 1.34. The horizontal line in the box represents the median price change, while the upper and lower edges of the box represent the 25th and 75th percentiles, respectively. The short horizontal lines represent the 10th and 90th percentiles, respectively. Here again, the box and whiskers plots look somewhat alike. The median and 10th and 90th percentiles are very similar. However, the distribution of the negative price changes is somewhat more compact than the distribution of positive prices, as indicated by the smaller box for the negative price changes. Although suggestive, it is likely that these price changes are symmetrical around the 0 and thus the underlying time series is probably independent and potentially does not contain serial dependencies. But, it is important to note that we should also examine this time series with the other methods described below, as they make different assumptions about the underlying time series.

The differential spectrum for our asymmetric data with $1.00 bin size is shown in Figure 1.35 and the summary statistics are shown in Table 1.8a-b. The cumulative probability density function is shown in Figure 1.36. The distribution of positive and negative price changes does look less symmetrical and this is confirmed when the data are plotted in the scatter diagram format as shown in Figure 1.37 and the box and whiskers plot in Figure 1.38, respectively. Again, these plots are suggestive.

Figure 1.34

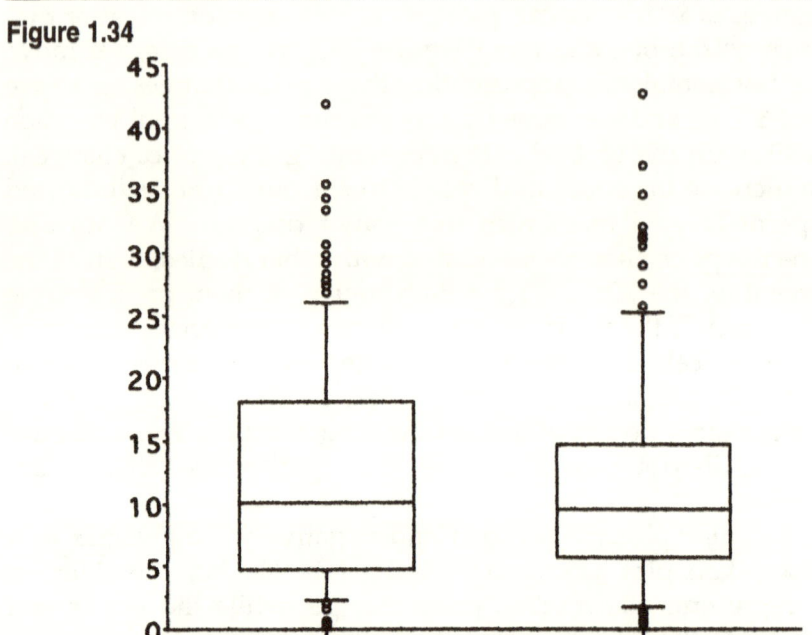

Figure 1.35

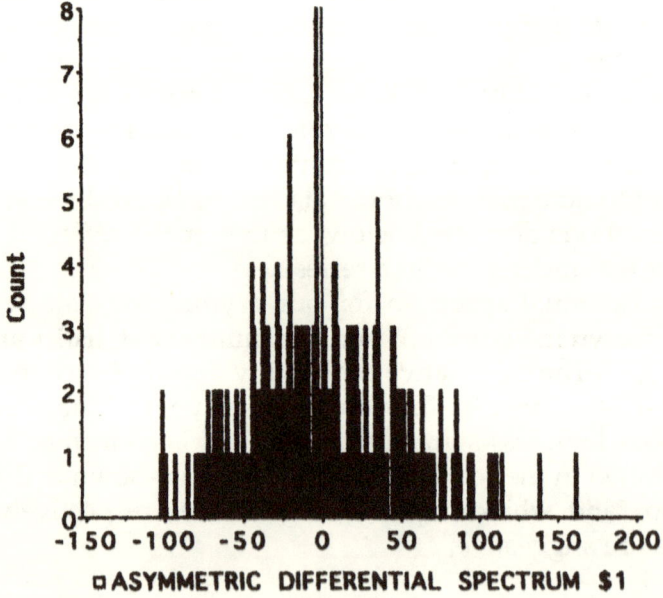

Table 1.8a-b

X_1: *Asymmetric Differential Spectrum +*

Mean	Std. Dev.	Std. Error	Variance	Coef. Var.	Count
40.295	31.145	2.904	969.998	77.293	115

Minimum	Maximum	Range	Sum	Sum of Sqr.	# Missing
1	161.7	160.7	4633.87	297299.375	112

# < 10th %	10th %	25th %	50th %	75th %	90th %
11	7.2	17.7	34.9	56.4	85.6

# > 90th %	Mode	Geo. Mean	Har. Mean	Kurtosis	Skewness
11	•	27.7	13.979	1.587	1.198

X_2: *Asymmetric Differential Spectrum –*

Mean	Std. Dev.	Std. Error	Variance	Coef. Var.	Count
34.446	26.072	2.252	679.73	75.688	134

Minimum	Maximum	Range	Sum	Sum of Sqr.	# Missing
0	103	103	4615.77	249399.087	93

# < 10th %	10th %	25th %	50th %	75th %	90th %
12	3.9	13.6	28.65	50.4	72.24

# > 90th %	Mode	Geo. Mean	Har. Mean	Kurtosis	Skewness
13	•	•	•	–.08	.807

We can also use a formal statistical method to compare the relative distribution of positive and negative price changes. The Chi-Square statistic[3] can be used to compare two distributions and

3. Croxton, F. E. *Elementary Statistics: With Applications in Medicine and the Biological Sciences.* Dover, 1953 provides an excellent introduction to Chi-Square and its uses.

Figure 1.36

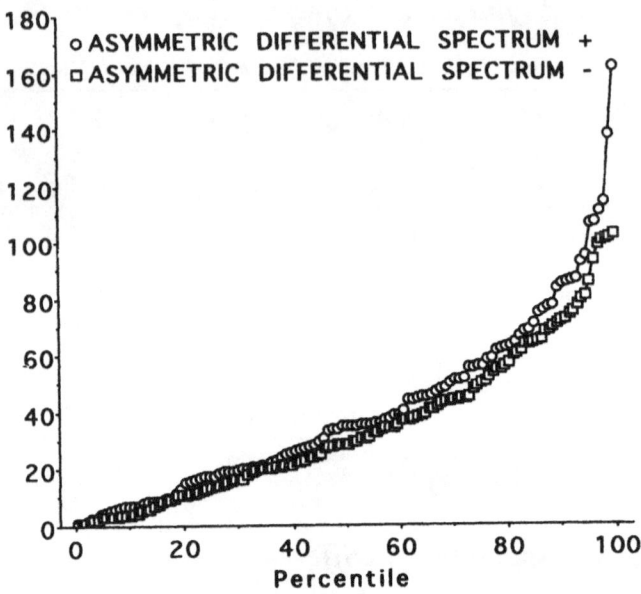

Figure 1.37

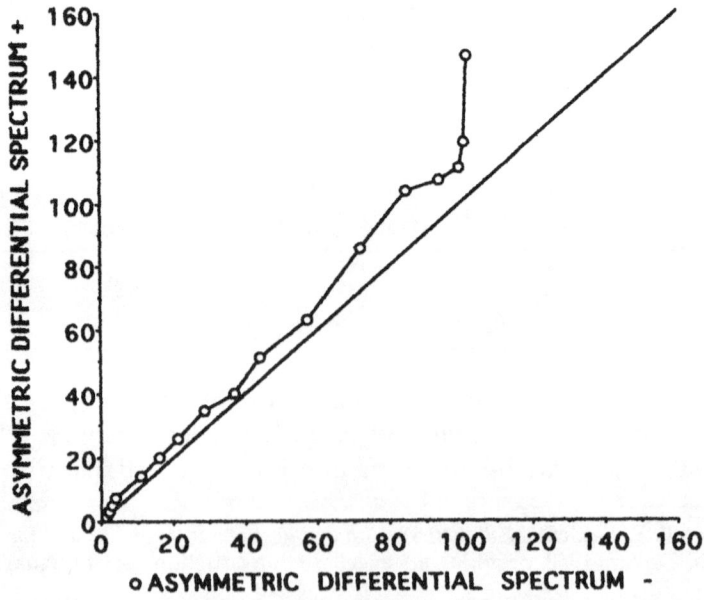

Figure 1.38

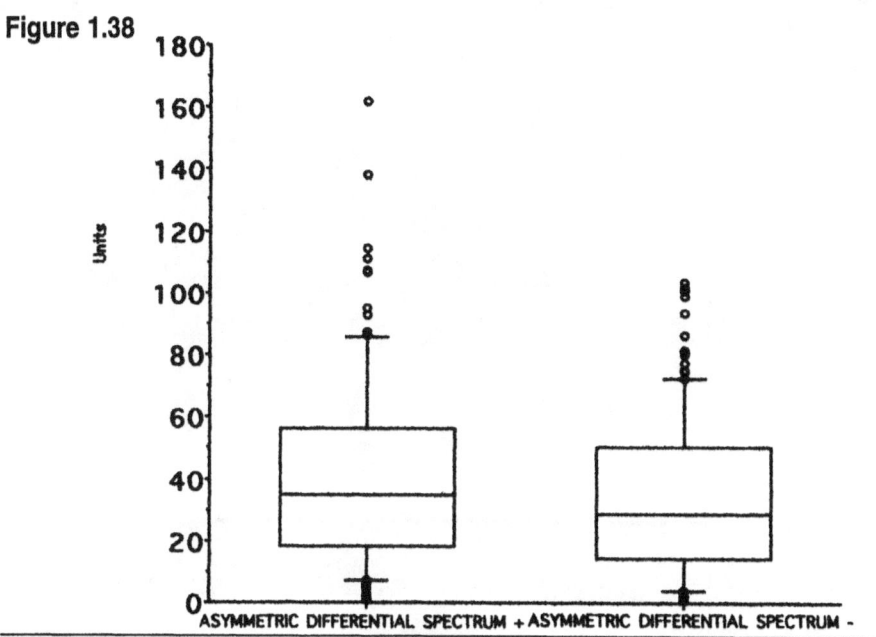

ASYMMETRIC DIFFERENTIAL SPECTRUM + ASYMMETRIC DIFFERENTIAL SPECTRUM -

Figure 1.39

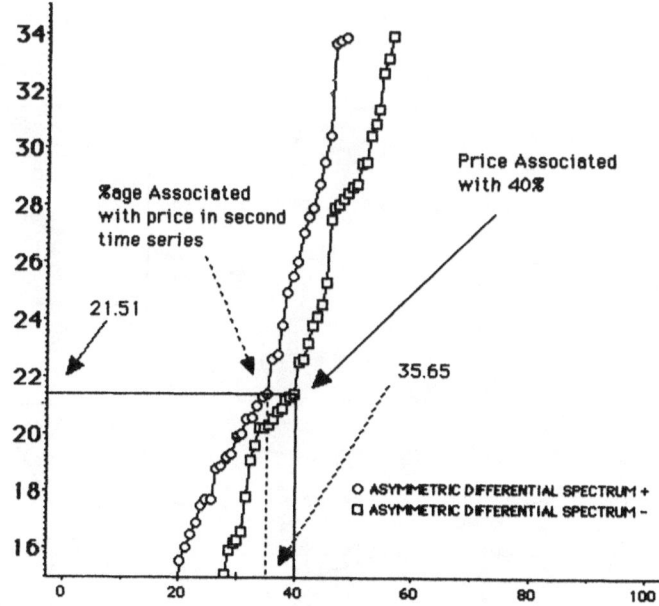

%age Associated with price in second time series

Price Associated with 40%

21.51

35.65

○ ASYMMETRIC DIFFERENTIAL SPECTRUM +
□ ASYMMETRIC DIFFERENTIAL SPECTRUM -

Figure 1.40

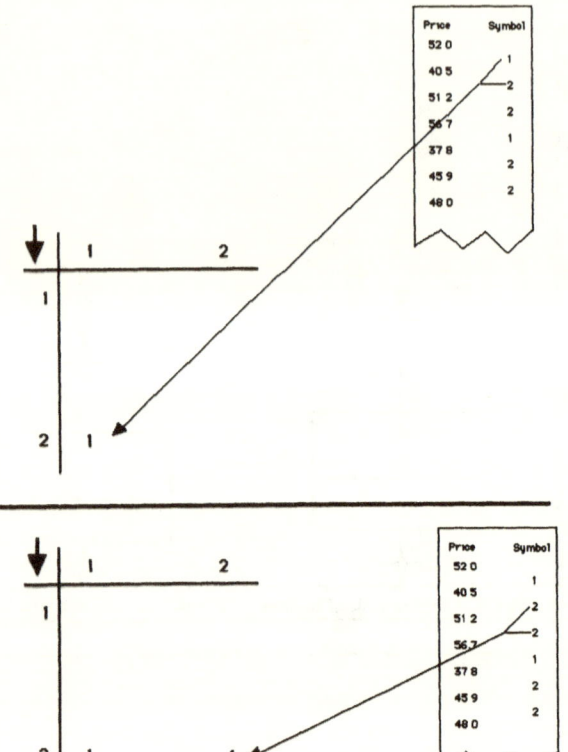

Figure 1.41

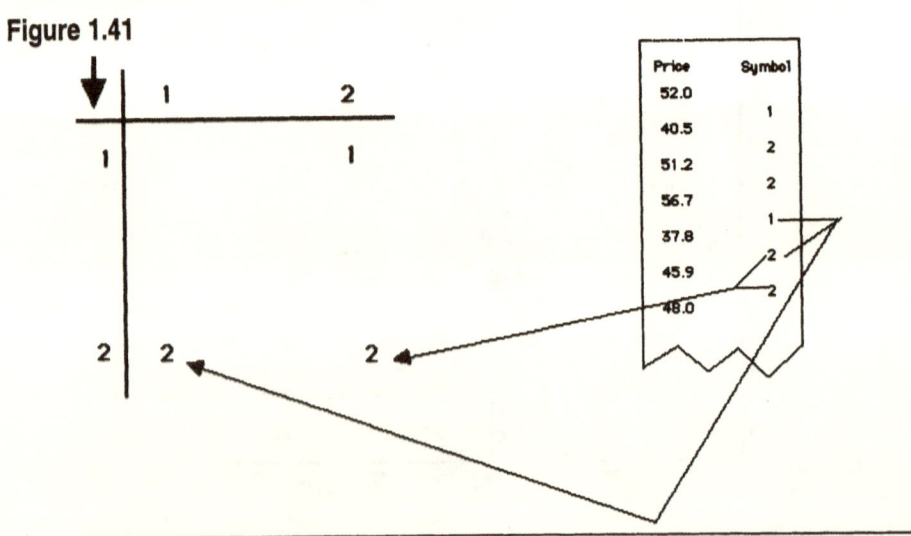

determine if they are the same or different. We will compare the two probability density functions using the quantile approach. We will use the decile points; that is, at 10, 20, 30, ... 90 percentiles, as the quantiles for these calculations. The first step is illustrated in Figure 1.39. This figure is a *blowup* of a portion of Figure 1.36. In this illustration, we determine the price associated with the 40th percentile by drawing a vertical line from the horizontal axis to the cumulative probability density function for the negative half of the differential spectrum. At the level that the vertical line touches the cumulative probability density function, we draw a line horizontal to the vertical axis. The point where the horizontal line intersects with the vertical line represents the price associated with the 40th percentile. In this case, that price is $21.51. We then find that price on the second probability density function (i.e., the positive half of the differential spectrum). We do this by following this horizontal line back to the second probability density function. This is indicated by the dashed vertical line on Figure 1.41. We then determine the percentile associated with the price. In this case, it is 35.65. We repeat this process at each decile; that is, at 10, 20, 30, 40, 50, 60, 70, 80, and 90th percentile.

The Chi-Square statistic has the following form (Equation 1.1):

$$X^2 = \sum_{1}^{n} \frac{[(O)bserved - (E)xpected]^2}{(E)xpected}$$

To calculate the Chi-Square statistic, we will need a matrix that will have seven columns, labeled Percentile, (E)xpected, (O)bserved, (O–E), (O–E)^2, (O–E)^2 / E, and Sum, as shown in Table 1.9. The matrix will have nine active rows, labeled 10, 20, ... 90th percentile. The (E)xpected values are all 10%, because we used the decile points as the basis of this calculation. The (O)bserved values (Table 1.9, column 3) are probabilities drawn from second cumulative probability density function (see Figure 1.41) and are shown in Table 1.10, column 4. The Chi-Square is 9.05, as shown in Table 1.9, column 7. The degrees of freedom for this calculation are n–1, where n = the number of deciles (9). Therefore the degrees of freedom are 9–1 = 8. A table of critical values for the Chi-Square statistic is shown in Appendix A. We enter the Chi-Square table at 8 degrees of freedom and note that the critical value is 15.51 (p = 0.05). This value is larger than our calculated Chi-Square, which strongly suggests that the positive

Table 1.9

	A	B	C	D	E	F	G
1		DS -	DS + (O)				
2	PERCENTILE	(E)XPECTED	(O)BSERVED	(O-E)	(O-E)^2	((O-E)^2)/E	SUM
3							
4	10	10.00	13.92	3.92	15.37	1.54	
5	20	10.00	5.22	-4.78	22.85	2.28	3.82
6	30	10.00	12.17	2.17	4.71	0.47	4.29
7	40	10.00	8.7	-1.30	1.69	0.17	4.46
8	50	10.00	13.91	3.91	15.29	1.53	5.99
9	60	10.00	5.22	-4.78	22.85	2.28	8.28
10	70	10.00	12.71	2.71	7.34	0.73	9.01
11	80	10.00	9.56	-0.44	0.19	0.02	9.03
12	90	10.00	9.57	-0.43	0.18	0.02	9.05
13							
14						CHI-SQUARE	9.05

Table 1.10

Percentile Associated with Prices at Deciles in 1st CPDF

	A	B	C	D
1	%ile in 1st CPDF	Price	%ile in 2nd CPDF	Difference
2	10	3.9	4.34	
3	20	10.99	18.26	13.92
4	30	16.27	23.48	5.22
5	40	21.51	35.65	12.17
6	50	28.65	44.35	8.7
7	60	37.02	58.26	13.91
8	70	44.33	63.48	5.22
9	80	57.18	75.65	12.17
10	90	72.24	85.21	9.56
11	100	101.74	94.78	9.57

and negative Delta-1 price changes are not significantly different from each other and thus the underlying time series is probably independent and does not contain serial dependencies. If our calculated Chi-Square value had been higher than the tabled value, then this would indicate that the positive and negative Delta-1 price changes were significantly different from each other and thus the underlying time series probably contained serial dependencies.

Based on these results, however, the asymmetric time series would not be a good candidate for using conventional technical analysis methods to attempt to develop a trading strategy. This is because these methods assume that the time series does contain serial dependencies. But, as noted above, it is appropriate to test the time series with other methods for detecting dependence.

Relative Price Change

While valuable as a screening tool, the differential spectrum does not allow us to determine what types of serial dependencies are present. On the other hand, the relative price change (RPC) method allows us to determine if serial dependencies exist in our time series, but it also allows us to describe them. An additional advantage of the relative price change method is that it allows us to determine the duration of a *temporal window* during which the prices or price changes are not independent (see below). The relative price change method makes minimal assumptions about the underlying time series.

We will translate the raw prices contained in our time series into a series of arbitrary symbols based on an unvarying rule. We will examine sequential pairs of prices and determine if the first price in the pair is larger than the second, in which case we will record a 1. If the first price is smaller than the second, we will record a 2. Thus a 2 represents a price increase, while a 1 represents a price decrease. This is illustrated in Figure 1.40, top right. The first price in our symmetrical time series is $52.00 and the second is $40.50. The first is larger than the second, so we encode this sequential relationship as a 1. The second price in this time series is $40.50 and the third is $51.20, so we encode this relationship as a 2. We continue this process by examining price3 and 4, price4 and 5, etc. until we reach the last pair of prices. We have created a new time series that consists of a long series of 1s and 2s.

We will now collect these symbols (i.e., 1s and 2s) into a series of transition matrices. This is a relatively simple and straightforward process. For example, the digram transition matrix, as shown in Figure 1.40, top left, allows us to specify how often a 1 is followed by a 1 or a 2 and how often a 2 is followed by a 1 or a 2. We will construct a matrix with four cells as shown in the figure. By convention, the horizontal axis represents the first symbol in a pair and the vertical axis, the second. The small arrow in the upper left hand corner of the matrix means 'followed by'. In our time series, a 1 is followed by a 2, so we increase the count in Cell-1,2 (a 1 followed by a 2) from 0 to 1, as indicated by the 1 in that in Cell-1,2 in the matrix. The next pair of symbols consists of a 2 followed by a 2, so we increase the count in Cell-2,2 from 0 to 1, as indicated by the small 1 in Cell-2,2 (see Figure 1.40, bottom left). The third pair of symbols consists of a 2 followed by a 1, so we increase the count in Cell-2,1 from 0 to 1 (Figure 1.41-no arrow). The next pair of symbols is a 1 followed by a 2, so we increase the count in Cell-1,2 from 1 to 2 as shown in Figure 1.41, left. The next pair of symbols is a 2 followed by a 2, so we increase the count in Cell-2,2 from 1 to 2, as shown in the figure. We continue this process until we reach the final pair of symbols.

Now we want to compare the digram transition matrix we just generated with one that was generated under the assumption of independence. If we do not allow ties to occur, then the matrix generated under the assumption of independence is relatively easy to construct and does not depend on the probability density function of the original time series.[4] On the other hand, if we allow ties to occur, the matrix generated under the assumption of independence does depend on the underlying probability density function of the time series and the calculations become somewhat complex and involved. The formulas for calculating the frequency of occurrence of the nine possible digrams (1,1; 1,0; 1,2; 0,1; 0,0; 0,2; 2,1; 2,0; 2,2, where a 0 is defined as a tie) are shown in the appendix to this chapter.[5]

---

4. One of my former associates, Dr. Heinz von Foerster, a well-known cyberneticist, provided the first proof of this phenomenon for digrams through pentagrams in: Saxena, K., von Foerster, H., and Wolf, D. "Two Theorems Regarding Interspike Interval Histograms," Report #722, 1971/72, pp. 127–41, Biological Computer Laboratory, University of Illinois, Urbana. The other former colleagues generalized von Foerster's results to higher order n-grams in Brudno, S., and Marczynski, T. "Temporal Patterns, Their Distribution and Redundancy in Trains of Spontaneous Neuronal Spike Intervals of the Feline Hippocampus Studies with a Non-parametric Technique." *Brain Research* 1977, vol. 126, pp. 65–89.

5. The formulas for determining the probability of occurrence of each of the nine digrams under the assumption of independence has been described in: Sherry, C. J. and Marczynski, T. J. "A New Analysis of Neuronal Interspike Intervals Based on Inequality Tests." *Intern. J. Neurosci.* 1972, vol. 3, pp. 259–70. We have not as yet generalized these formulas for higher order patterns. We have, however, developed a shuffling technique. It is described in: Sherry, C. J., Barrow, D. L., and Klemm, W. R. "Serial dependencies and Markov properties of neuronal interspike intervals from rat cerebellum." *Brain Res Bull.* 1982, vol. 8, pp. 169–82.

A tie is defined as a price that is followed by a price of the same magnitude. For example, the first price in our symmetric time series is $52.00. If the second price was also $52.00, then we have a tie. We can resolve ties in some arbitrary manner, such as flipping a coin. If the number of ties is small (< 10% of all comparisons) and if we resolve them in an arbitrary manner, then the probability of a 1,2 and a 2,1 is 0.333 and the probability of a 1,1 and a 2,2 is 0.1667.

We can convert these probabilities into frequencies by multiplying them by the total number of pairs in our sample. We can then compare the two transition matrices using Chi-Square. The Chi-Square calculations for the symmetric time series are shown in Table 1.11, while the calculations for the asymmetric time series are shown in Table 1.12. The Chi-Square is 1.03 in Table 1.11 and 5.07 in Table 1.12. Neither of these Chi-Square values are statistically significant. This means that the two matrices were similar and thus the underlying time series did not contain serial dependencies. This confirms the findings of the differential spectrum. **Normally, we would stop the analysis at this point. But, for the purpose of illustration, we will continue the analysis.**

We now want to determine how often each digram is followed by a 1 or a 2. We collect our symbol time series into a trigram transition matrix as shown in Figure 1.42. For example, the first digram is 1,2 and it is followed by a 2, so we increase the Cell-1,2,2 in our

Table 1.11

	A	B	C	D	E	F	G	H	I
1	n	Digram	Expec Prob	(E)xpec Freq	(O)bser Freq	(E-O)	(E-O)^2	((E-O)^2)/E	SUM
2	248	11	0.1667	41.3416	47	-5.6584	32.01749056	0.774461815	
3		12	0.3333	82.6584	80	2.6584	7.06709056	0.085497548	0.859959363
4		21	0.3333	82.6584	80	2.6584	7.06709056	0.085497548	0.945456912
5		22	0.1667	41.3416	41	0.3416	0.11669056	0.002822594	1.03095446
6									
7									
8								Chi-Square	1.033777054

Table 1.12

	A	B	C	D	E	F	G	H	I
1	n	Digram	Expec Prob	(E)xpec Freq	(O)bser Freq	(E-O)	(E-O)^2	((E-O)^2)/E	SUM
2	248	11	0.1667	41.3416	53	-11.6584	135.9182906	3.287688202	
3		12	0.3333	82.6584	81	1.6584	2.75029056	0.033272971	3.320961173
4		21	0.3333	82.6584	81	1.6584	2.75029056	0.033272971	3.354234144
5		22	0.1667	41.3416	33	8.3416	69.58229056	1.683105892	3.387507115
6									
7									
8								Chi-Square	5.070613008

Figure 1.42

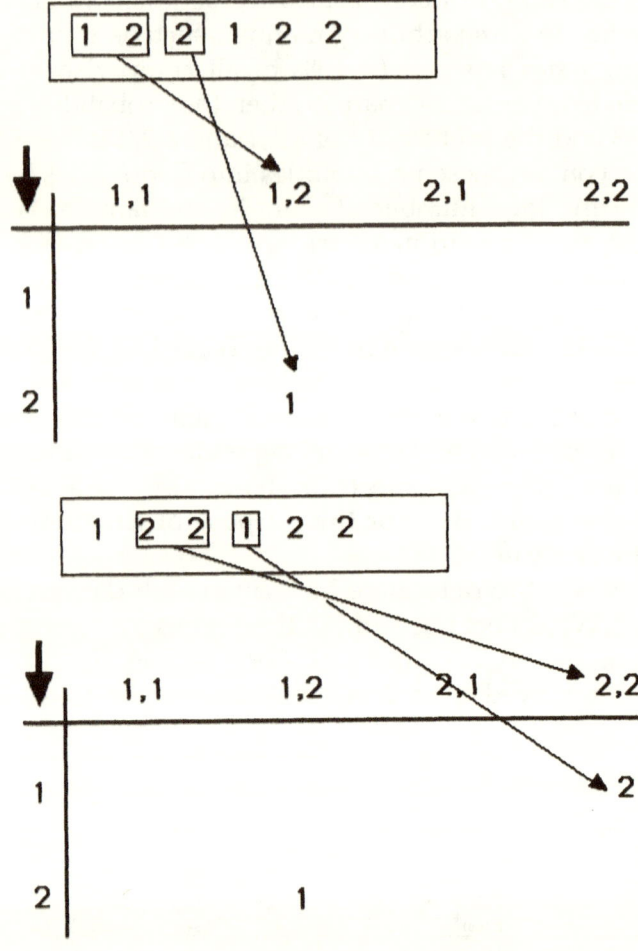

trigram transition matrix from 0 to 1, as indicated by the small 1 in that cell (Figure 1.42 top). The second digram is 2,2 which is followed by a 1, so we increase the count in Cell-2,2,1 from 0 to 1, as indicated by the small 2 in the figure (bottom).

The trigram transition matrix for the symmetric time series is shown in Table 1-13 and the matrix for the asymmetric is shown in Table 1.14. The probabilities associated with the independence case are shown in column 3 in each matrix. The Chi-Square values are also not significant, which confirms that these time series are independent

Table 1.13

	A	B	C	D	E	F	G	H	I
1	n	Tngram	Expec Prob	(E)xpec Freq	(O)bser Freq	(E-O)	(E-O)^2	((E-O)^2)/E	SUM
2	247	111	0.04167	10.29249	15	-4 70751	22.1606504	2.15308933	
3		112	0.125	30.875	32	-1.125	1.265625	0.0409919	2.19408123
4		121	0.20833	51.45751	47	4.45751	19.8693954	0.38613208	2.58021331
5		122	0.125	30.875	33	-2.125	4.515625	0.14625506	2.72646837
6		211	0.125	30.875	32	-1.125	1.265625	0.0409919	2.76746028
7		212	0.20833	51.45751	47	4.45751	19.8693954	0.38613208	3.15359236
8		221	0.125	30.875	33	-2.125	4.515625	0.14625506	3.29984742
9		222	0.04167	10.29249	8	2.29249	5.2555104	0.51061603	
10									
11								Chi-Square	3 81046345

Table 1.14

	A	B	C	D	E	F	G	H	I
1	n	Tngram	Expec Prob	(E)xpec Freq	(O)bser Freq	(E-O)	(E-O)^2	((E-O)^2)/E	SUM
2	247	111	0.04167	10.29249	17	-6.70751	44.9906904	4.37121536	
3		112	0.125	30.875	36	-5.125	26.265625	0.8507085	5.22192386
4		121	0.20833	51.45751	56	-4.54249	20.6342154	0.40099522	5.62291908
5		122	0.125	30.875	25	5.875	34.515625	1.11791498	6.74083406
6		211	0.125	30.875	36	-5.125	26.265625	0.8507085	7.59154256
7		212	0.20833	51.45751	45	6.45751	41.6994354	0.81036637	8.40190893
8		221	0.125	30.875	24	6.875	47.265625	1.53087045	9.93277937
9		222	0.04167	10.29249	8	2.29249	5.2555104	0.51061603	
10									
11								Chi-Square	10.4433954

and probably do not contain serial dependencies. **Normally, when our Chi-Square calculations yield a nonsignificant result, we would stop the analysis at that point. But, for the purpose of illustration, we will continue the analysis.**

The tetragram transition matrices are generated in the same manner. The transition matrices for the symmetric and asymmetric tetragrams are shown in Tables 1.15 and 1.16, respectively. The theoretical probabilities are shown in column 3 of these tables.

The methods used for determining the presence and duration of a *temporal window* will be described below.

Category Price Change

We may want to determine if our time series has specific types of dependencies. For example, we might want to determine if high prices tend to follow high prices or if low prices tend to follow low prices. We can do this by dividing time series into approximately equal quarters, so that each quarter contains about the same number of prices. We can do this by collecting our time series into a frequency histogram (see Figure 1.43, top). Then we divide the fre-

Table 1.15

	A	B	C	D	E	F	G	H	I
1	n	Tetragram	Expected P	(E)xpected f	(O)bser f	(E-O)	(E-O)^2	((E-O)^2)/E	Sum
2	246	1111	0.00833	2.049	6	-3.951	15.609	7.617	
3		1112	0.03333	8.199	9	-0.801	0.641	0.078	7.695
4		1121	0.075	18.450	17	1.450	2.102	0.114	7.809
5		1122	0.05	12.300	15	-2.700	7.290	0.593	8.402
6		1211	0.075	18.450	22	-3.550	12.603	0.683	9.085
7		1212	0.13333	32 799	24	8.799	77.426	2.361	11.446
8		1221	0.096166	23.657	28	-4.343	18.863	0.797	12.243
9		1222	0.03333	8.199	5	3.199	10.235	1.248	13.491
10		2111	0.03333	8.199	9	-0.801	0.641	0.078	13.570
11		2112	0.09166	22.548	23	-0.452	0.204	0.009	13.579
12		2121	0.13333	32.799	30	2.799	7.835	0.239	13.817
13		2122	0.075	18.450	17	1.450	2.102	0.114	13.931
14		2211	0.05	12.300	10	2 300	5.290	0.430	14.362
15		2212	0.075	18.450	23	-4.550	20.703	1.122	15.484
16		2221	0.03333	8.199	5	3.199	10.235	1.248	16.732
17		2222	0.00833	2.049	3	-0.951	0.904	0.441	17.173
18									
19									
20				N=	246			Chi-Square	17.173

Table 1.16

	A	B	C	D	E	F	G	H	I
1	n	Tetragram	Expected P	(E)xpected f	(O)bser f	(E-O)	(E-O)^2	((E-O)^2)/E	Sum
2	246	1111	0.00833	2.049	5	-2.951	8.707	4.249	
3		1112	0.03333	8.199	12	-3.801	14.446	1.762	6.011
4		1121	0.075	18.450	23	-4.550	20.703	1.122	7.133
5		1122	0.05	12.300	13	-0.700	0.490	0.040	7.173
6		1211	0.075	18.450	23	-4.550	20.703	1.122	8.295
7		1212	0.13333	32.799	33	-0.201	0.040	0.001	8.296
8		1221	0.096166	23.657	17	6.657	44.313	1.873	10.170
9		1222	0.03333	8.199	7	1.199	1.438	0.175	10.345
10		2111	0.03333	8.199	12	-3.801	14.446	1.762	12.107
11		2112	0.09166	22 548	24	-1.452	2.107	0.093	12.200
12		2121	0.13333	32.799	33	-0.201	0.040	0.001	12.201
13		2122	0.075	18.450	12	6.450	41.602	2.255	14.456
14		2211	0.05	12 300	12	0.300	0.090	0.007	14.464
15		2212	0.075	18.450	12	6.450	41.602	2.255	16.719
16		2221	0.03333	8.199	7	1.199	1.438	0.175	16.894
17		2222	0.00833	2.049	1	1.049	1.101	0.537	17.431
18									
19									
20				N=	246			Chi-Square	17.431

quency histogram into quarters. We then examine the original time series and encode it into arbitrary symbols, 1, 2, 3, and 4, which represents the four quarters in our frequency histogram. For example, the first price in our symmetric time series is $52.00, which falls into the third quarter, so we encode it as a 3, as shown in Figure 1.43, bottom, right. The next price is $40.50 and it falls in the first quarter, so we encode it as a 1. We repeat this process with each price in our time series and we end up with a new time series, in which the prices are encoded as 1s, 2s, etc. We can then use this new time series to generate a digram transition matrix. Our first symbol is a 3 and it is

Figure 1.43

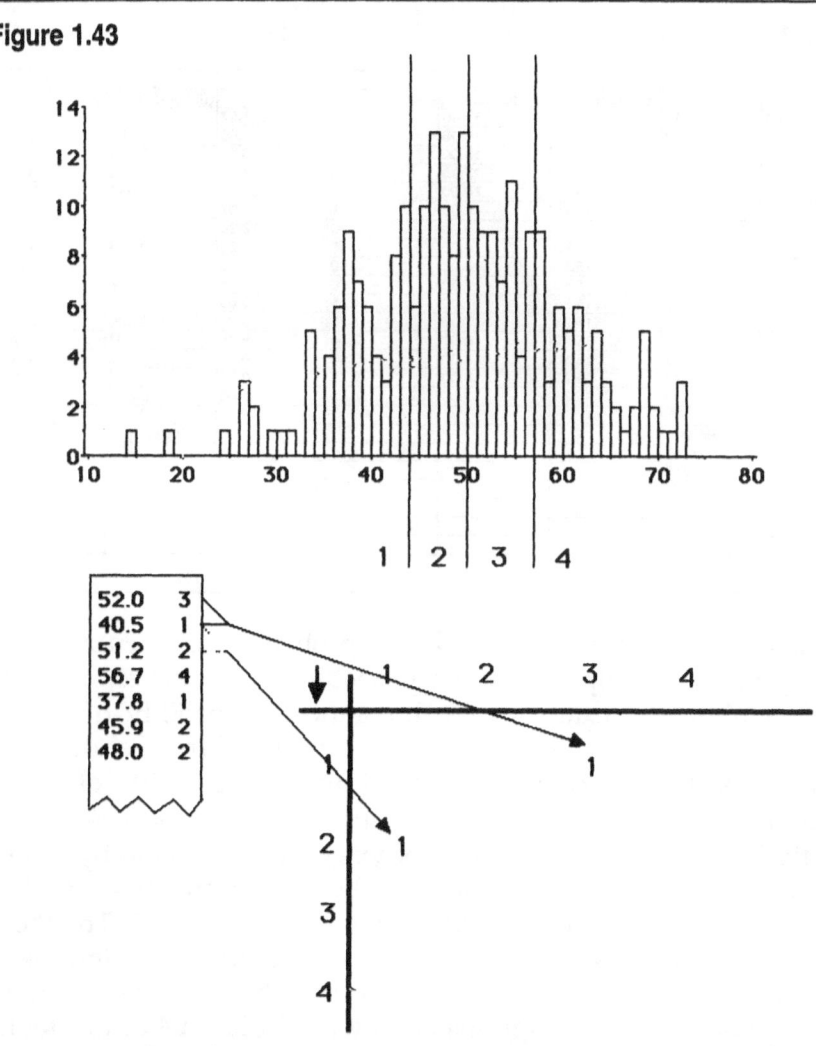

followed by a 1, so we increase the count in Cell-3,1 from 0 to 1, as indicated by the small 1 in that cell. Next, a 1 is followed by a 2, so we increase the count in Cell-1,2 from 0 to 1. We continue this process until we reach the last pair of symbols.

We can then compare this digram transition matrix with one generated under the assumption of independence. It is important to note that each price is used only once in generating our arbitrary symbols (See Figure 1.43, bottom left). If the symbols are distributed independently, then basic probability theory says that the probability

Table 1.17

	A	B	C	D	E	F	G	H
1	n	Digrams	(E)xpected Freq	(O)bserved Freq	(E-O)	(E-O)^2	((E-O)^2)/E	Sum
2	248	11	15.75	17.00	-1.25	1.57	0.099	
3		12	15.75	14.00	1.75	3.06	0.194	0.294
4		13	15.50	15.00	0.50	0.25	0.016	0.310
5		14	15.50	17.00	-1.50	2.25	0.145	0.455
6		21	15.75	10.00	5.75	33.05	2.099	2.554
7		22	15.75	18.00	-2.25	5.07	0.322	2.875
8		23	15.50	19.00	-3.50	12.26	0.791	3.666
9		24	15.50	15.00	0.50	0.25	0.016	3.682
10		31	15.50	18.00	-2.50	6.25	0.404	4.086
11		32	15.50	16.00	-0.50	0.25	0.016	4.102
12		33	15.25	11.00	4.25	18.09	1.186	5.288
13		34	15.25	17.00	-1.75	3.05	0.200	5.488
14		41	15.50	18.00	-2.50	6.25	0.404	5.892
15		42	15.50	15.00	0.50	0.25	0.016	5.908
16		43	15.25	16.00	-0.75	0.56	0.037	5.944
17		44	15.25	13.00	2.25	5.08	0.333	
18								
19	Symbols	Probabilities					Chi-Square	6.277
20	1	0.252						
21	2	0.252						
22	3	0.248						
23	4	0.248						
24								
25	n	248						

of occurrence of a digram is simply the probability of occurrence of the first symbol multiplied by the probability of occurrence of the second symbol. For example, the theoretical probability of occurrence of the digram 1,1 under the assumption of independence is simply the probability of a 1 multiplied by the probability of a 1 or 0.252×0.252 (see Table 1.17, lower left) which equals 0.0635. The theoretical frequency of occurrence of this digram is found by multiplying this probability by the total number of digrams (n=248). The theoretical frequency of occurrence of this digram is 15.75. The theoretical frequency of occurrence of each of the other 15 digrams is shown in column 3 of Table 1.17, while the observed frequency of occurrence of each of the digrams is shown in column 4 of this table. The table also contains the Chi-Square calculations for comparing the observed frequencies and expected frequencies (based on the assumption of independence). The Chi-Square is not significant, so when the prices are parsed in this manner, their distribution is independent. The summary statistics for the four quarters are shown in Table 1.18a-d.

The Category Price Change method can be used to parse the time series in other ways. For example, we might determine if the distribution of high and low prices shows serial dependencies. We can divide the time series into unequal thirds, so that a 1 represents

Table 1.18a-b

X_1: 1–Symmetric

Mean	Std. Dev.	Std. Error	Variance	Coef. Var.	Count
36.162	5.844	.736	34.152	16.161	63

Minimum	Maximum	Range	Sum	Sum of Sqr.	# Missing
14	42.8	28.8	2278.2	84501.5	163

# < 10th %	10th %	25th %	50th %	75th %	90th %
6	26.94	34.1	37.6	39.875	42.2

# > 90th %	Mode	Geo. Mean	Har. Mean	Kurtosis	Skewness
5	•	35.56	34.745	2.84	−1.613

X_2: 2–Symmetric

Mean	Std. Dev.	Std. Error	Variance	Coef. Var.	Count
46.302	1.858	.234	3.451	4.012	63

Minimum	Maximum	Range	Sum	Sum of Sqr.	# Missing
43.1	49.3	6.2	2917	135275.68	163

# < 10th %	10th %	25th %	50th %	75th %	90th %
6	43.58	44.775	46.5	47.775	48.92

# > 90th %	Mode	Geo. Mean	Har. Mean	Kurtosis	Skewness
6	46.5	46.265	46.228	−1.129	−.086

the prices in the lowest 10% of the distribution and a 3 represents the prices in the highest 10% of the distribution, while a 2 represents the middle 80% of the distribution, as shown in Figure 1.44. The frequency of occurrence of each of the nine digrams is shown in column 3 of Table 1.19, while the frequency of occurrence of these digrams under the assumption of independence is shown in column 2 in this

Table 1.18c-d

X_3: 3–Symmetric

Mean	Std. Dev.	Std. Error	Variance	Coef. Var.	Count
52.621	2.052	.261	4.212	3.9	62

Minimum	Maximum	Range	Sum	Sum of Sqr.	# Missing
49.5	56.5	7	3262.5	171932.85	164

# < 10th %	10th %	25th %	50th %	75th %	90th %
5	49.9	50.8	52.6	54.2	55.52

# > 90th %	Mode	Geo. Mean	Har. Mean	Kurtosis	Skewness
6	•	52.582	52.543	−1.123	.153

X_4: 4–Symmetric

Mean	Std. Dev.	Std. Error	Variance	Coef. Var.	Count
62.782	4.909	.623	24.094	7.818	62

Minimum	Maximum	Range	Sum	Sum of Sqr.	# Missing
56.7	74.8	18.1	3892.5	245849.65	164

# < 10th %	10th %	25th %	50th %	75th %	90th %
6	57.14	58.9	61.3	66.9	69.97

# > 90th %	Mode	Geo. Mean	Har. Mean	Kurtosis	Skewness
6	•	62.599	62.421	−.612	.657

table. The Chi-Square calculations suggest that the high and low prices are independently distributed. The summary statistics for each third are shown in Table 1.20 a-c.

We can use this method to determine if Delta-1 price changes are distributed independently or if they show serial dependencies. We will divide the Delta-1 price changes for our symmetric time

Figure 1.44

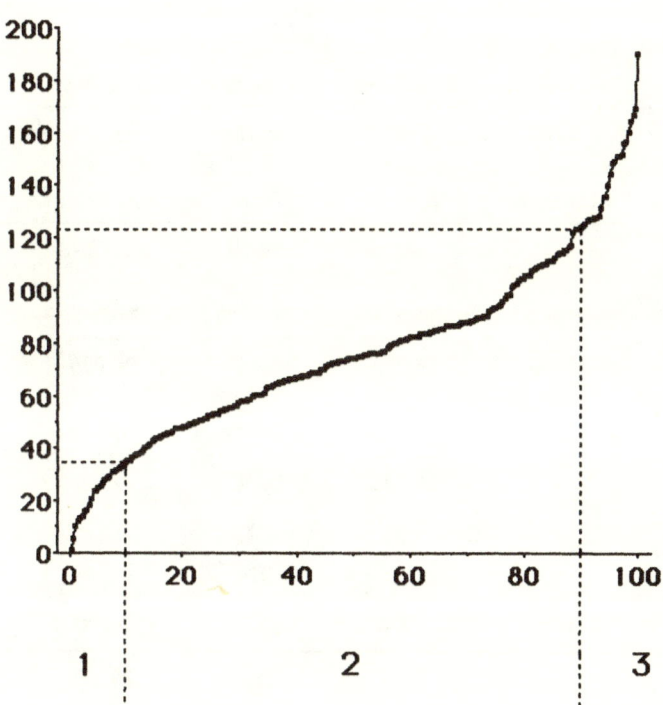

Table 1.19

	A	B	C	D	E	F	G
1	Symbols	Frequency	Probability				
2	1	25	0.1000				
3	2	200	0.8000				
4	3	25	0.1000				
5							
6	Sum	250					
7							
8							
9							
10	Digrams	(E)xpected	(O)bserved	(E-O)	(E-O)^2	((E-O)^2)/E	Sum
11	11	2.50	3	-0.50	0.250	0.100	
12	12	20.00	20	0.00	0.000	0.000	0.100
13	13	2.50	2	0.50	0.250	0.100	0.200
14	21	20.00	19	1.00	1.000	0.050	0.250
15	22	160.00	161	-1.00	1.000	0.006	0.256
16	23	20.00	19	1.00	1.000	0.050	0.306
17	31	2.50	3	-0.50	0.250	0.100	0.406
18	32	20.00	18	2.00	4.000	0.200	0.606
19	33	2.50	4	-1.50	2.250	0.900	
20							
21						Chi-Square	1.506

Table 1.20a-b

X_1: 1–Column 1

Mean	Std. Dev.	Std. Error	Variance	Coef. Var.	Count
22.676	9.407	1.881	88.488	41.483	25

Minimum	Maximum	Range	Sum	Sum of Sqr.	# Missing
1	34.7	33.7	566.9	14978.73	201

# < 10th %	10th %	25th %	50th %	75th %	90th %
2	10.6	15.275	24.7	30.825	33

# > 90th %	Mode	Geo. Mean	Har. Mean	Kurtosis	Skewness
2	•	19.175	11.107	−.578	−.629

X_2: 2–Column 1

Mean	Std. Dev.	Std. Error	Variance	Coef. Var.	Count
74.819	21.829	1.544	476.498	29.176	200

Minimum	Maximum	Range	Sum	Sum of Sqr.	# Missing
35.1	123.5	88.4	14963.73	1214389.089	26

# < 10th %	10th %	25th %	50th %	75th %	90th %
20	46.1	57.5	74.4	87.8	108.1

# > 90th %	Mode	Geo. Mean	Har. Mean	Kurtosis	Skewness
20	•	71.55	68.217	−.689	.234

series into thirds, such that a 1 represents the lowest 10% of the price changes, a 3 represents the highest 10% of price changes, and a 2 represents the middle 80% of the price changes, as shown in Figure 1.45.

The frequency of occurrence of each of the nine digrams under the assumption of independence is shown in column 2, Table 1.21

Table 1.20c

X_3: 3 – Column 1

Mean	Std. Dev.	Std. Error	Variance	Coef. Var.	Count
144.356	17.217	3.443	296.422	11.927	25

Minimum	Maximum	Range	Sum	Sum of Sqr.	# Missing
124.2	189.7	65.5	3608.9	528080.49	201

# < 10th %	10th %	25th %	50th %	75th %	90th %
2	126.2	127.625	144.1	155.525	166.7

# > 90th %	Mode	Geo. Mean	Har. Mean	Kurtosis	Skewness
2	•	143.411	142.506	–.052	.708

Figure 1.45

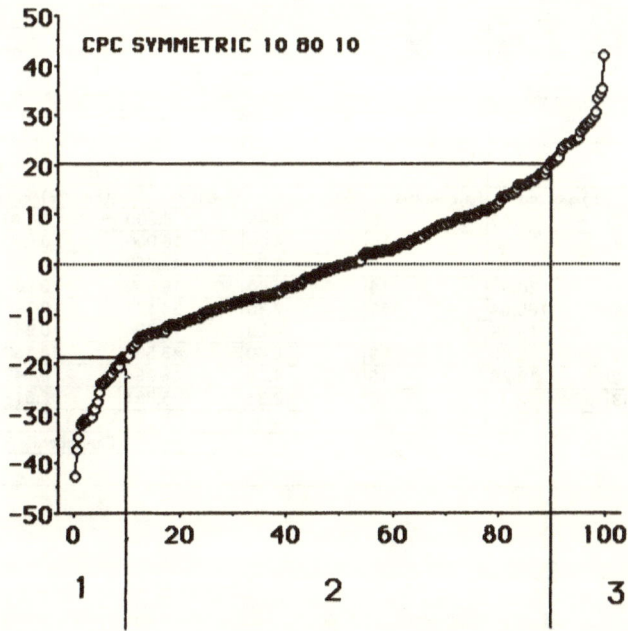

and the observed frequency of these digrams is shown in column 3. The Chi-Square calculation is statistically significant and thus suggests that serial dependencies occur when the price changes are parsed in this manner. The summary statistics for the three thirds are shown in Table 1.22.

If we examine Table 1.21, we will note that an intermediate price change follows an intermediate price change (i.e., 2,2) about as often as one might expect if the price changes were independent. A small price change is never followed by a small price change (i.e., the observed frequency of occurrence of a 1,1 is 0). A large price change also does not follow a large price change (i.e., a 3,3). But, a large price change follows a small price change about three times more often than would occur if the price changes were distributed independently.

We will now use these techniques to analyze *real* time series.

Table 1.21

	A	B	C	D	E	F	G
	A	**B**	**C**	**D**	**E**	**F**	**G**
1	Symbols	Frequency	Probability				
2	1	25	0.1000				
3	2	200	0.8032				
4	3	24	0.0964				
5							
6	Sum	249					
7							
8							
9							
10	Digrams	(E)xpected	(O)bserved	(E-O)	(E-O)^2	((E-O)^2)/E	Sum
11	11	2.49	0	2.49	6.200	2.490	
12	12	20.00	16	4.00	16.000	0.800	3.290
13	13	2.40	9	-6.60	43.560	18.150	21.440
14	21	20.00	16	4.00	16.000	0.800	22.240
15	22	160.64	168	-7.36	54.132	0.337	22.577
16	23	19.28	15	4.28	18.294	0.949	23.526
17	31	2.40	9	-6.60	43.560	18.150	41.676
18	32	19.28	15	4.28	18.294	0.949	42.625
19	33	2.31	0	2.31	5.351	2.313	
20							
21						Chi-Square	44.938

Figure 1.22a-b

X_1: 1–Category Price Change 10 80 10

Mean	Std. Dev.	Std. Error	Variance	Coef. Var.	Count
−26.9	6.3	1.3	39.1	−23.3	25

Minimum	Maximum	Range	Sum	Sum of Sqr.	# Missing
−42.7	−18.8	23.9	−671.3	18963.3	175

# < 10th %	10th %	25th %	50th %	75th %	90th %
2	−34.7	−31.2	−25.8	−21.8	−18.9

# > 90th %	Mode	Geo. Mean	Har. Mean	Kurtosis	Skewness
2	•	•	•	−.2	−.6

X_2: 2–Category Price Change 10 80 10

Mean	Std. Dev.	Std. Error	Variance	Coef. Var.	Count
.1	9.9	.7	97.8	18838.6	200

Minimum	Maximum	Range	Sum	Sum of Sqr.	# Missing
−18.7	20.7	39.4	10.5	19466.3	0

# < 10th %	10th %	25th %	50th %	75th %	90th %
20	−12.6	−8.1	−.3	8.2	14.1

# > 90th %	Mode	Geo. Mean	Har. Mean	Kurtosis	Skewness
20	•	•	•	−1	.2

Table 1.22c

X₃: 3–Category Price Change 10 80 10

Mean	Std. Dev.	Std. Error	Variance	Coef. Var.	Count
27.4	5	1	25.5	18.4	24

Minimum	Maximum	Range	Sum	Sum of Sqr.	# Missing
20.8	41.9	21.1	656.5	18543.6	176

# < 10th %	10th %	25th %	50th %	75th %	90th %
2	21.5	24.1	26.2	29.6	34.5

# > 90th %	Mode	Geo. Mean	Har. Mean	Kurtosis	Skewness
2	24.1	26.9	26.6	1.1	1.1

Standard & Poor's 500 Stock Index   2

The Standard and Poor's 500 Stock Index is important for a number of reasons. First, it is used as a gauge of the stock market in general. It is a market-weighted (i.e., market price/share * number of shares outstanding) index of 500 leading stocks. It measures the percentage change in market value (10) compared to the base period (the years 1941–1943). The stocks fall into four major categories: 400 industrials, 40 utilities, 20 transportations, and 40 financials. These companies are not always the largest in their sectors, but they represent large capitalization stocks with blue chip qualities. The only stock market indicator that is quoted more often is the Dow Jones Industrial Average.

The S&P 500 Stock Index is one of the 12 time series that make up the National Bureau of Economic Research and the U. S. Department of Commerce's Composite Index of Leading Economic Indicators (see Chapter 8 for a discussion of this indicator and its other components). Changes in the indicator lead (come before) changes in the aggregate business cycle (i.e., periods of inflation and recession) by 9–12 months. It is important to keep in mind that the indicator predicts the direction, but not the magnitude, of changes and that the lead times are not necessarily constant. Some economists (and the politicians they influence) believe that they can *manage* the economy to maximize employment, while minimizing the effects of the peaks and troughs of inflation and recession. These economic *planners* seem to believe that these manipulations are essentially independent of each other, especially if they are separated in time by more than a few months. That is, they seem to believe that if they make a manipulation today (i.e., increase or decrease money supply), this manipulation will have little or no impact on manipulations they make 3, 6, or 12 months or more from now.

The S&P 500 Stock Index is also used as a surrogate for the market as a whole when calculating β, which is used to compare the performance of a specific stock with the market as a whole. It is also commonly used to calculate α, which is a stock's unsystematic return (i.e., its average return that is independent of the market's return). Last, but certainly not least, the S&P 500 is itself traded on both the options and futures markets.

The closing 'price' for the S&P 500 for 04/21/82 to 09/30/93 is shown in Figure 2.1. The frequency histogram of 'prices' is shown in Figure 2.2, where the bin size is $1.00. It is important to note the difference between these two plots. The plot in Figure 2.1 is a sequential plot; that is, the prices are presented in the order in which they occurred. On the other hand, the plot in Figure 2.2 is a frequency histogram. It tells us how often a price or price change occurred, but not the order in which they occurred. The cumulative probability density function shown in Figure 2.3 a, is also a nonsequential plot. But it can provide us with some valuable information. For example, the plot is essentially a straight line, running from the lower left corner of the plot (origin) to the upper right corner. This suggests that the various prices tended to occur equally often. This is confirmed by looking at Figure 2.2. The count in most of the bins is between 5 and 15, except for the bins around $170, which occur more often. The cumulative probability density plot reflects this by the 'bump' that occurs in the plot just above the parallel horizontal lines labeled 'sd' and '25th percentile'. Further, we know by Tchebysheff's Theorem that the mean (μ) $\pm$ 1 standard deviation (sd) contains 68% of the prices. The mean and the median are close together, which suggests that the distribution of prices is symmetrical. We can make this plot even more useful by transforming the prices into 'z' or standard scores. We do this by subtracting each individual price from the mean price and dividing the result by the standard deviation (see summary statistics in Table 2.1 a). A cumulative probability density plot of the 'z' scores for the S&P 500 is shown in Figure 2.3 b.

Figure 2.1

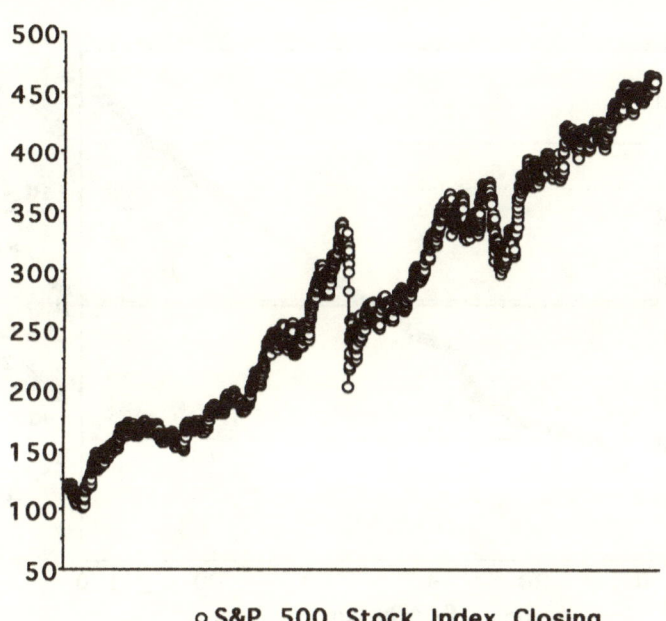

o S&P 500 Stock Index Closing

Figure 2.2

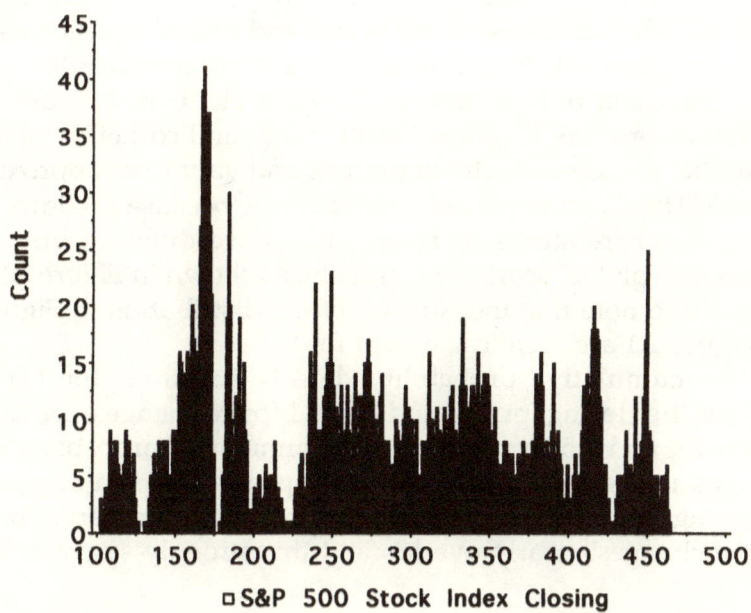

□ S&P 500 Stock Index Closing

Figure 2.3a

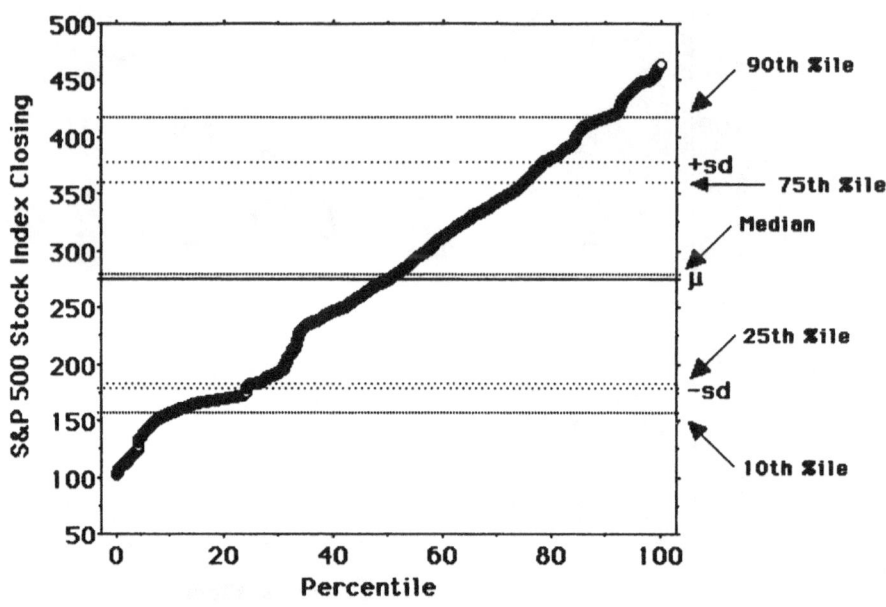

The 'z' scores have a mean of '0' and a standard deviation of 1 (see summary statistics in Table 2.1 b). The shape of the cumulative probability density functions of the prices and the 'z' transformation of prices are identical. You can see this if you compare the plots in Figure 2.3 a and b. It is confirmed by the fact that the kurtosis and skewness scores are identical (lower right hand corner of Table 2.1a-b). But the 'z' scores are dimensionless and vary from approximately +3 to –3. This is an important characteristic because it allows you to directly compare stocks or futures that have different price scales. We can also plot 'z' scores sequentially, as shown in Figure 2.3 c. It is important to note that the 'shapes' of the distribution in Figure 2.3 c and Figure 2.1 are identical, except for the scale.

The cumulative probability density functions for Half1 and Half2 for the closing 'prices' and Delta-1 'price changes' are shown in Figures 2.4 and 2.5 respectively. The cumulative probability density functions in Figure 2.4 are essentially parallel, which suggests that the closing 'prices' of the S&P 500 are probably stationary. The Delta-1 'price changes' clearly overlap and this strongly suggests that the

Table 2.1a

X_1: S&P 500 Stock Index Closing

Mean	Std. Dev.	Std. Error	Variance	Coef. Var.	Count
278.9	99.4	1.8	9879.6	35.6	2896

Minimum	Maximum	Range	Sum	Sum of Sqr.	# Missing
102.4	464.1	361.7	807835.3	253945951.9	0

# < 10th %	10th %	25th %	50th %	75th %	90th %
290	156.7	182.9	274.5	360.4	417.4

# > 90th %	Mode	Geo. Mean	Har. Mean	Kurtosis	Skewness
290	•	260	240.7	−1.2	.1

Figure 2.3b

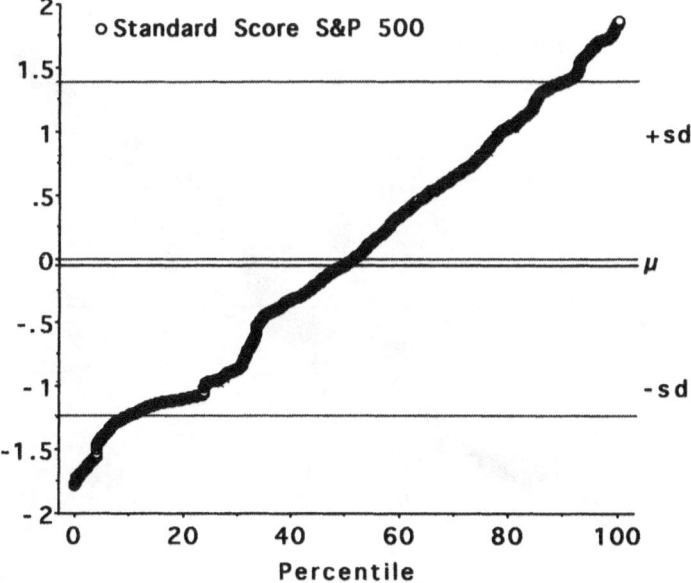

Table 2.1b

X_1: *Standard Score*

Mean	Std. Dev.	Std. Error	Variance	Coef. Var.	Count
–3.349E–6	1	.019	1	–2.986E7	2896

Minimum	Maximum	Range	Sum	Sum of Sqr.	# Missing
–1.776	1.863	3.639	–.01	2895.002	0

# < 10th %	10th %	25th %	50th %	75th %	90th %
290	–1.23	–.966	–.045	.819	1.393

# > 90th %	Mode	Geo. Mean	Har. Mean	Kurtosis	Skewness
290	•	•	•	–1.193	.102

Figure 2.3c

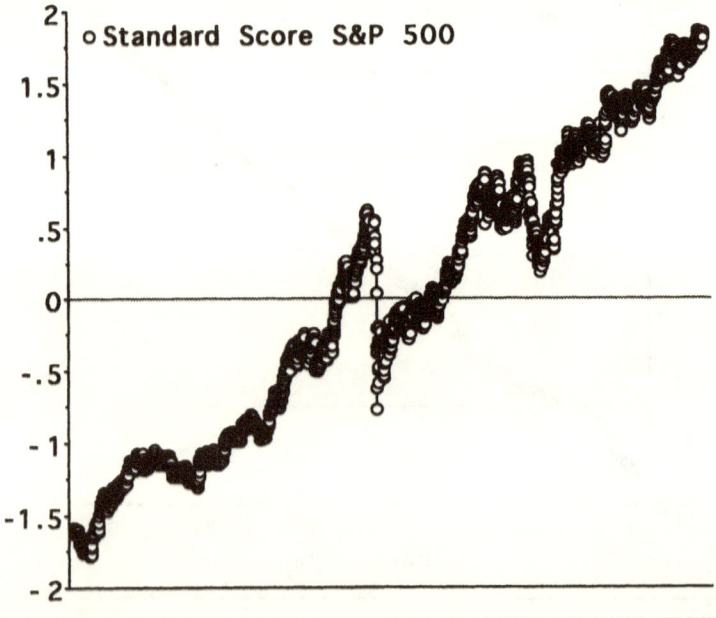

Figure 2.4

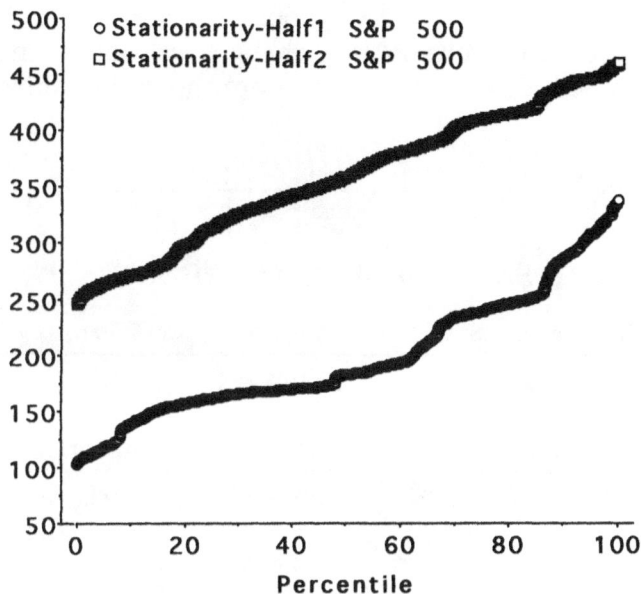

Figure 2.5

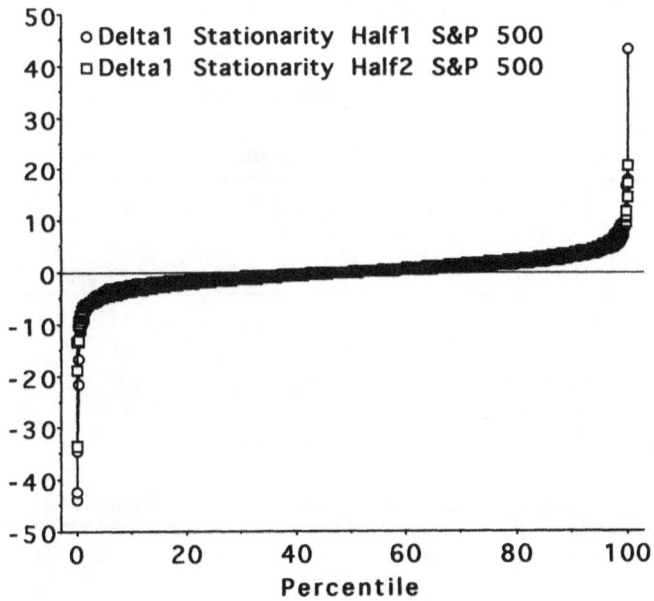

Delta-1 'price changes' are stationary. The summary statistics are shown in Tables 2.2 a-b and 2.3 a-b, respectively.

The Odd and Even cumulative probability density functions for the closing 'prices' and Delta-1 'price changes' are shown in Figures 2.6 and 2.7, while the summary statistics are shown in Tables 2.4 a-b

Table 2.2a-b

X_1: Stationarity–Half1 S&P 500

Mean	Std. Dev.	Std. Error	Variance	Coef. Var.	Count
198.1	55.8	1.5	3108.2	28.1	1448

Minimum	Maximum	Range	Sum	Sum of Sqr.	# Missing
102	341.2	239.2	286821	61311273.6	0

# < 10th %	10th %	25th %	50th %	75th %	90th %
144	138.2	161.2	182.8	239.5	288.6

# > 90th %	Mode	Geo. Mean	Har. Mean	Kurtosis	Skewness
145	•	190.6	183.6	–.4	.6

X_2: Stationarity–Half2

Mean	Std. Dev.	Std. Error	Variance	Coef. Var.	Count
359.6	59.9	1.6	3584.1	16.6	1448

Minimum	Maximum	Range	Sum	Sum of Sqr.	# Missing
244.8	464.1	219.3	520754.5	192468918.2	0

# < 10th %	10th %	25th %	50th %	75th %	90th %
145	271.2	314.1	359.9	412.6	422.9

# > 90th %	Mode	Geo. Mean	Har. Mean	Kurtosis	Skewness
145	•	354.5	349.2	–1.1	–.1

Table 2.3a-b

X_1: Delta-1 Stationarity Half1 S&P 500

Mean	Std. Dev.	Std. Error	Variance	Coef. Var.	Count
.1	3.4	.1	11.8	3795.1	1448

Minimum	Maximum	Range	Sum	Sum of Sqr.	# Missing
−43.9	43.3	87.2	131.1	17100.7	0

# < 10th %	10th %	25th %	50th %	75th %	90th %
145	−2.4	−1.1	.2	1.3	3

# > 90th %	Mode	Geo. Mean	Har. Mean	Kurtosis	Skewness
145	•	•	•	60.8	−2.2

X_2: Delta-1 Stationarity Half2

Mean	Std. Dev.	Std. Error	Variance	Coef. Var.	Count
.1	3.3	.1	10.6	2216.3	1447

Minimum	Maximum	Range	Sum	Sum of Sqr.	# Missing
−33.7	20.6	54.3	213.1	15429.5	1

# < 10th %	10th %	25th %	50th %	75th %	90th %
145	−3.5	−1.6	.1	2	3.8

# > 90th %	Mode	Geo. Mean	Har. Mean	Kurtosis	Skewness
145	.8	•	•	10.7	−.6

and 2.5 a-b, respectively. These strongly suggest that both the 'prices' and Delta-1 'price changes' are random.

The differential spectrum is shown in Figure 2.8, where the bin size is $1.00. The summary statistics are shown in Table 2.6. The differential spectrum looks asymmetrical. But, if we parse it into positive and negative halves as shown in Table 2.7 a-b, the two

Figure 2.6

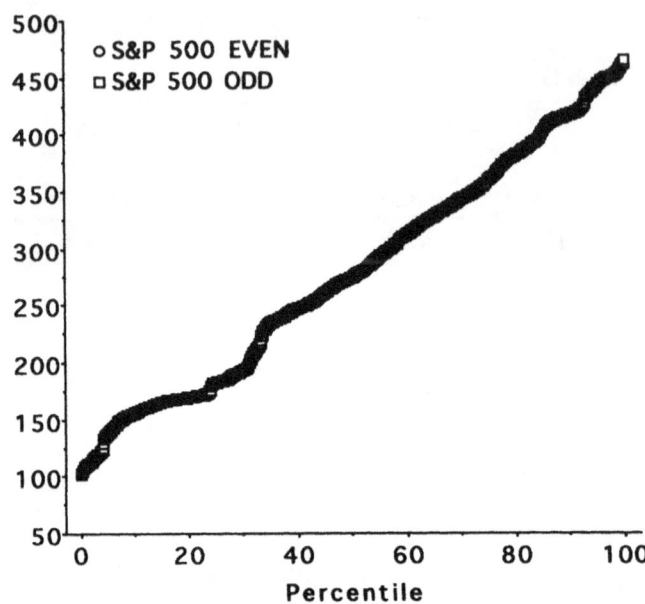

Figure 2.7

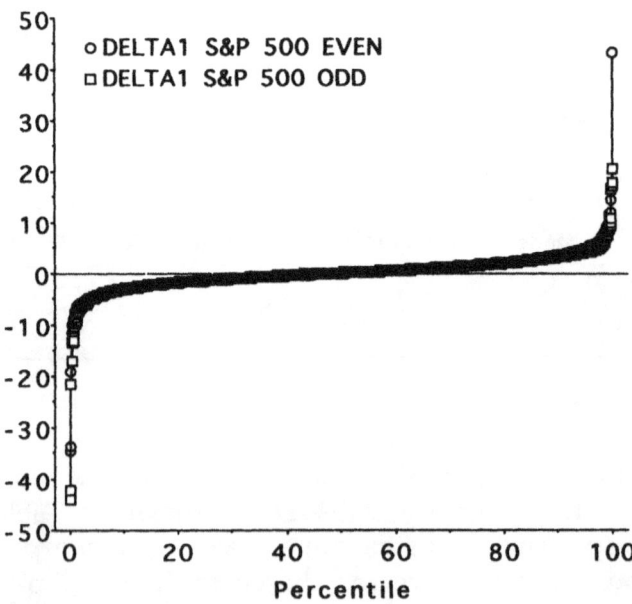

Table 2.4a-b

X_1: S&P 500 Even

Mean	Std. Dev.	Std. Error	Variance	Coef. Var.	Count
278.9	99.3	2.6	9868	35.6	1450

Minimum	Maximum	Range	Sum	Sum of Sqr.	# Missing
103.5	463.2	359.7	404395	127081741.5	0

# < 10th %	10th %	25th %	50th %	75th %	90th %
145	156.8	183.2	274.1	359.9	417.2

# > 90th %	Mode	Geo. Mean	Har. Mean	Kurtosis	Skewness
145	170	260	240.7	−1.2	.1

X_2: S&P 500 Odd

Mean	Std. Dev.	Std. Error	Variance	Coef. Var.	Count
278.8	99.3	2.6	9864.3	35.6	1450

Minimum	Maximum	Range	Sum	Sum of Sqr.	# Missing
102	464.1	362.1	404306.6	127027071.4	0

# < 10th %	10th %	25th %	50th %	75th %	90th %
145	156.4	182.8	274	359.9	417.2

# > 90th %	Mode	Geo. Mean	Har. Mean	Kurtosis	Skewness
145	•	259.9	240.6	−1.2	.1

halves appear to be very similar (we use the absolute value of the negative 'price changes' so the two sets of values can be compared easily). This is confirmed by the positive and negative cumulative probability density plots shown in Figure 2.9 and the box and whiskers plots shown in Figure 2.10.

Table 2.5a-b

X_1: Delta-1 S&P 500 Even

Mean	Std. Dev.	Std. Error	Variance	Coef. Var.	Count
.2	3.4	.1	11.7	1691.7	1449

Minimum	Maximum	Range	Sum	Sum of Sqr.	# Missing
−34.5	43.3	77.8	293.5	17057.3	1

# < 10th %	10th %	25th %	50th %	75th %	90th %
145	−3	−1.4	.1	1.7	3.7

# > 90th %	Mode	Geo. Mean	Har. Mean	Kurtosis	Skewness
145	−.8	•	•	32.7	.3

X_2: Delta-1 S&P 500 Odd

Mean	Std. Dev.	Std. Error	Variance	Coef. Var.	Count
.1	3.3	.1	11.1	5467.8	1450

Minimum	Maximum	Range	Sum	Sum of Sqr.	# Missing
−43.9	20.6	64.6	88.3	16074	0

# < 10th %	10th %	25th %	50th %	75th %	90th %
145	−2.9	−1.3	.1	1.6	3.3

# > 90th %	Mode	Geo. Mean	Har. Mean	Kurtosis	Skewness
145	.2	•	•	42.7	−3.1

The findings of symmetry with the differential spectrum would tend to suggest that there are no serial dependencies in the S&P 500 time series. However, this is not confirmed by the digram, trigram, tetragram, and pentagram relative price change transition matrices shown in Tables 2.8 to 2.10, respectively. Each of these matrices shows that the S&P 500 time series significantly diverges from inde-

Figure 2.8

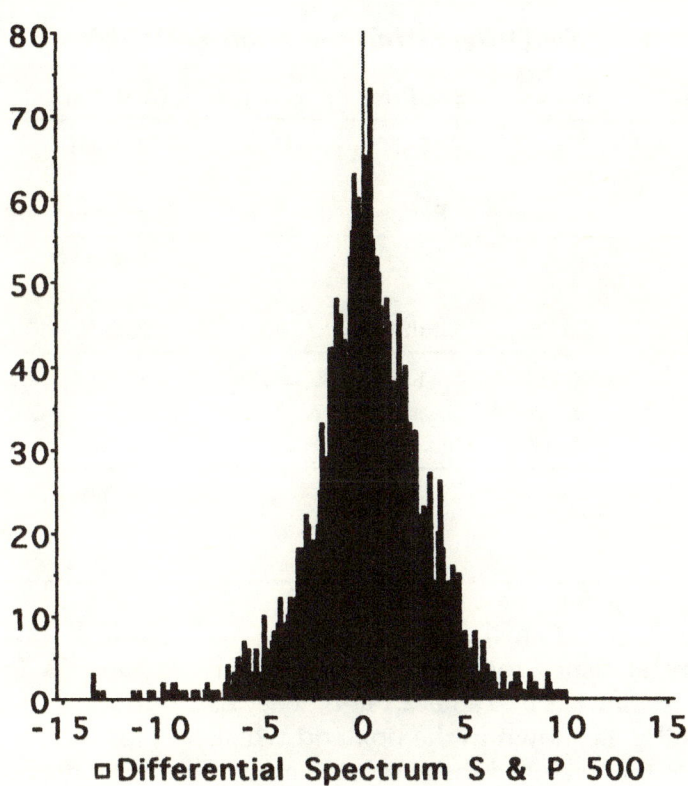

□ Differential Spectrum S & P 500

pendence, which strongly suggests that the S&P 500 Stock Index contains significant serial dependencies.

Or we can parse our Delta-1 'price changes' into three unequal
thirds and encode the lowest 10% (–43.9 to –2.9) as a '1', the middle
80% (–3.0 to +3.0) as a '2', and the upper 10% (3.4 to 43.3) as a '3', as
shown in Figure 2.11. The summary statistics for each of the thirds
are shown in Table 2.11 a-c. The digram transition matrix, shown in
Table 2.12, diverges from independence.

We can also collect price data within a day. For example, we can
subtract the opening 'price' from the closing 'price' and form the
Close-Open time series. The summary statistics for this new time
series are shown in Table 2.13. We can then parse this Close-Open
time series and if the opening 'price' is higher than the closing 'price',
we encode this relationship as a '1'. On the other hand, if the opening

Table 2.6

X_1: *Differential Spectrum S&P 500*

Mean	Std. Dev.	Std. Error	Variance	Coef. Var.	Count
.119	3.351	.062	11.226	2818.361	2895

Minimum	Maximum	Range	Sum	Sum of Sqr.	# Missing
−43.93	43.29	87.22	344.17	32530.229	1

# < 10th %	10th %	25th %	50th %	75th %	90th %
289	−2.98	−1.34	.13	1.65	3.44

# > 90th %	Mode	Geo. Mean	Har. Mean	Kurtosis	Skewness
289	•	•	•	38.486	−1.438

'price' is lower than the closing 'price', we encode it as a '2'. The summary statistics for the 1s ('price' decrease) and 2s ('price' increase) are shown in Table 2.14a-b. The distribution of 1s and 2s is very similar as shown in the box and whiskers plot shown in Figure 2.12 (note that we used the absolute value of the 1s to allow us to plot the two distributions on the same axis). Or we can parse this time series into approximately equal quarters and encode the lowest quarter (−$64.80 to −$1.20) as a '1', the next quarter (−$1.199 to $0.10) as a '2', the next quarter ($0.101 to $1.40) as a '3', and the highest quarter ($1.401 to $46.20) as a '4'. The summary statistics for each of these quarters are shown in Table 2.15 a-d. The digram category price change Chi-Square calculations (Table 2.16, column 7) are not significant. Thus the Close-Open time series is independent and does not contain serial dependencies.

We can also generate a High-Low time series, where we subtract the daily low 'price' from the daily high 'price' and divide the resulting time series into approximately equal quarters. The summary statistics for the High-Low time series are shown in Table 2.17. We can parse this time series into roughly equal quarters and encode the lowest quarter ($0.50 to $2.10) as a '1', the next quarter ($2.101 to $3.00) as a '2', the next quarter ($3.001 to $4.30) as a '3', and the

Table 2.7a-b

X_1: + *Differential Spectrum S&P 500*

Mean	Std. Dev.	Std. Error	Variance	Coef. Var.	Count
2.1	2.3	.1	5.5	111.6	1530

Minimum	Maximum	Range	Sum	Sum of Sqr.	# Missing
1.0E–2	43.3	43.3	3220.7	15216.8	0

# < 10th %	10th %	25th %	50th %	75th %	90th %
152	.2	.6	1.5	2.9	4.5

# > 90th %	Mode	Geo. Mean	Har. Mean	Kurtosis	Skewness
153	•	1.2	.4	69.4	5.5

X_2: – *Differential Spectrum S&P 500*

Mean	Std. Dev.	Std. Error	Variance	Coef. Var.	Count
2.1	2.9	.1	8.3	136.2	1364

Minimum	Maximum	Range	Sum	Sum of Sqr.	# Missing
0	43.9	43.9	2876.5	17313.4	166

# < 10th %	10th %	25th %	50th %	75th %	90th %
131	.2	.6	1.4	2.7	4.5

# > 90th %	Mode	Geo. Mean	Har. Mean	Kurtosis	Skewness
136	•	•	•	85.4	7.3

highest quarter ($4.301 to $72.20) as a '4'. The summary statistics for each of the quarters are shown in Table 2.18 a-d. The High-Low Chi-Square calculations are shown in table 2.19. The Chi-Square is 497.14. We could look this value up in a Table of crucial Chi-Square values. However, if the number of degrees of freedom is > 10, then we can use Equation 2.1 to determine if the calculated Chi-Square is

Figure 2.9

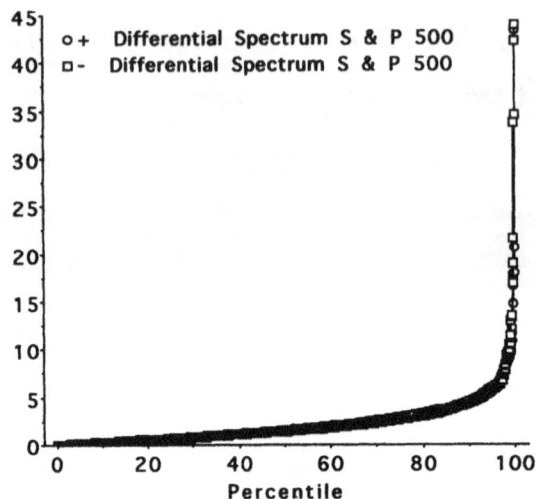

Figure 2.10

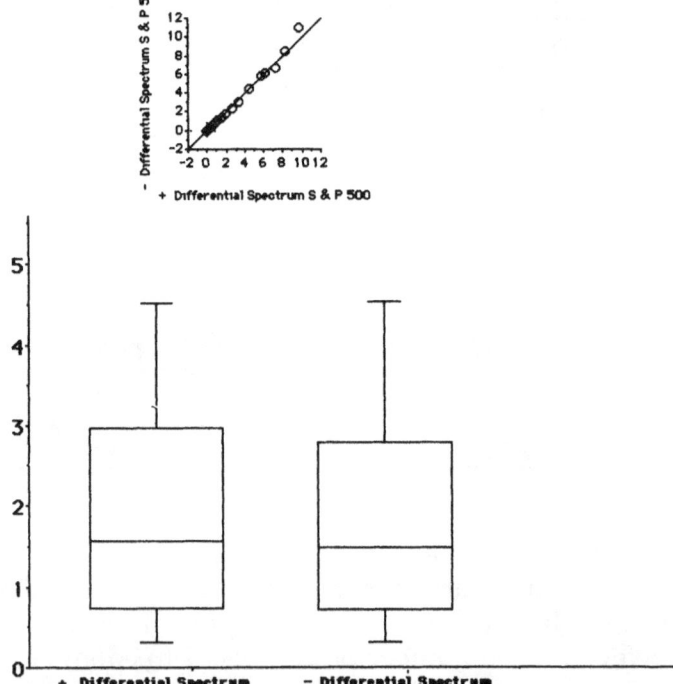

Table 2.8

	A	B	C	D	E	F	G	H	I
1	n	Digram	Expec Prob	(E)xpec Freq	(O)bser Freq	(E-O)	(E-O)^2	((E-O)^2)/E	SUM
2	2894	11	0.1667	482.43	581.00	-98.57	9716.08	20.14	
3		12	0.3333	964.57	782.00	182.57	33331.88	34.56	54.70
4		21	0.3333	964.57	783.00	181.57	32967.74	34.18	88.87
5		22	0.1667	482.43	748.00	-265.57	70527.53	146.19	123.05
6									
7									
8					2894.00			Chi-Square	269.25

Table 2.9a

	A	B	C	D	E	F	G	H	I
1	n	Tngram	Expec Prob	(E)xpec Freq	(O)bser Freq	(E-O)	(E-O)^2	((E-O)^2)/E	SUM
2	2893	111	0.04167	120.55	263.00	-142.45	20291.63	168.32	
3		112	0.125	361.63	317.00	44.63	1991.39	5.51	173.83
4		121	0.20833	602.70	401.00	201.70	40682.36	67.50	241.33
5		122	0.125	361.63	381.00	-19.38	375.39	1.04	242.37
6		211	0.125	361.63	318.00	43.63	1903.14	5.26	247.63
7		212	0.20833	602.70	465.00	137.70	18960.93	31.46	279.09
8		221	0.125	361.63	382.00	-20.38	415.14	1.15	280.24
9		222	0.04167	120.55	366.00	-245.45	60245.06	499.75	
10									
11					2893.00			Chi-Square	779.99

Table 2.9b

	A	B	C	D	E	F	G	H	I
1	n	Tetragram	Expected P	(E)xpected P	(O)bser P	(E-O)	(E-O)^2	((E-O)^2)/E	Sum
2	2892	1111	0.00833	24.09	114.00	-89.91	8083.74	335.56	
3		1112	0.03333	96.39	149.00	-52.61	2767.77	28.71	364.27
4		1121	0.075	216.90	158.00	58.90	3469.21	15.99	380.27
5		1122	0.05	144.60	159.00	-14.40	207.36	1.43	381.70
6		1211	0.075	216.90	147.00	69.90	4886.01	22.53	404.23
7		1212	0.13333	385.59	254.00	131.59	17316.02	44.91	449.14
8		1221	0.096166	278.11	192.00	86.11	7415.29	26.66	475.80
9		1222	0.03333	96.39	189.00	-92.61	8576.55	88.98	564.78
10		2111	0.03333	96.39	149.00	-52.61	2767.77	28.71	593.49
11		2112	0.09166	265.08	168.00	97.08	9424.67	35.55	629.04
12		2121	0.13333	385.59	243.00	142.59	20332.01	52.73	681.77
13		2122	0.075	216.90	222.00	-5.10	26.01	0.12	681.89
14		2211	0.05	144.60	171.00	-26.40	696.96	4.82	686.71
15		2212	0.075	216.90	211.00	5.90	34.81	0.16	686.87
16		2221	0.03333	96.39	190.00	-93.61	8762.76	90.91	777.78
17		2222	0.00833	24.09	176.00	-151.91	23076.54	957.92	1735.70
18									
19					2892.00				
20								Chi-Square	1735.70

significantly different from 0. In this case, the degrees of freedom (df) is the number of digrams (16) – 1 or 15. Equation 2.1 has the following form:

$$Z = \{\sqrt{(2)} * (\text{Chi–Square})\} - \{\sqrt{(2)} * (df - 1)\};$$

Table 2.10

	N	PENTAGRAM	(EXPECTED P)	(E)XPECTED f	(O)BSERVED f	(O-E)	(O-E)^2	((O-E)^2)/E	SUM
2	2891	11111	0.00139	4.018	43	38.98	1519.56	378.14	
3		11112	0.00694	20.064	71	50.94	2594.52	129.32	507.46
4		11121	0.01944	56.201	69	12.80	163.81	2.91	510.37
5		11122	0.01389	40.156	80	39.84	1587.55	39.53	549.91
6		11211	0.02639	76.293	45	-31.29	979.28	12.84	562.74
7		11212	0.04861	140.532	113	-27.53	757.98	5.39	568.14
8		11221	0.03611	104.394	74	-30.39	923.80	8.85	576.98
9		11222	0.01389	40.156	85	44.84	2010.99	50.08	627.06
10		12111	0.01944	56.201	71	14.80	219.01	3.90	630.96
11		12112	0.0555	160.451	76	-84.45	7131.89	44.45	675.41
12		12121	0.08472	244.926	131	-113.93	12979.02	52.99	728.40
13		12122	0.04861	140.532	123	-17.53	307.35	2.19	730.59
14		12211	0.03611	104.394	81	-23.39	547.28	5.24	735.83
15		12212	0.0555	160.451	111	-49.45	2445.35	15.24	751.07
16		12221	0.02639	76.293	98	21.71	471.17	6.18	757.25
17		12222	0.00694	20.064	91	70.94	5031.98	250.80	1008.05
18		21111	0.00694	20.064	71	50.94	2594.52	129.32	1137.37
19		21112	0.02639	76.293	78	1.71	2.91	0.04	1137.40
20		21121	0.0555	160.451	89	-71.45	5105.17	31.82	1169.22
21		21122	0.03611	104.394	79	-25.39	644.86	6.18	1175.40
22		21211	0.04861	140.532	102	-38.53	1484.68	10.56	1185.96
23		21212	0.08472	244.926	141	-103.93	10800.51	44.10	1230.06
24		21221	0.0555	160.451	118	-42.45	1802.04	11.23	1241.29
25		21222	0.01944	56.201	104	47.80	2284.74	40.65	1281.94
26		22111	0.01389	40.156	78	37.84	1432.17	35.67	1317.61
27		22112	0.03611	104.394	92	-12.39	153.61	1.47	1319.08
28		22121	0.04861	140.532	112	-28.53	814.05	5.79	1324.87
29		22122	0.02639	76.293	99	22.71	515.59	6.76	1331.63
30		22211	0.01389	40.156	90	49.84	2484.43	61.87	1393.50
31		22212	0.01944	56.201	100	43.80	1918.35	34.13	1427.63
32		22221	0.00694	20.064	92	71.94	5174.85	257.92	1685.56
33		22222	0.00139	4.018	84	79.98	6397.04	1591.90	
34				n	2891				
35							Chi-Square		3277.45967

Figure 2.11

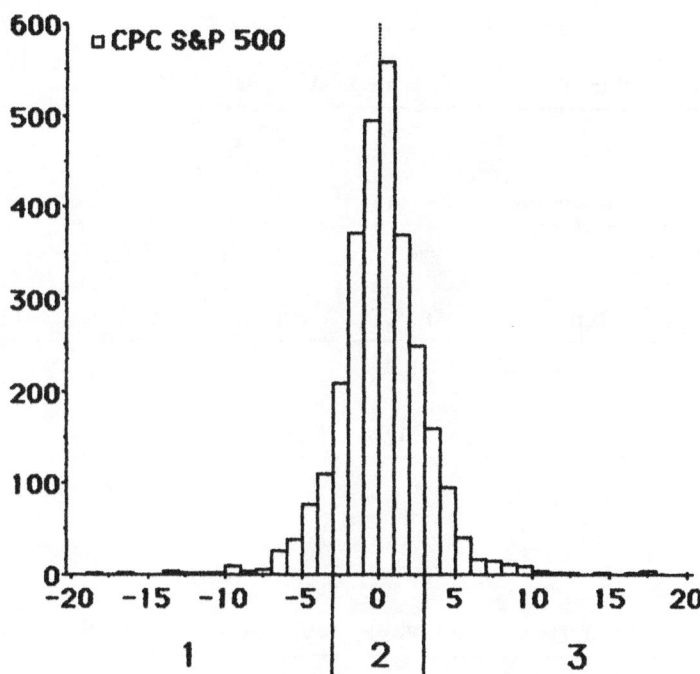

Table 2.11a

X_1: 1 – CPC S&P 500

Mean	Std. Dev.	Std. Error	Variance	Coef. Var.	Count
–5.5	4.7	.3	21.9	–84.8	288

Minimum	Maximum	Range	Sum	Sum of Sqr.	# Missing
–43.9	–2.9	41	–1590.7	15075.1	2026

# < 10th %	10th %	25th %	50th %	75th %	90th %
29	–8.1	–5.8	–4.3	–3.5	–3.1

# > 90th %	Mode	Geo. Mean	Har. Mean	Kurtosis	Skewness
29	•	•	•	36.9	–5.6

Table 2.11b-c

X_2: 2 – CPC S&P 500

Mean	Std. Dev.	Std. Error	Variance	Coef. Var.	Count
.1	1.5	3.2E–2	2.3	1065.7	2314

Minimum	Maximum	Range	Sum	Sum of Sqr.	# Missing
–3	3.4	6.4	330.9	5418.2	0

# < 10th %	10th %	25th %	50th %	75th %	90th %
230	–1.9	–1	.1	1.2	2.3

# > 90th %	Mode	Geo. Mean	Har. Mean	Kurtosis	Skewness
227	•	•	•	–.7	.1

X_3: 3 – CPC S&P 500

Mean	Std. Dev.	Std. Error	Variance	Coef. Var.	Count
5.5	3.3	.2	11.2	61	293

Minimum	Maximum	Range	Sum	Sum of Sqr.	# Missing
3.4	43.3	39.9	1604	12036.9	2021

# < 10th %	10th %	25th %	50th %	75th %	90th %
28	3.6	3.8	4.5	5.8	8.4

# > 90th %	Mode	Geo. Mean	Har. Mean	Kurtosis	Skewness
29	4.3	5	4.8	57.8	6.2

If $z' > 1.96$, it means that the Chi-Square is significantly different from 0. In this case our $z = 23.17$, which is highly significant. Therefore, the High-Low time series does contain significant serial dependencies.

Table 2.12

	A	B	C	D	E	F	G
1	Symbols	Frequency	Probability				
2	1	288	0.1000				
3	2	2314	0.7993				
4	3	293	0.1012				
5							
6	Sum	2895					
7							
8							
9							
10	Digrams	(E)xpected	(O)bserved	(E-O)	(E-O)^2	((E-O)^2)/E	Sum
11	11	28.95	38	-9.05	81.902	2.829	
12	12	231.40	192	39.40	1552.360	6.709	9.538
13	13	29.30	58	-28.70	823.690	28.112	37.650
14	21	231.40	208	23.40	547.560	2.366	40.016
15	22	1849.60	1895	-45.40	2061.035	1.114	41.131
16	23	234.20	210	24.20	585.523	2.500	43.631
17	31	29.30	42	-12.70	161.290	5.505	49.135
18	32	234.20	226	8.20	67.200	0.287	49.422
19	33	29.65	25	4.65	21.662	0.730	
20							
21						Chi-Square	50.153

Table 2.13

X_1: S&P 500 Close–Open

Mean	Std. Dev.	Std. Error	Variance	Coef. Var.	Count
.1	3	.1	8.9	3327.7	2896

Minimum	Maximum	Range	Sum	Sum of Sqr.	# Missing
−64.8	46.2	111.1	259.8	25822.7	0

# < 10th %	10th %	25th %	50th %	75th %	90th %
290	−2.6	−1.2	.1	1.4	3

# > 90th %	Mode	Geo. Mean	Har. Mean	Kurtosis	Skewness
290	•	•	•	102	−2.7

The correlation coefficient for the comparison of S&P 500 High-Low and S&P 500 Close-Open is 0.053 (i.e., the square root of r^2), as shown in the scatter diagram in Figure 2.13.

Table 2.14a-b

X_1: 1–S&P 500 Close–Open

Mean	Std. Dev.	Std. Error	Variance	Coef. Var.	Count
−1.8	2.6	.1	6.8	−143.7	1395

Minimum	Maximum	Range	Sum	Sum of Sqr.	# Missing
−64.8	0	64.8	−2527.4	14032.2	106

# < 10th %	10th %	25th %	50th %	75th %	90th %
139	−3.9	−2.3	−1.2	−.5	−.2

# > 90th %	Mode	Geo. Mean	Har. Mean	Kurtosis	Skewness
137	−.3	•	•	256.9	−12

X_2: 2–S&P 500 Close–Open

Mean	Std. Dev.	Std. Error	Variance	Coef. Var.	Count
1.9	2.1	.1	4.4	113.1	1501

Minimum	Maximum	Range	Sum	Sum of Sqr.	# Missing
1.0E−2	46.2	46.2	2787.2	11790.5	0

# < 10th %	10th %	25th %	50th %	75th %	90th %
150	.2	.6	1.3	2.6	4

# > 90th %	Mode	Geo. Mean	Har. Mean	Kurtosis	Skewness
150	.6	1.1	.4	138.2	7.8

We can also determine if the daily volume contains serial dependencies. The cumulative probability density function for volume is shown in Figure 2.14 and the summary statistics are shown in Table 2.20. We can parse the Volume time series into approximately equal quarters as shown in Figure 2.14. The summary statistics for each of the quarters are shown in Table 2.21 a-d. The Volume time

Figure 2.12

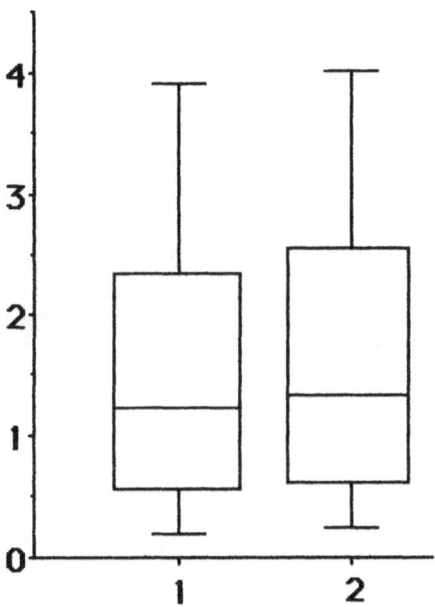

series contains significant serial dependencies as shown by the category price change digram transition matrix in Table 2.22.

We can also use the category price change method to determine if there are serial dependencies in daily closing 'price'. For example, we can parse the data into six unequal classes, where the lowest class contains the lowest 10% of the price changes and is encoded as a '1', the next as a '2', etc. The summary statistics for each of the six classes are shown in Table 2.23 a-f. The category price change transition matrices and Chi-Square calculations are shown in Table 2.24. The Chi-Square is not statistically significant, so when the closing 'price' time series is parsed in this manner, it does not contain serial dependencies.

We can also parse this time series so that the lowest 10% of the 'price' changes are contained in the first class and are encoded as a '1', the middle 80% as a '2', and the upper 10% as a 3. The summary statistics for these three classes are shown in Table 2.25 a-c. The transition matrices and Chi-Square calculations are shown in Table 2.26. The Chi-Square is statistically significant and thus when the closing

Table 2.15a-b

X_1: 1–S&P 500 Close–Open

Mean	Std. Dev.	Std. Error	Variance	Coef. Var.	Count
–3	3.2	.1	10.1	–105.5	714

Minimum	Maximum	Range	Sum	Sum of Sqr.	# Missing
–64.8	–1.2	63.6	–2156.9	13750.6	31

# < 10th %	10th %	25th %	50th %	75th %	90th %
71	–5.2	–3.4	–2.3	–1.6	–1.4

# > 90th %	Mode	Geo. Mean	Har. Mean	Kurtosis	Skewness
68	–1.6	•	•	205.3	–11.7

X_2: 2–S&P 500 Close–Open

Mean	Std. Dev.	Std. Error	Variance	Coef. Var.	Count
–.5	.4	1.3E–2	.1	–74.7	745

Minimum	Maximum	Range	Sum	Sum of Sqr.	# Missing
–1.2	.1	1.3	–367.2	281.8	0

# < 10th %	10th %	25th %	50th %	75th %	90th %
74	–1	–.8	–.5	–.2	–1.0E–2

# > 90th %	Mode	Geo. Mean	Har. Mean	Kurtosis	Skewness
73	–.3	•	•	–1.1	–.2

'price' time series is parsed in this manner, it contains serial dependencies.

The Half1 and Half2 cumulative probability density functions for the closing 'price' and Delta-1 'price changes' for the S&P 500 Stock Index for 09/03/92 to 09/30/93 are shown in Figures 2.15 and 2.16, respectively, while the summary statistics are shown in Tables

Table 2.15c-d

X_3: 3 – S&P 500 Close–Open

Mean	Std. Dev.	Std. Error	Variance	Coef. Var.	Count
.7	.4	1.4E–2	.1	52.6	721

Minimum	Maximum	Range	Sum	Sum of Sqr.	# Missing
.1	1.4	1.3	519.7	478.1	24

# < 10th %	10th %	25th %	50th %	75th %	90th %
72	.2	.4	.7	1	1.3

# > 90th %	Mode	Geo. Mean	Har. Mean	Kurtosis	Skewness
69	.6	.6	.5	–1.2	.1

X_4: 4 – S&P 500 Close–Open

Mean	Std. Dev.	Std. Error	Variance	Coef. Var.	Count
3.2	2.4	.1	5.8	76.2	716

Minimum	Maximum	Range	Sum	Sum of Sqr.	# Missing
1.4	46.2	44.8	2264.1	11312.2	29

# < 10th %	10th %	25th %	50th %	75th %	90th %
72	1.6	1.9	2.6	3.8	4.8

# > 90th %	Mode	Geo. Mean	Har. Mean	Kurtosis	Skewness
72	2	2.8	2.5	146	9.2

2.27 a-b and 2.28 a-b. The Delta-1 'price changes' appear to be stationary, as evidenced by the fact that the cumulative probability density functions overlap. The 'prices' may be stationary.

The Odd and Even cumulative probability density functions for the same time period for the 'prices' and Delta-1 'price changes' are

Table 2.16

	A	B	C	D	E	F	G	H
1	Symbols	Frequency	Probability					
2	1	714	0.2466					
3	2	744	0.2570					
4	3	721	0.2491					
5	4	716	0.2473					
6	Sum	2895						
7								
8								
9								
10	Digrams	(E)xpected	(O)bserved	(E-O)	(E-O)^2	((E-O)^2)/E	Sum	
11	11	176.095	170.000	6.095	37.153	0.211		
12	12	183.494	152.000	31.494	991.891	5.406	5.617	
13	13	177.822	179.000	-1.178	1.388	0.008	5.624	
14	14	176.589	213.000	-36.411	1325.790	7.508	13.132	
15	21	183.494	178.000	5.494	30.187	0.165	5.789	
16	22	191.204	204.000	-12.796	163.734	0.856	6.645	
17	23	185.293	187.000	-1.707	2.913	0.016	6.661	
18	24	184.008	175.000	9.008	81.149	0.441	7.102	
19	31	177.822	182.000	-4.178	17.458	0.098	6.759	
20	32	185.293	203.000	-17.707	313.528	1.692	8.451	
21	33	179.565	172.000	7.565	57.231	0.319	8.770	
22	34	178.320	164.000	14.320	205.058	1.150	9.920	
23	41	176.589	184.000	-7.411	54.929	0.311	10.231	
24	42	184.008	186.000	-1.992	3.967	0.022	10.252	
25	43	178.320	182.000	-3.680	13.543	0.076	10.328	
26	44	177.083	164.000	13.083	171.171	0.967	11.295	
27								
28		n=	2895			Chi-Square	11.2950104	
29								
30		2895						
31						4.75289604	5.47722558	
32								
33								
34								
35								
36								
37								
38								
39								
40								
41								
42								
43								
44								
45								
46								
47								
48								
49								
50								
51								-0.72432953

clearly random as shown by the overlap in Figures 2.17 and 2.18 and the summary statistics in Tables 2.29a-b and 2.30a-b, respectively.

The digram and trigram relative price change matrices show that the S&P has significant serial dependencies, as shown by statistically significant Chi-Square values in Tables 2.31 and 2.32. The posi-

Table 2.17

X_1: S&P 500 High–Low

Mean	Std. Dev.	Std. Error	Variance	Coef. Var.	Count
3.5	3	.1	8.7	83.4	2896

Minimum	Maximum	Range	Sum	Sum of Sqr.	# Missing
.5	72.2	71.8	10251.1	61492.9	0

# < 10th %	10th %	25th %	50th %	75th %	90th %
288	1.5	2.1	3	4.3	5.9

# > 90th %	Mode	Geo. Mean	Har. Mean	Kurtosis	Skewness
290	•	3	2.6	194.2	10.3

tive and negative 'price changes' are similar, as shown in Table 2.33 a-b.

The closing 'prices' for the monthly S&P 500 Stock Index (the form used in the Composite Index of Leading Economic Indicators) for January 1948 to November 1992 is shown in Figure 2.19, the cumulative probability density function is shown in Figure 2.20, and the 'price' frequency histogram is shown in Figure 2.21, while the summary statistics are shown in Table 2.34.

The Half1 and Half2 cumulative probability density functions for the monthly Delta-1 'price changes' are shown in Figure 2.22. The cumulative probability density functions are different, as shown by the box and whiskers plots shown in Figure 2.23 and is confirmed by the compare percentiles in the upper half of this figure and the summary statistics shown in Table 2.35 a-b. Since this time series is not stationary, we will not analyze it further.

Table 2.18a-b

X_1: 1 – S&P 500 High–Low

Mean	Std. Dev.	Std. Error	Variance	Coef. Var.	Count
1.6	.4	1.3E–2	.1	22.8	773

Minimum	Maximum	Range	Sum	Sum of Sqr.	# Missing
.5	2.1	1.6	1211.1	1996.4	0

# < 10th %	10th %	25th %	50th %	75th %	90th %
74	1.1	1.3	1.6	1.9	2

# > 90th %	Mode	Geo. Mean	Har. Mean	Kurtosis	Skewness
75	1.7	1.5	1.5	–.6	–.5

X_2: 2 – S&P 500 High–Low

Mean	Std. Dev.	Std. Error	Variance	Coef. Var.	Count
2.5	.3	9.8E–3	.1	10.2	701

Minimum	Maximum	Range	Sum	Sum of Sqr.	# Missing
2.1	3	.9	1781.8	4575.9	72

# < 10th %	10th %	25th %	50th %	75th %	90th %
68	2.2	2.3	2.5	2.8	2.9

# > 90th %	Mode	Geo. Mean	Har. Mean	Kurtosis	Skewness
70	•	2.5	2.5	–1.2	1.5E–2

Table 2.18c-d

X₃: 3 – S&P 500 High–Low

Mean	Std. Dev.	Std. Error	Variance	Coef. Var.	Count
3.6	.4	1.4E–2	.1	10.4	697

Minimum	Maximum	Range	Sum	Sum of Sqr.	# Missing
3	4.3	1.3	2519.2	9203.9	76

# < 10th %	10th %	25th %	50th %	75th %	90th %
70	3.1	3.3	3.6	3.9	4.1

# > 90th %	Mode	Geo. Mean	Har. Mean	Kurtosis	Skewness
68	3.1	3.6	3.6	–1.2	.1

X₄: 4 – S&P 500 High–Low

Mean	Std. Dev.	Std. Error	Variance	Coef. Var.	Count
6.5	4.5	.2	20.4	69	725

Minimum	Maximum	Range	Sum	Sum of Sqr.	# Missing
4.3	72.2	67.9	4739	45716.7	48

# < 10th %	10th %	25th %	50th %	75th %	90th %
72	4.5	4.8	5.5	6.8	8.9

# > 90th %	Mode	Geo. Mean	Har. Mean	Kurtosis	Skewness
72	5.1	6	5.8	113.6	9.4

Table 2.19

	A	B	C	D	E	F	G	H
1	Symbols	Frequency	Probability					
2	1	773	0.2670					
3	2	701	0.2421					
4	3	696	0.2404					
5	4	725	0.2504					
6	Sum	2895						
7								
8								
9								
10	Digrams	(E)xpected	(O)bserved	(E-O)	(E-O)^2	((E-O)^2)/E	Sum	
11	11	206.400	408.000	-201.600	40642.421	196.911		
12	12	187.175	183.000	4.175	17.435	0.093	197.004	
13	13	185.840	114.000	71.840	5161.045	27.771	224.775	
14	14	193.584	68.000	125.584	15771.282	81.470	306.245	
15	21	187.175	215.000	-27.825	774.204	4.136	228.911	
16	22	169.741	183.000	-13.259	175.794	1.036	229.947	
17	23	168.531	169.000	-0.469	0.220	0.001	229.948	
18	24	175.553	134.000	41.553	1726.625	9.835	239.784	
19	31	185.840	105.000	80.840	6535.173	35.166	265.114	
20	32	168.531	181.000	-12.469	155.487	0.923	266.036	
21	33	167.328	194.000	-26.672	711.369	4.251	270.288	
22	34	174.301	216.000	-41.699	1738.847	9.976	280.264	
23	41	193.584	44.000	149.584	22375.303	115.585	395.849	
24	42	175.553	154.000	21.553	464.518	2.646	398.495	
25	43	174.301	220.000	-45.699	2088.443	11.982	410.476	
26	44	181.563	307.000	-125.437	15734.431	86.661	497.137	
27								
28	Sum	2895	2895			Chi-Square	497.137414	
29								
30								
31						31.5321238	8.36660027	
32								
33								
34								
35								
36								
37								
38								
39								
40								
41								
42								
43								
44								
45								
46								
47								
48								
49								
50								
51								23.1655235

Figure 2.13

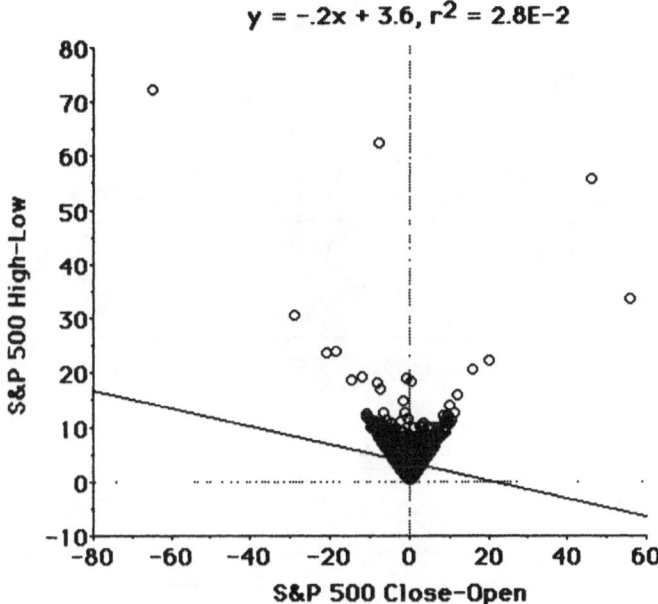

Figure 2.14

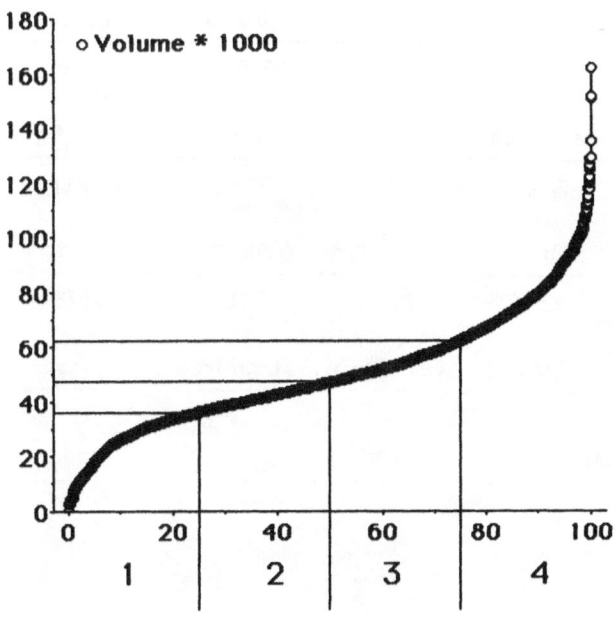

Table 2.20

X_1: *Volume*

Mean	Std. Dev.	Std. Error	Variance	Coef. Var.	Count
50286.9	21484.6	399.2	461589635.1	42.7	2896

Minimum	Maximum	Range	Sum	Sum of Sqr.	# Missing
0	162020	162020	145630993	8.7E12	0

# < 10th %	10th %	25th %	50th %	75th %	90th %
290	26212.3	36337	47313	62482.5	79593.7

# > 90th %	Mode	Geo. Mean	Har. Mean	Kurtosis	Skewness
290	•	•	•	.8	.7

Table 2.21a

X_1: *1 – Volume*

Mean	Std. Dev.	Std. Error	Variance	Coef. Var.	Count
25952.9	8682.6	322.7	75388188	33.5	724

Minimum	Maximum	Range	Sum	Sum of Sqr.	# Missing
0	36335	36335	18789877	5.4E11	0

# < 10th %	10th %	25th %	50th %	75th %	90th %
72	12208.2	20781	28413	32925	34869.2

# > 90th %	Mode	Geo. Mean	Har. Mean	Kurtosis	Skewness
72	•	•	•	–.1	–.9

Table 2.21b-c

X_2: 2 – Volume

Mean	Std. Dev.	Std. Error	Variance	Coef. Var.	Count
41680.8	3157.7	117.4	9970861.5	7.6	724

Minimum	Maximum	Range	Sum	Sum of Sqr.	# Missing
36339	47305	10966	30176892	1.3E12	0

# < 10th %	10th %	25th %	50th %	75th %	90th %
72	37344	38980.5	41589	44478	46103.6

# > 90th %	Mode	Geo. Mean	Har. Mean	Kurtosis	Skewness
72	•	41561.3	41441.9	−1.2	.1

X_3: 3 – Volume

Mean	Std. Dev.	Std. Error	Variance	Coef. Var.	Count
53968.9	4310.1	160.2	18576920.8	8	724

Minimum	Maximum	Range	Sum	Sum of Sqr.	# Missing
47321	62480	15159	39073463	2.1E12	0

# < 10th %	10th %	25th %	50th %	75th %	90th %
72	48311.4	50218	53277.5	57533.5	60370.5

# > 90th %	Mode	Geo. Mean	Har. Mean	Kurtosis	Skewness
72	•	53798.6	53630.2	−1.1	.3

Table 2.21d

X_4: 4 – Volume

Mean	Std. Dev.	Std. Error	Variance	Coef. Var.	Count
79545.2	14366.5	533.9	206395487.1	18.1	724

Minimum	Maximum	Range	Sum	Sum of Sqr.	# Missing
62485	162020	99535	57590761	4.7E12	0

# < 10th %	10th %	25th %	50th %	75th %	90th %
72	64774.2	68431.5	76111.5	87381.5	99138.8

# > 90th %	Mode	Geo. Mean	Har. Mean	Kurtosis	Skewness
72	•	78396.7	77369.4	3.3	1.5

Table 2.22

	A	B	C	D	E	F	G	H
1	Symbols	Frequency	Probability					
2	1	723	0 2497					
3	2	724	0 2501					
4	3	724	0 2501					
5	4	724	0 2501					
6	Sum	2895						
7								
8								
9								
10	Digrams	(E)xpected	(O)bserved	(E-O)	(E-O)^2	((E-O)^2)/E	Sum	
11	11	180.563	489.000	-308 437	95133 572	526.873		
12	12	180.812	167 000	13 812	190.783	1 055	527 928	
13	13	180.812	60 000	120.812	14595 645	80 723	608.651	
14	14	180.812	7 000	173 812	30210 763	167 083	775 734	
15	21	180.812	170.000	10 812	116 909	0 647	609 297	
16	22	181 063	309 000	-127 937	16367 998	90.400	699.697	
17	23	181.063	211 000	-29 937	896 253	4 950	704 647	
18	24	181 063	34 000	147 063	21627 385	119 447	824 094	
19	31	180.812	61 000	119 812	14355 020	79 392	784 039	
20	32	181 063	208.000	-26 937	725 628	4 008	788.046	
21	33	181 063	304 000	-122.937	15113 624	83 472	871 518	
22	34	181 063	151.000	30.063	903 755	4 991	876 509	
23	41	180.812	3 000	177 812	31617 262	174 862	1051 372	
24	42	181 063	40 000	141 063	19898 635	109 899	1161 271	
25	43	181 063	149 000	32 063	1028 005	5.678	1166 948	
26	44	181 063	532.000	-350 937	123157 114	680.191	1847 140	
27								
28	Sum=	2895	2895			Chi-Square	1847 13956	
29								
30								
31					Sqrt((Chi-Square)*(2))	60 78058169		
32					Sqrt((df-1)*(2))	8 366600265		
33								
34				Z=		52 41398142		
35								
36								
37								
38								
39								
40								
41								
42								
43								
44								
45								
46								
47								
48								
49								
50								
51								52 4139814

Table 2.23a-b

X_1: 1

Mean	Std. Dev.	Std. Error	Variance	Coef. Var.	Count
−5.4	5.4	.3	28.7	−99.6	290

Minimum	Maximum	Range	Sum	Sum of Sqr.	# Missing
−80.9	−2.8	78.1	−1558.8	16666.4	474

# < 10th %	10th %	25th %	50th %	75th %	90th %
29	−7.7	−5.7	−4.3	−3.5	−3

# > 90th %	Mode	Geo. Mean	Har. Mean	Kurtosis	Skewness
27	−3.6	•	•	137	−10.4

X_2: 2

Mean	Std. Dev.	Std. Error	Variance	Coef. Var.	Count
−1.9	.4	2.2E−2	.2	−23.6	415

Minimum	Maximum	Range	Sum	Sum of Sqr.	# Missing
−2.8	−1.2	1.6	−779.5	1545.3	349

# < 10th %	10th %	25th %	50th %	75th %	90th %
41	−2.5	−2.2	−1.8	−1.5	−1.3

# > 90th %	Mode	Geo. Mean	Har. Mean	Kurtosis	Skewness
41	−1.8	•	•	−1	−.4

Table 2.23c-d

X_3: 3

Mean	Std. Dev.	Std. Error	Variance	Coef. Var.	Count
−.5	.4	1.4E−2	.1	−74.2	713

Minimum	Maximum	Range	Sum	Sum of Sqr.	# Missing
−1.2	.1	1.3	−359.4	280.8	51

# < 10th %	10th %	25th %	50th %	75th %	90th %
71	−1.1	−.8	−.5	−.2	2.0E−3

# > 90th %	Mode	Geo. Mean	Har. Mean	Kurtosis	Skewness
71	0	•	•	−1.1	−.2

X_4: 4

Mean	Std. Dev.	Std. Error	Variance	Coef. Var.	Count
.7	.4	1.4E−2	.2	54.6	764

Minimum	Maximum	Range	Sum	Sum of Sqr.	# Missing
.1	1.5	1.4	560.2	533	0

# < 10th %	10th %	25th %	50th %	75th %	90th %
74	.2	.4	.7	1.1	1.3

# > 90th %	Mode	Geo. Mean	Har. Mean	Kurtosis	Skewness
75	.3	.6	.5	−1.2	.2

Table 2.23e-f

X_5: 5

Mean	Std. Dev.	Std. Error	Variance	Coef. Var.	Count
2.3	.5	2.6E–2	.3	23.1	429

Minimum	Maximum	Range	Sum	Sum of Sqr.	# Missing
1.5	3.4	1.9	981.1	2363.5	335

# < 10th %	10th %	25th %	50th %	75th %	90th %
38	1.6	1.8	2.2	2.7	3.1

# > 90th %	Mode	Geo. Mean	Har. Mean	Kurtosis	Skewness
41	•	2.2	2.2	–1	.4

X_6: 6

Mean	Std. Dev.	Std. Error	Variance	Coef. Var.	Count
5.3	3	.2	8.9	56.5	284

Minimum	Maximum	Range	Sum	Sum of Sqr.	# Missing
3.4	41.9	38.5	1498.1	10417.6	480

# < 10th %	10th %	25th %	50th %	75th %	90th %
28	3.6	3.9	4.5	5.6	7.7

# > 90th %	Mode	Geo. Mean	Har. Mean	Kurtosis	Skewness
28	5	4.9	4.7	79.5	7.3

Table 2.24

	A	B	C	D	E	F	G	H
1	Symbols	Frequency	Probability					
2	1	290	0.1002					
3	2	414	0.1431					
4	3	713	0.2464					
5	4	764	0.2640					
6	5	429	0.1482					
7	6	284	0.0981					
8	Sum	2894						
9								
10								
11								
12	Digrams	(E)xpected	(O)bserved	(E-O)	(E-O)^2	((E-O)^2)/E	Sum	
13	11	29.060	50.000	-20.940	438.478	15.089		
14	12	41.486	31.000	10.486	109.953	2.650	17.739	
15	13	71.448	49.000	22.448	503.905	7.053	24.792	
16	14	76.558	69.000	7.558	57.129	0.746	25.538	
17	15	42.989	56.000	-13.011	169.288	3.938	29.476	
18	16	28.459	35.000	-6.541	42.786	1.503	30.979	
19	21	41.486	41.000	0.486	0.236	0.006	24.797	
20	22	59.225	57.000	2.225	4.949	0.084	24.881	
21	23	101.998	88.000	13.998	195.942	1.921	26.802	
22	24	109.294	115.000	-5.706	32.562	0.298	27.100	
23	25	61.370	61.000	0.370	0.137	0.002	27.102	
24	26	40.628	52.000	-11.372	129.334	3.183	30.286	
25	31	71.448	61.000	10.448	109.157	1.528	28.330	
26	32	101.998	99.000	2.998	8.988	0.088	28.418	
27	33	175.663	176.000	-0.337	0.114	0.001	28.419	
28	34	188.228	212.000	-23.772	565.105	3.002	31.421	
29	35	105.694	104.000	1.694	2.868	0.027	31.448	
30	36	69.970	61.000	8.970	80.454	1.150	32.598	
31	41	76.558	70.000	6.558	43.013	0.562	33.160	
32	42	109.294	119.000	-9.706	94.212	0.862	34.022	
33	43	188.228	220.000	-31.772	1009.456	5.363	39.385	
34	44	201.692	189.000	12.692	161.081	0.799	40.183	
35	45	113.254	102.000	11.254	126.644	1.118	41.302	
36	46	74.974	64.000	10.974	120.438	1.606	42.908	
37	51	42.989	39.000	3.989	15.912	0.370	43.278	
38	52	61.370	68.000	-6.630	43.951	0.716	43.994	
39	53	105.694	107.000	-1.306	1.707	0.016	44.010	
40	54	113.254	115.000	-1.746	3.050	0.027	44.037	
41	55	63.594	62.000	1.594	2.541	0.040	44.077	
42	56	42.100	38.000	4.100	16.806	0.399	44.476	
43	61	28.459	29.000	-0.541	0.293	0.010	44.487	
44	62	40.628	41.000	-0.372	0.139	0.003	44.490	
45	63	69.970	73.000	-3.030	9.183	0.131	44.621	
46	64	74.974	63.000	11.974	143.387	1.912	46.534	
47	65	42.100	44.000	-1.900	3.612	0.086	46.620	
48	66	27.870	34.000	-6.130	37.576	1.348	47.968	
49								
50	Sum=	2894	2894			Chi-Square	47.9678976	
51								
52								
53						9.79468199	8.36660027	1.42808172

Table 2.25a-b

X_1: 1

Mean	Std. Dev.	Std. Error	Variance	Coef. Var.	Count
−5.4	5.3	.3	28.5	−99.7	292

Minimum	Maximum	Range	Sum	Sum of Sqr.	# Missing
−80.9	−2.8	78.2	−1564.4	16682.1	2027

# < 10th %	10th %	25th %	50th %	75th %	90th %
29	−7.7	−5.7	−4.3	−3.5	−3

# > 90th %	Mode	Geo. Mean	Har. Mean	Kurtosis	Skewness
29	−3.6	•	•	137.7	−10.4

X_2: 2

Mean	Std. Dev.	Std. Error	Variance	Coef. Var.	Count
.2	1.4	2.9E−2	2	803.7	2319

Minimum	Maximum	Range	Sum	Sum of Sqr.	# Missing
−2.8	3.4	6.2	408	4707	0

# < 10th %	10th %	25th %	50th %	75th %	90th %
232	−1.7	−.8	.1	1.1	2.2

# > 90th %	Mode	Geo. Mean	Har. Mean	Kurtosis	Skewness
232	.3	•	•	−.6	.1

Table 2.25c

X_3: 3

Mean	Std. Dev.	Std. Error	Variance	Coef. Var.	Count
5.3	3	.2	8.9	56.5	284

Minimum	Maximum	Range	Sum	Sum of Sqr.	# Missing
3.4	41.9	38.5	1498.1	10417.6	2035

# < 10th %	10th %	25th %	50th %	75th %	90th %
28	3.6	3.9	4.5	5.6	7.7

# > 90th %	Mode	Geo. Mean	Har. Mean	Kurtosis	Skewness
28	5	4.9	4.7	79.5	7.3

Table 2.26

	A	B	C	D	E	F	G
1	Symbols	Frequency	Probability				
2	1	292	0.1000				
3	2	2318	0.8010				
4	3	284	0.0981				
5							
6	Sum	2894					
7							
8							
9							
10	Digrams	(E)xpected	(O)bserved	(E-O)	(E-O)^2	((E-O)^2)/E	Sum
11	11	28.94	50	-21.06	443.524	15.326	
12	12	231.80	207	24.80	615.040	2.653	17.979
13	13	28.40	35	-6.60	43.560	1.534	19.513
14	21	231.80	213	18.80	353.440	1.525	21.038
15	22	1856.64	1890	-33.36	1112.709	0.599	21.637
16	23	227.47	215	12.47	155.620	0.684	22.321
17	31	28.40	29	-0.60	0.360	0.013	22.334
18	32	227.47	221	6.47	41.923	0.184	22.518
19	33	27.87	34	-6.13	37.576	1.348	
20							
21	Sum=	2888.80234	2894			Chi-Square	23.866
22							
23							
24					6.90885984	4	

Figure 2.15

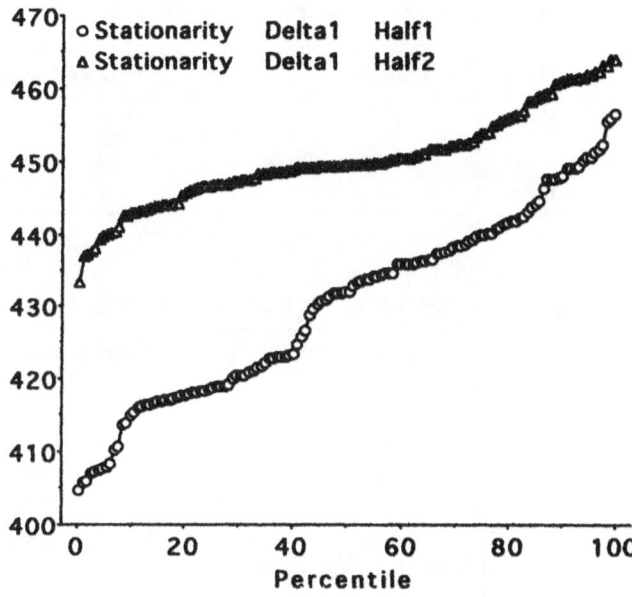

Figure 2.16

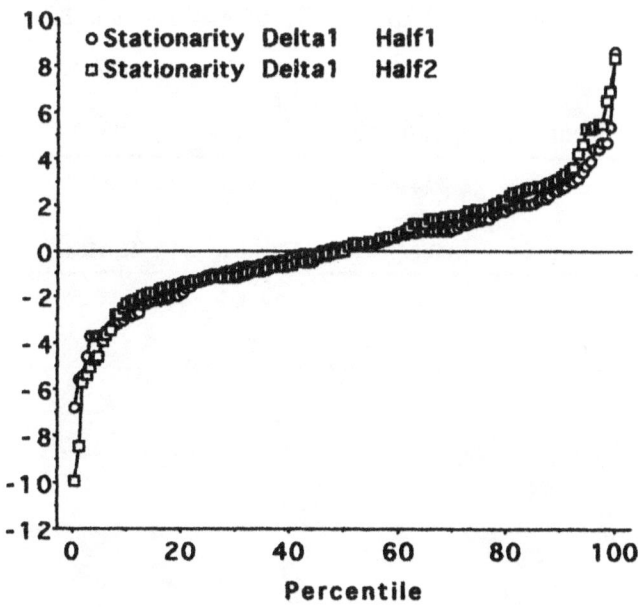

Table 2.27a-b

X_1: *Stationarity Delta-1 Half1*

Mean	Std. Dev.	Std. Error	Variance	Coef. Var.	Count
430.3	13	1.1	169.9	3	136

Minimum	Maximum	Range	Sum	Sum of Sqr.	# Missing
404.7	456.6	51.9	58514.8	25199275.5	136

# < 10th %	10th %	25th %	50th %	75th %	90th %
14	414.9	418.6	432.1	439.8	448

# > 90th %	Mode	Geo. Mean	Har. Mean	Kurtosis	Skewness
14	•	430.1	429.9	−1	−2.6E−2

X_2: *Stationarity Delta-1 Half2*

Mean	Std. Dev.	Std. Error	Variance	Coef. Var.	Count
450.2	6.5	.6	42.2	1.4	136

Minimum	Maximum	Range	Sum	Sum of Sqr.	# Missing
433.2	464.1	30.9	61233	27575408	136

# < 10th %	10th %	25th %	50th %	75th %	90th %
14	442.9	446.7	449.4	453.7	460.9

# > 90th %	Mode	Geo. Mean	Har. Mean	Kurtosis	Skewness
14	•	450.2	450.1	−.2	.2

Table 2.28a-b

X_1: *Stationarity Delta-1 Half1*

Mean	Std. Dev.	Std. Error	Variance	Coef. Var.	Count
.1	2.3	.2	5.2	4266.9	135

Minimum	Maximum	Range	Sum	Sum of Sqr.	# Missing
−6.8	8.5	15.3	7.2	700.1	137

# < 10th %	10th %	25th %	50th %	75th %	90th %
13	−2.9	−1.1	.1	1.3	2.7

# > 90th %	Mode	Geo. Mean	Har. Mean	Kurtosis	Skewness
13	.8	•	•	1.2	.1

X_2: *Stationarity Delta-1 Half2*

Mean	Std. Dev.	Std. Error	Variance	Coef. Var.	Count
.2	2.7	.2	7.3	1184.3	135

Minimum	Maximum	Range	Sum	Sum of Sqr.	# Missing
−9.9	8.3	18.2	30.9	991.7	137

# < 10th %	10th %	25th %	50th %	75th %	90th %
13	−2.3	−1.2	−4.0E−2	1.8	3.2

# > 90th %	Mode	Geo. Mean	Har. Mean	Kurtosis	Skewness
13	−.5	•	•	1.8	−.2

Figure 2.17

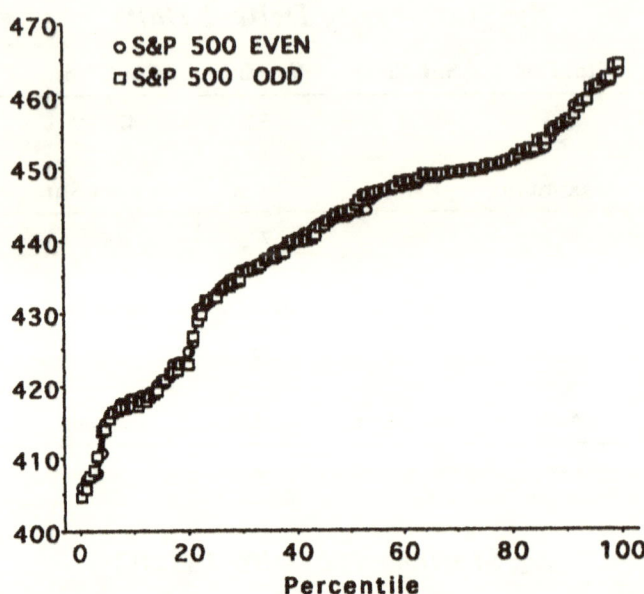

Figure 2.18

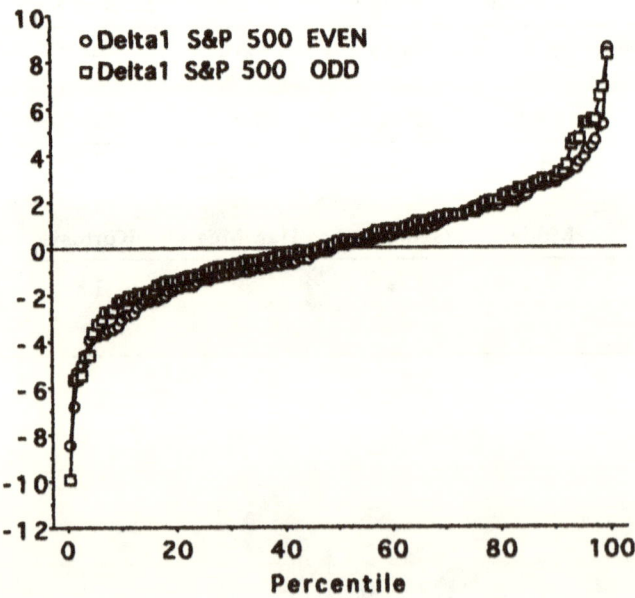

Table 2.29a-b

X_1: S&P 500 Even

Mean	Std. Dev.	Std. Error	Variance	Coef. Var.	Count
440.2	14.3	1.2	203.2	3.2	136

Minimum	Maximum	Range	Sum	Sum of Sqr.	# Missing
406	463.2	57.2	59873.2	26386235.6	0

# < 10th %	10th %	25th %	50th %	75th %	90th %
14	418.2	432.3	443.5	449.9	456

# > 90th %	Mode	Geo. Mean	Har. Mean	Kurtosis	Skewness
14	461.3	440	439.8	–.4	–.7

X_2: S&P 500 Odd

Mean	Std. Dev.	Std. Error	Variance	Coef. Var.	Count
440.3	14.5	1.2	210	3.3	136

Minimum	Maximum	Range	Sum	Sum of Sqr.	# Missing
404.7	464.1	59.5	59874.6	26388447.9	0

# < 10th %	10th %	25th %	50th %	75th %	90th %
14	417.1	432.1	444	450	456.2

# > 90th %	Mode	Geo. Mean	Har. Mean	Kurtosis	Skewness
14	•	440	439.8	–.5	–.6

Table 2.30a-b

X$_1$: Delta-1 S&P 500 Even

Mean	Std. Dev.	Std. Error	Variance	Coef. Var.	Count
−1.1E−2	2.5	.2	6	−22708.1	136

Minimum	Maximum	Range	Sum	Sum of Sqr.	# Missing
−8.4	8.5	17	−1.5	813.3	0

# < 10th %	10th %	25th %	50th %	75th %	90th %
14	−3	−1.3	.1	1.7	2.8

# > 90th %	Mode	Geo. Mean	Har. Mean	Kurtosis	Skewness
13	•	•	•	1.2	−.1

X$_2$: Delta-1 S&P 500 Odd

Mean	Std. Dev.	Std. Error	Variance	Coef. Var.	Count
.3	2.6	.2	6.5	805.6	135

Minimum	Maximum	Range	Sum	Sum of Sqr.	# Missing
−9.9	8.3	18.2	42.8	888.9	1

# < 10th %	10th %	25th %	50th %	75th %	90th %
13	−2.2	−1.1	.3	1.6	3

# > 90th %	Mode	Geo. Mean	Har. Mean	Kurtosis	Skewness
13	.3	•	•	2.1	−.1

Table 2.31

	A	B	C	D	E	F	G	H	I
1	n	Digram	Expec Prob	(E)xpec Freq	(O)bser Freq	(E-O)	(E-O)^2	((E-O)^2)/E	SUM
2	270	11	0.1667	45.01	59.00	-13.99	195.75	4.35	
3		12	0.3333	89.99	71.00	18.99	360.66	4.01	8.36
4		21	0.3333	89.99	71.00	18.99	360.66	4.01	12.36
5		22	0.1667	45.01	69.00	-23.99	575.57	12.79	16.37
6									
7									
8					270.00			Chi-Square	29.16

Table 2.32

	A	B	C	D	E	F	G	H	I
1	n	Trigram	Expec Prob	(E)xpec Freq	(O)bser Freq	(E-O)	(E-O)^2	((E-O)^2)/E	SUM
2	269	111	0.04167	11.21	24.00	-12.79	163.60	14.60	
3		112	0.125	33.63	34.00	-0.38	0.14	0.00	14.60
4		121	0.20833	56.04	34.00	22.04	485.80	8.67	23.27
5		122	0.125	33.63	37.00	-3.38	11.39	0.34	23.61
6		211	0.125	33.63	34.00	-0.38	0.14	0.00	23.61
7		212	0.20833	56.04	37.00	19.04	362.55	6.47	30.08
8		221	0.125	33.63	37.00	-3.38	11.39	0.34	30.42
9		222	0.04167	11.21	32.00	-20.79	432.26	38.56	
10									
11					269.00			Chi-Square	68.98

Table 2.33a

X_1: S&P 500 RPC 1

Mean	Std. Dev.	Std. Error	Variance	Coef. Var.	Count
−1.8	1.7	.1	2.8	−94.5	131

Minimum	Maximum	Range	Sum	Sum of Sqr.	# Missing
−9.9	−3.0E−2	9.9	−233	782	9

# < 10th %	10th %	25th %	50th %	75th %	90th %
13	−3.7	−2.3	−1.2	−.7	−.2

# > 90th %	Mode	Geo. Mean	Har. Mean	Kurtosis	Skewness
13	−.5	•	•	5.2	−2

Table 2.33b

X_2: S&P 500 RPC 2

Mean	Std. Dev.	Std. Error	Variance	Coef. Var.	Count
2	1.7	.1	2.8	84.6	140

Minimum	Maximum	Range	Sum	Sum of Sqr.	# Missing
.1	8.5	8.5	274.4	920.3	0

# < 10th %	10th %	25th %	50th %	75th %	90th %
14	.3	.8	1.5	2.7	4.4

# > 90th %	Mode	Geo. Mean	Har. Mean	Kurtosis	Skewness
14	•	1.3	.7	2.6	1.5

Figure 2.19

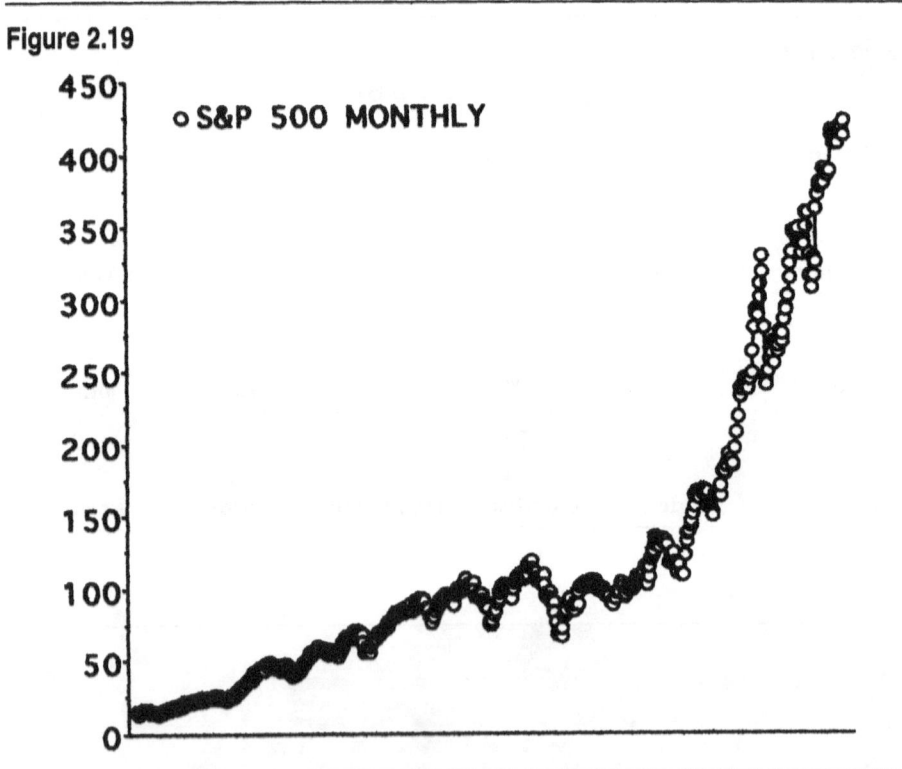

o S&P 500 MONTHLY

Figure 2.20

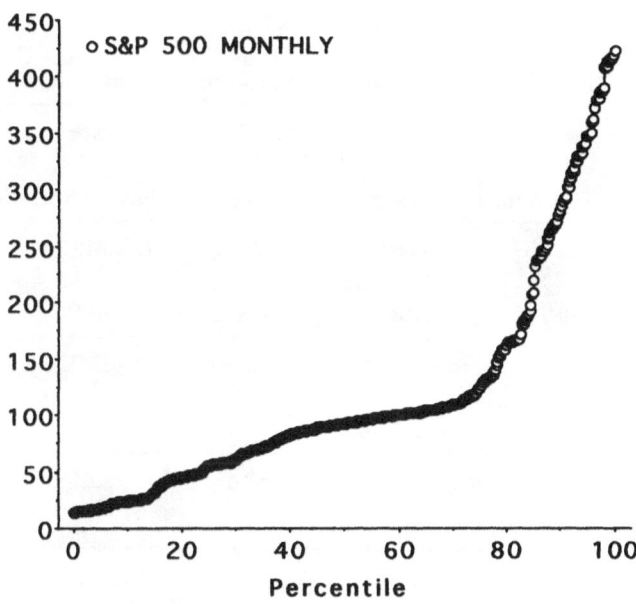

Figure 2.21

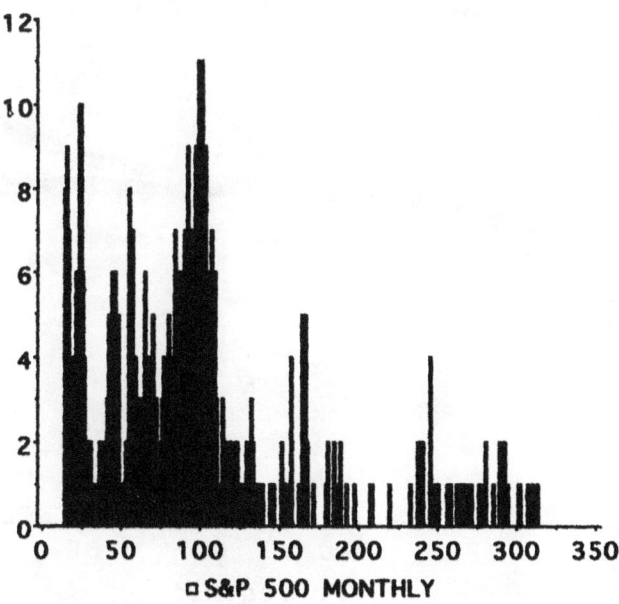

Table 2.34

X_1: *S&P 500 Monthly*

Mean	Std. Dev.	Std. Error	Variance	Coef. Var.	Count
116.3	97.3	4.2	9461.4	83.6	539

Minimum	Maximum	Range	Sum	Sum of Sqr.	# Missing
14	422.8	408.9	62709.6	12386121.4	1

# < 10th %	10th %	25th %	50th %	75th %	90th %
54	24	54.9	92.5	123.7	283.6

# > 90th %	Mode	Geo. Mean	Har. Mean	Kurtosis	Skewness
54	•	84.5	59.6	1.6	1.6

Figure 2.22

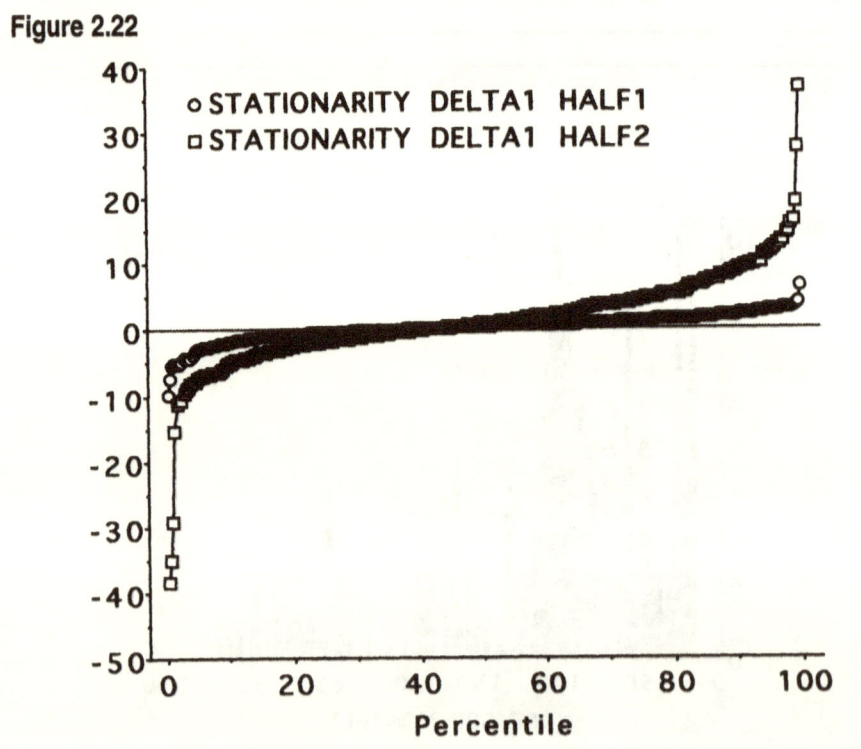

Figure 2.23

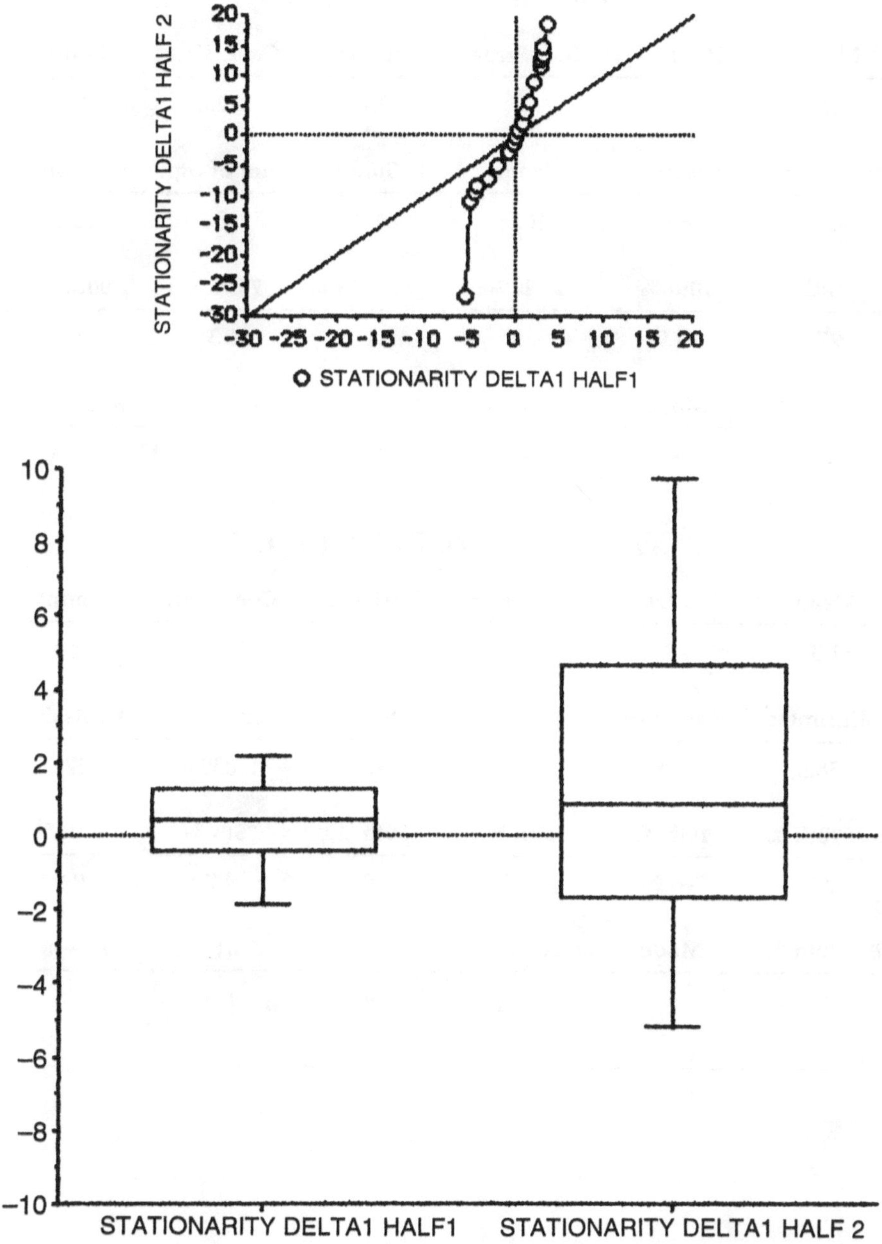

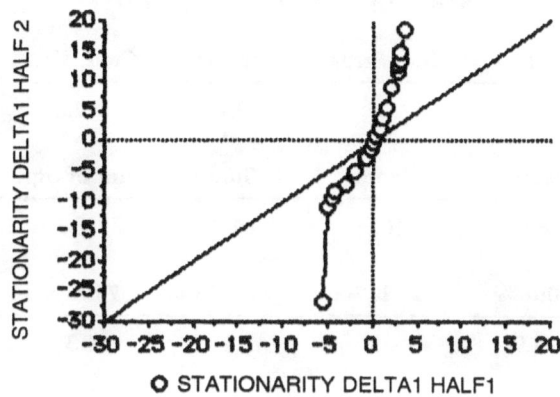

Table 2.35a-b

X_1: *Stationarity Delta-1 Half1*

Mean	Std. Dev.	Std. Error	Variance	Coef. Var.	Count
.2	1.9	.1	3.4	812.2	268

Minimum	Maximum	Range	Sum	Sum of Sqr.	# Missing
−9.9	6.6	16.5	61.2	933.3	272

# < 10th %	10th %	25th %	50th %	75th %	90th %
27	−1.9	−.5	.4	1.3	2.2

# > 90th %	Mode	Geo. Mean	Har. Mean	Kurtosis	Skewness
27	•	•	•	4.7	−1.3

X_2: *Stationarity Delta-1 Half2*

Mean	Std. Dev.	Std. Error	Variance	Coef. Var.	Count
1.3	7.1	.4	50.2	554.5	268

Minimum	Maximum	Range	Sum	Sum of Sqr.	# Missing
−38.5	36.8	75.3	342.4	13839.8	272

# < 10th %	10th %	25th %	50th %	75th %	90th %
27	−5.2	−1.8	.9	4.7	9.1

# > 90th %	Mode	Geo. Mean	Har. Mean	Kurtosis	Skewness
27	•	•	•	8.9	−.6

General Motors, Chrysler, Ford

General Motors operates in a single industry and consists of the manufacture, assembly, and sale of automobiles and trucks, as well as related parts and accessories. A plot of the closing prices for GM for 01/02/70 to 09/30/93 is shown in Figure 3.1. The frequency histogram of closing prices and Delta-1 price changes is shown in Figures 3.2 a and b, respectively. The cumulative probability density function for closing prices is shown in Figure 3.3. The summary statistics for closing prices are shown in Table 3.1. The summary statistics for Delta-1 price changes are shown in Table 3.2.

A box and whiskers plot for the daily prices for the years 1970 to 1992 is shown in Figure 3.4. Remember that the lower and upper margins of the box and whiskers plot represent the prices that correspond with the 25th and 75th percentiles, while the upper and lower parallel lines represent the 10th and 90th percentiles. The line inside the box represents the 50th (median) percentile. When we examine Figure 3.4, we will note that there are significant trends in our price data. On a year-to-year basis, the prices tend to be skewed and the direction of the skew (i.e., the line representing the median is not in the middle of the box) varies from year to year. This is confirmed by the cumulative probability density functions for the first and second halves of our GM price data, as shown in Figure 3.5. These cumulative probability density functions do not overlap, but they are essen-

Figure 3.1

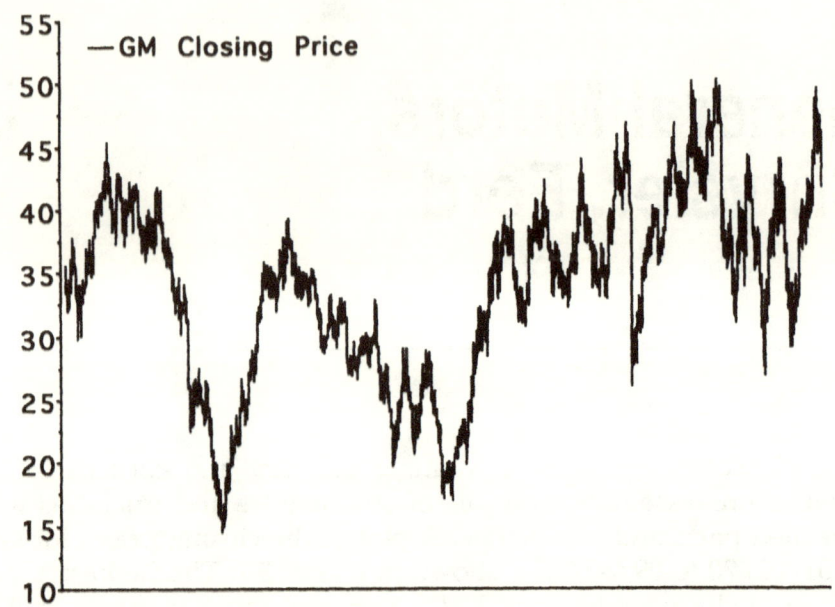

Figure 3.2a

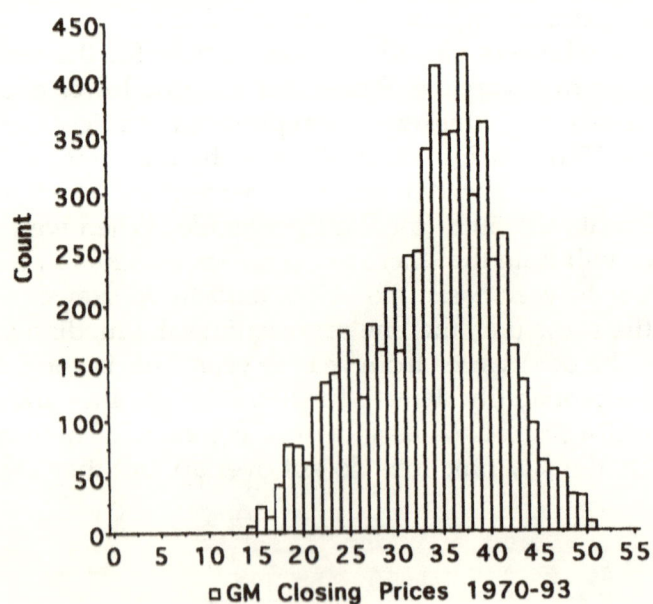

Figure 3.2b

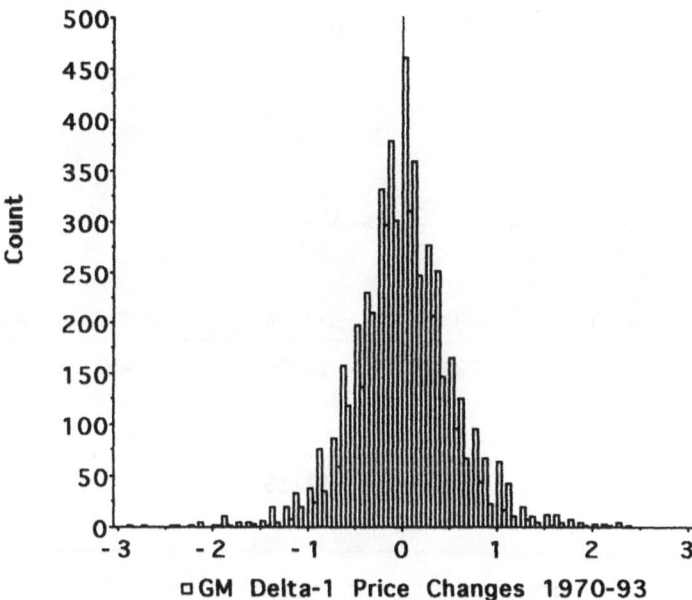

□ GM Delta-1 Price Changes 1970-93

Figure 3.3

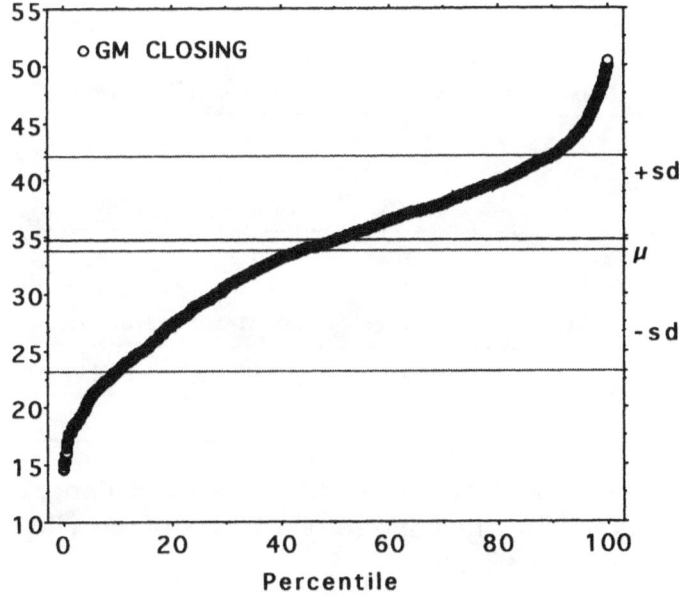

Table 3.1

X_1: GM Closing Price

Mean	Std. Dev.	Std. Error	Variance	Coef. Var.	Count
33.808	7.111	.092	50.572	21.035	6004

Minimum	Maximum	Range	Sum	Sum of Sqr.	# Missing
14.562	50.375	35.812	202984.812	7166149.105	0

# < 10th %	10th %	25th %	50th %	75th %	90th %
588	23.25	29.062	34.75	38.938	42.125

# > 90th %	Mode	Geo. Mean	Har. Mean	Kurtosis	Skewness
591	34	32.978	32.055	−.403	−.36

Table 3.2

X_1: GM Delta-1 Price Change

Mean	Std. Dev.	Std. Error	Variance	Coef. Var.	Count
.001	.537	.007	.288	52637.746	6003

Minimum	Maximum	Range	Sum	Sum of Sqr.	# Missing
−6.938	5.5	12.438	6.125	1731.281	1

# < 10th %	10th %	25th %	50th %	75th %	90th %
467	−.625	−.312	0	.312	.625

# > 90th %	Mode	Geo. Mean	Har. Mean	Kurtosis	Skewness
534	0	•	•	8.743	−.101

tially parallel, so it is likely that GM prices are stationary. The summary statistics for the first and second halves of our GM time series are shown in Table 3.3 a-b.

Figure 3.4

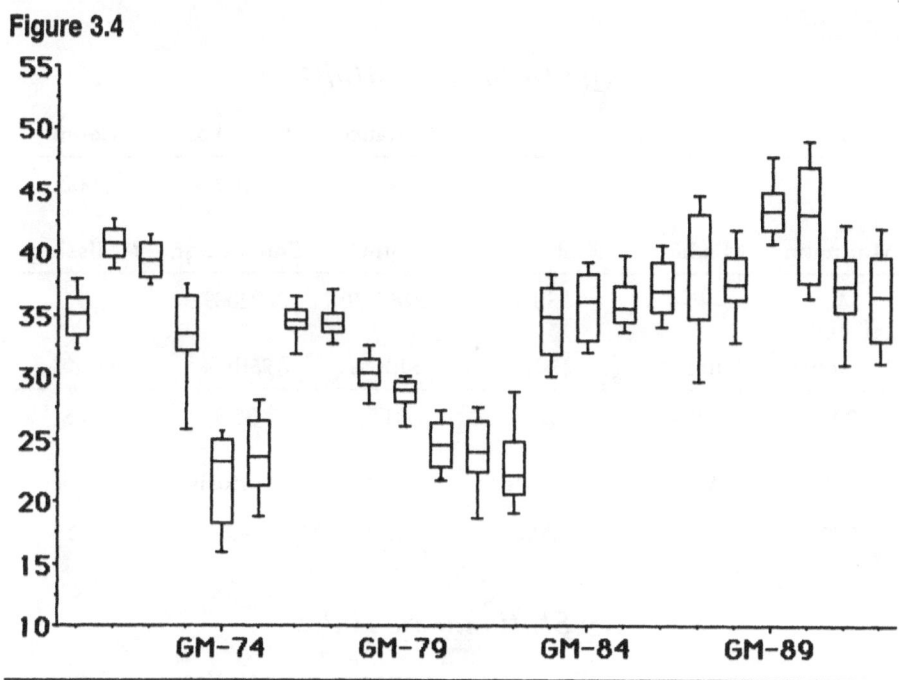

Figure 3.5

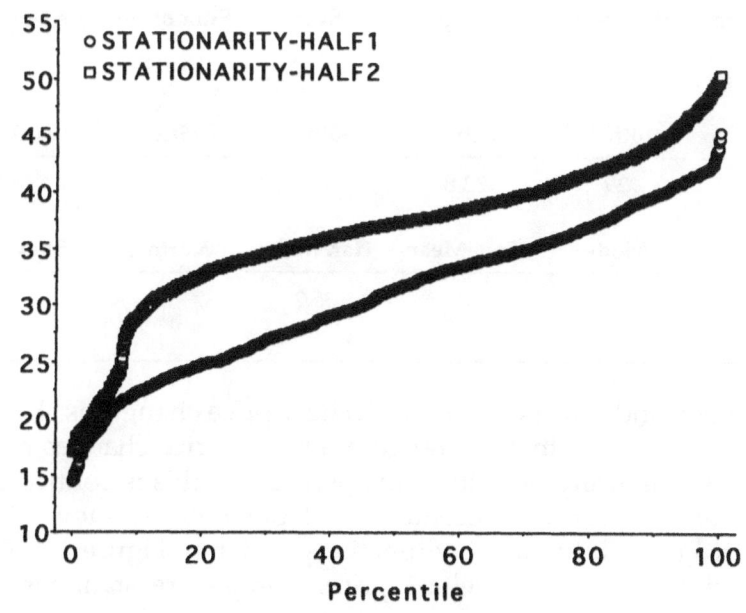

Table 3.3a-b

X_1: Stationarity–Half1

Mean	Std. Dev.	Std. Error	Variance	Coef. Var.	Count
30.9	6.6	.1	43	21.2	3002

Minimum	Maximum	Range	Sum	Sum of Sqr.	# Missing
14.6	45.4	30.8	92871.9	3002089.6	0

# < 10th %	10th %	25th %	50th %	75th %	90th %
300	22.4	25.7	31.5	35.9	39.8

# > 90th %	Mode	Geo. Mean	Har. Mean	Kurtosis	Skewness
298	34	30.2	29.4	–.8	–.2

X_2: Stationarity–Half2

Mean	Std. Dev.	Std. Error	Variance	Coef. Var.	Count
36.7	6.5	.1	41.7	17.6	3002

Minimum	Maximum	Range	Sum	Sum of Sqr.	# Missing
17.1	50.4	33.3	110112.9	4164059.5	0

# < 10th %	10th %	25th %	50th %	75th %	90th %
298	29	33.8	37.2	40.7	44.2

# > 90th %	Mode	Geo. Mean	Har. Mean	Kurtosis	Skewness
299	•	36	35.2	.8	–.8

The box and whiskers plot for Delta-1 price changes is shown in Figure 3.6. It is clear that the range of Delta-1 price changes has not changed dramatically over this time period and this is confirmed by the plot of the first and second half of our data as shown in the cumulative probability density functions shown in Figure 3.7. Therefore, it is clear that our Delta-1 price changes are stationary. The

Figure 3.6

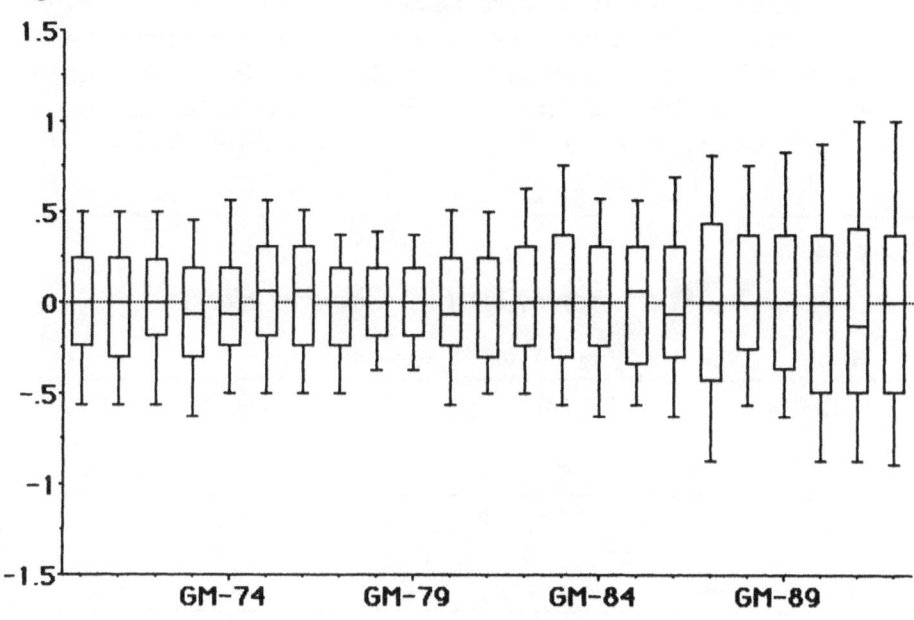

Figure 3.7

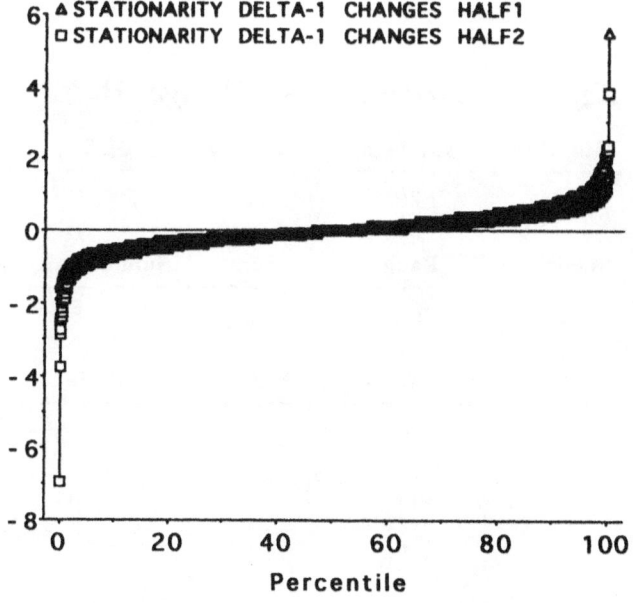

summary statistics for the first and second halves of our GM Delta-1 price changes are shown in Table 3.4 a-b.

GM prices parsed into two frequency histograms for the Odd and Even prices are shown in Figures 3.8 and 3.9. The bin size in these figures is $0.25, so the prices from $20.00 to $20.24 go into Bin-$20.00; prices from $20.25 to $20.49 go into Bin-$20.25, etc. The

---

Table 3.4a-b

X_1: *Stationarity Delta-1 Changes Half1*

Mean	Std. Dev.	Std. Error	Variance	Coef. Var.	Count
–6.0E–3	.4	7.6E–3	.2	–6996.3	3001

Minimum	Maximum	Range	Sum	Sum of Sqr.	# Missing
–1.8	5.5	7.3	–17.9	524.7	1

# < 10th %	10th %	25th %	50th %	75th %	90th %
282	–.5	–.2	0	.2	.5

# > 90th %	Mode	Geo. Mean	Har. Mean	Kurtosis	Skewness
242	0	•	•	10.7	.8

X_2: *Stationarity Delta-1 Changes Half2*

Mean	Std. Dev.	Std. Error	Variance	Coef. Var.	Count
8.0E–3	.6	1.2E–2	.4	7908.7	3001

Minimum	Maximum	Range	Sum	Sum of Sqr.	# Missing
–6.9	3.8	10.8	24.1	1206.6	1

# < 10th %	10th %	25th %	50th %	75th %	90th %
290	–.7	–.4	0	.4	.8

# > 90th %	Mode	Geo. Mean	Har. Mean	Kurtosis	Skewness
280	0	•	•	6.5	–.4

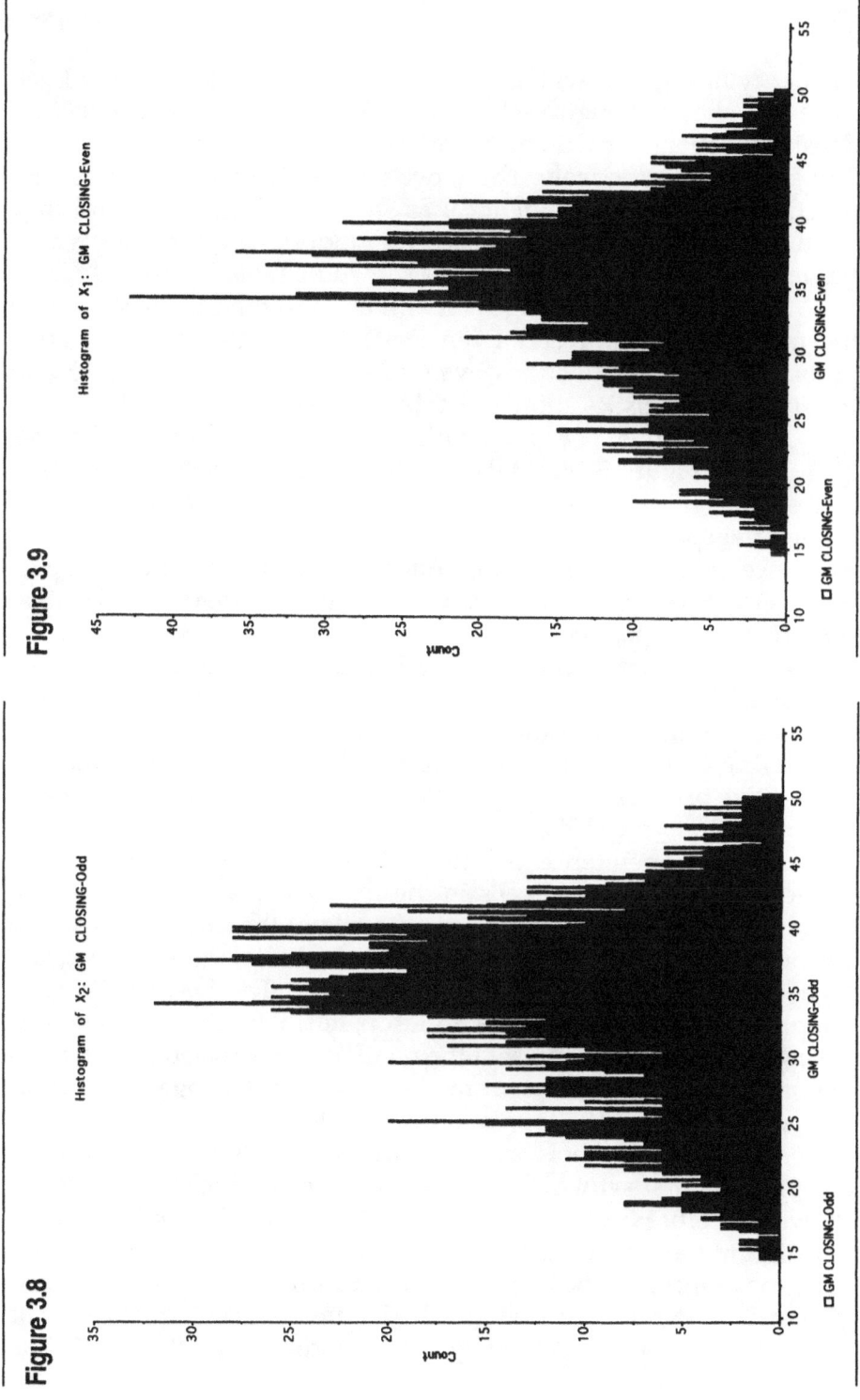

Figure 3.8

Histogram of X₂: GM CLOSING-Odd

Figure 3.9

Histogram of X₁: GM CLOSING-Even

frequency histograms with a bin width of $1.00 for the Odd and Even prices are shown in Figures 3.10 and 3.11. The cumulative probability density functions for Odd and Even GM prices are shown in Figures 3.12 and 3.13, respectively. These two cumulative probability density functions are essentially identical as shown in Figure 3.14 and thus the underlying GM price time series is random. The summary statistics for the even and odd prices are shown in Table 3.5 a-b.

The Delta-1 price changes for GM are also random as shown by the fact that the Odd and Even Delta-1 price change cumulative probability density functions overlap, as shown in Figure 3.15 and the summary statistics shown in Table 3.6 a-b.

The differential spectrum with a bin size of $1.00 for GM prices is shown in Figure 3.16, while the differential spectrum with a bin size of $0.50 is shown in Figure 3.17. The summary statistics for the differential spectrum are shown in Table 3.7. The negative and positive price changes are shown in Table 3.8 a-b. The differential spectrum is asymmetrical and this suggests that the underlying GM time series is not independent.

The relative price change digram, trigram, and tetragram transition matrices are shown in Tables 3.9 to 3.11. All these transition matrices are highly statistically significant, which indicates that the underlying GM time series contains serial dependencies that include at least five prices. Depending on your interest, it would be appropriate to continue to test the GM time series to determine if the diversion from independence continues beyond the pentagram level. However, the number of cells in the matrix and the computational effort increases sharply. Therefore, we will use the *temporal window* to test our GM relative price change time series to determine if there are higher order divergences from independence. The first 10 GM closing prices are shown in Figure 3.18 (top, left) with the Relative Price Change Symbols that represent the relationship between two sequential prices. We make a copy of the Relative Price Change symbols, as shown in Figure 3.18 (top, middle) and then we shift one of the copies down five symbols, as shown in Figure 3.18 (top, right). So we compare the first symbol and the fifth, the second and sixth, etc., as shown in Figure 3.18 (middle and bottom, left). We then construct our digram transition matrix. The first symbol is a '1' and the fifth is a '1', so we increase the count in Cell-1,1 from 0 to 1, as indicated by the small '1' in the transition matrix (Figure 3.18, bottom, right). The next pair is also a '1' followed by a '1', as indicated by the small '2' in

Figure 3.10

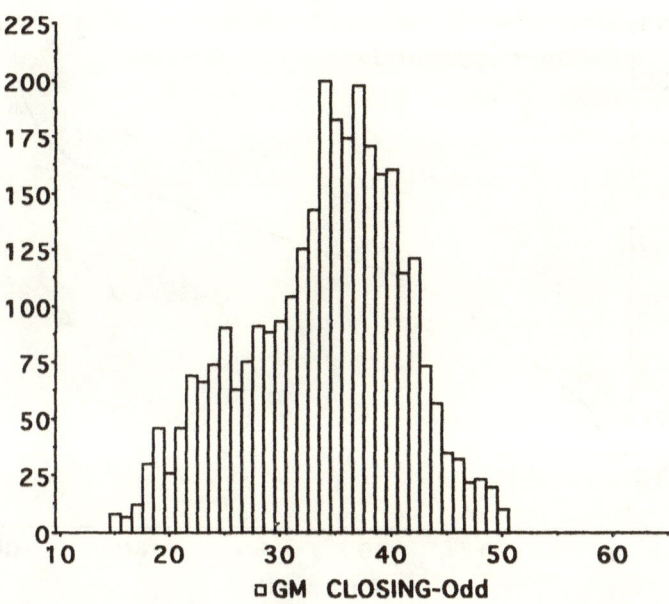

Figure 3.11

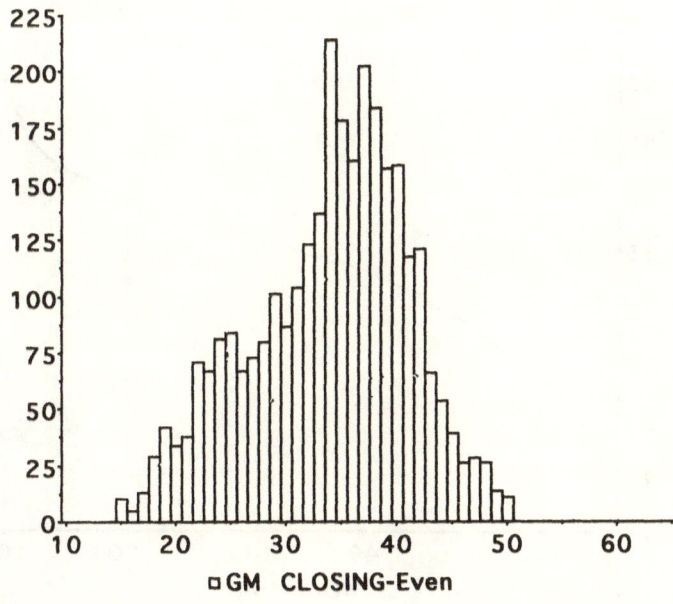

Figure 3.12

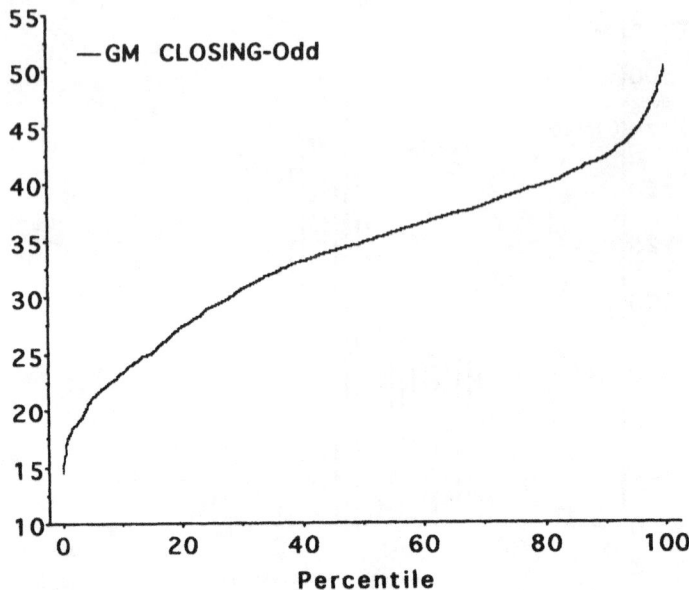

Figure 3.13

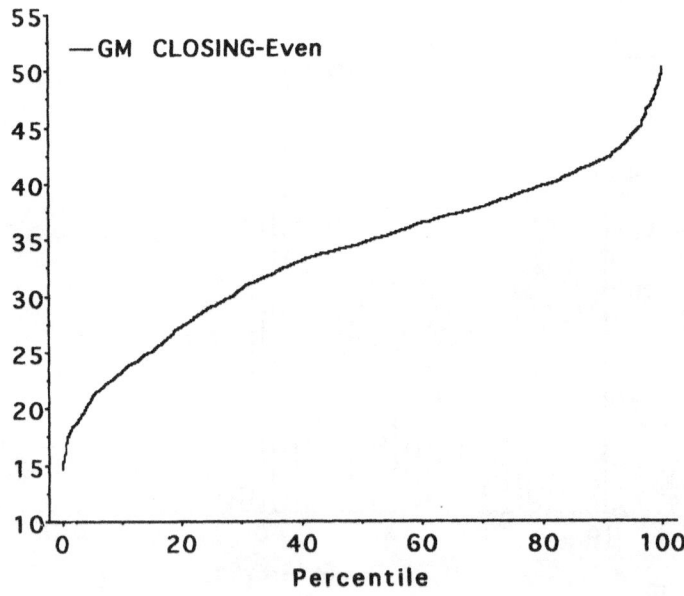

Figure 3.14

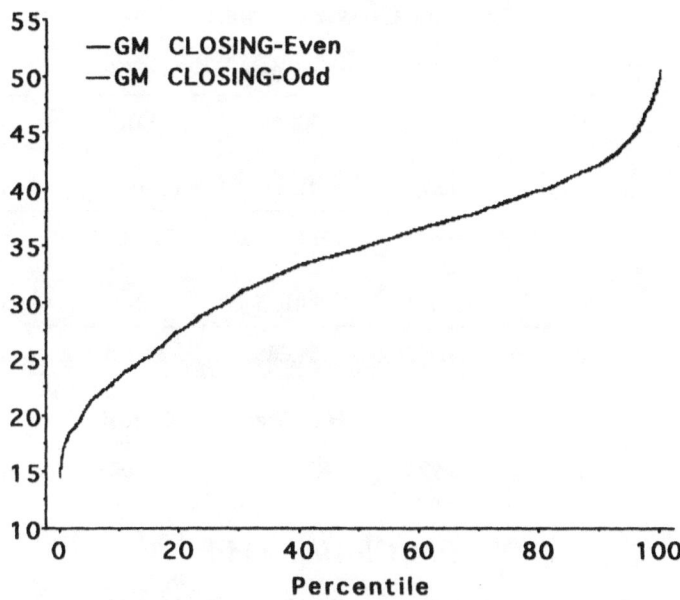

Figure 3.15

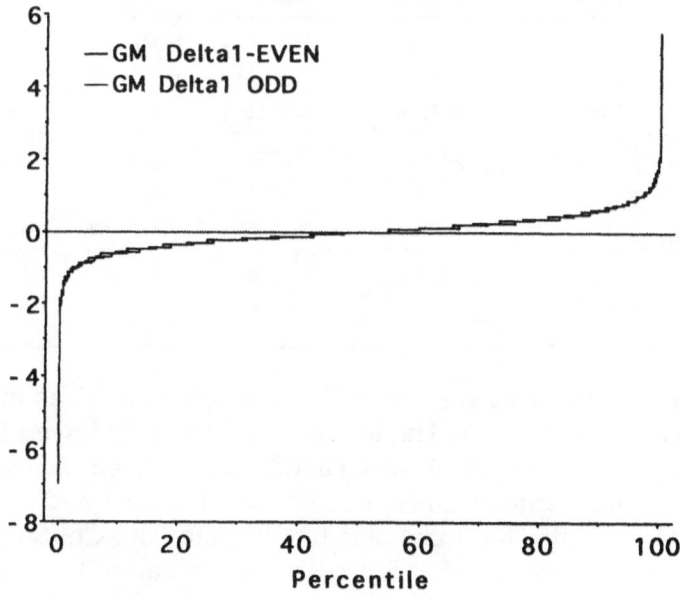

Table 3.5a-b

X₁: GM Closing–Even

Mean	Std. Dev.	Std. Error	Variance	Coef. Var.	Count
33.804	7.106	.13	50.499	21.022	3002

Minimum	Maximum	Range	Sum	Sum of Sqr.	# Missing
14.688	50.375	35.688	101479.562	3581960.926	0

# < 10th %	10th %	25th %	50th %	75th %	90th %
295	23.25	29.062	34.688	38.875	42.125

# > 90th %	Mode	Geo. Mean	Har. Mean	Kurtosis	Skewness
296	34	32.975	32.052	–.405	–.363

X₂: GM Closing–Odd

Mean	Std. Dev.	Std. Error	Variance	Coef. Var.	Count
33.815	7.119	.13	50.678	21.053	3002

Minimum	Maximum	Range	Sum	Sum of Sqr.	# Missing
14.562	50.25	35.688	101511.188	3584637.48	0

# < 10th %	10th %	25th %	50th %	75th %	90th %
300	23.294	29.062	34.75	39	42.125

# > 90th %	Mode	Geo. Mean	Har. Mean	Kurtosis	Skewness
295	36.25	32.983	32.059	–.403	–.356

the matrix. The third pair is also a '1' followed by a '1', as indicated by the small '3' in Cell-1,1. The fourth pair is a '1' followed by a '2', as indicated by the small '4' in Cell-1,2. The *temporal window-10* is generated in the same manner, but we use the first and tenth, the second and eleventh, the third and twelfth, etc. The Chi-Square calculations for *temporal window-5* are shown in Table 3.12. The Chi-

Table 3.6a-b

X_1: GM Delta-1–Even

Mean	Std. Dev.	Std. Error	Variance	Coef. Var.	Count
.013	.542	.01	.294	4238.805	3001

Minimum	Maximum	Range	Sum	Sum of Sqr.	# Missing
−6.938	5.5	12.438	38.375	881.891	1

# < 10th %	10th %	25th %	50th %	75th %	90th %
282	−.562	−.25	0	.312	.625

# > 90th %	Mode	Geo. Mean	Har. Mean	Kurtosis	Skewness
277	0	•	•	13.762	−.245

X_2: GM Delta-1 Odd

Mean	Std. Dev.	Std. Error	Variance	Coef. Var.	Count
−.011	.532	.01	.283	−5050.932	3002

Minimum	Maximum	Range	Sum	Sum of Sqr.	# Missing
−3.75	3.812	7.562	−31.625	850	0

# < 10th %	10th %	25th %	50th %	75th %	90th %
259	−.625	−.312	0	.25	.625

# > 90th %	Mode	Geo. Mean	Har. Mean	Kurtosis	Skewness
258	0	•	•	3.36	.048

Square values for *temporal window-10* through *temporal window-30* are shown in Table 3.13 a-c. These Chi-Square values are all highly significant, so it would be appropriate to continue to increase the size of the *temporal window* until the Chi-Square calculations are no longer significant.

Figure 3.16

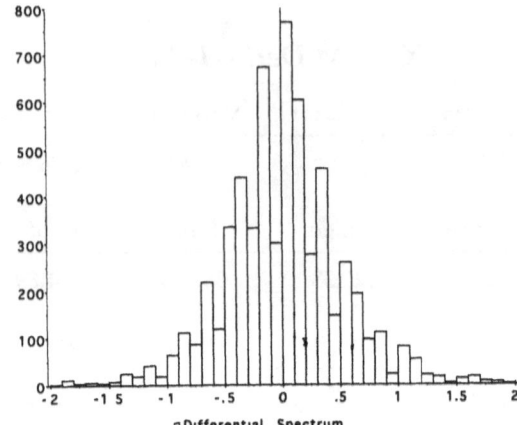

Figure 3.17

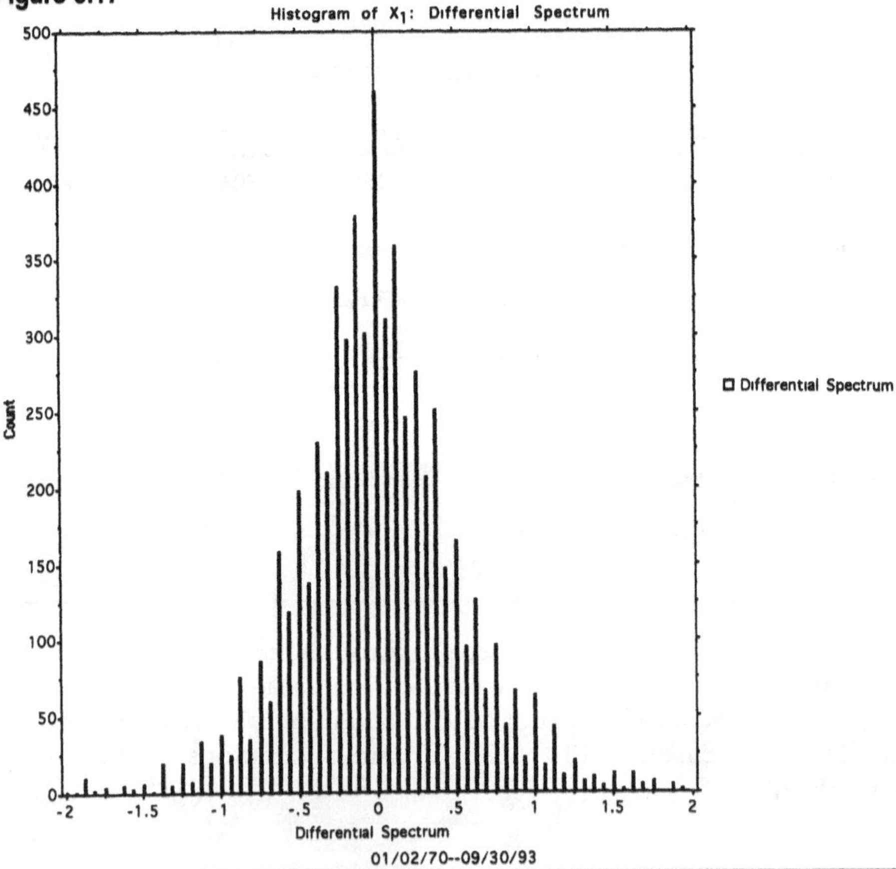

Histogram of X_1: Differential Spectrum

□ Differential Spectrum

Differential Spectrum

01/02/70--09/30/93

Table 3.7

X_1: *Differential Spectrum*

Mean	Std. Dev.	Std. Error	Variance	Coef. Var.	Count
.001	.537	.007	.288	52637.746	6003

Minimum	Maximum	Range	Sum	Sum of Sqr.	# Missing
−6.938	5.5	12.438	6.125	1731.281	1

# < 10th %	10th %	25th %	50th %	75th %	90th %
467	−.625	−.312	0	.312	.625

# > 90th %	Mode	Geo. Mean	Har. Mean	Kurtosis	Skewness
534	0	•	•	8.743	−.101

Table 3.8a

X_1: *1*

Mean	Std. Dev.	Std. Error	Variance	Coef. Var.	Count
−.352	.367	.006	.134	−104.161	3287

Minimum	Maximum	Range	Sum	Sum of Sqr.	# Missing
−6.938	0	6.938	−1156.688	848.512	0

# < 10th %	10th %	25th %	50th %	75th %	90th %
322	−.75	−.5	−.25	−.125	0

# > 90th %	Mode	Geo. Mean	Har. Mean	Kurtosis	Skewness
0	0	•	•	37.568	−3.526

We can use the category price change method to determine if small and large Delta-1 price changes are independently distributed. For example, we can divide our time series into six unequal parts,

Table 3.8b

X_2: 2

Mean	Std. Dev.	Std. Error	Variance	Coef. Var.	Count
.428	.377	.007	.142	87.948	2716

Minimum	Maximum	Range	Sum	Sum of Sqr.	# Missing
.062	5.5	5.438	1162.812	882.77	571

# < 10th %	10th %	25th %	50th %	75th %	90th %
0	.062	.188	.312	.562	.875

# > 90th %	Mode	Geo. Mean	Har. Mean	Kurtosis	Skewness
260	.125	.302	.206	17.254	2.645

Table 3.9

	A	B	C	D	E	F	G	H	I
1	n	Digram	Expec Prob	(E)xpec Freq	(O)bser Freq	(E-O)	(E-O)^2	((E-O)^2)/E	SUM
2	6002	11	0.1667	1000.5334	1869	-868.4666	754234.235	753.832141	
3		12	0.3333	2000.4666	1417	583.4666	340433.273	170.176934	924.009076
4		21	0.3333	2000.4666	1417	583.4666	340433.273	170.176934	1094.18601
5		22	0.1667	1000.5334	1299	-298.4666	89082.3113	89.0348201	1264.36294
6									
7					6002				
8								Chi-Square	1353.39776

Table 3.10

	A	B	C	D	E	F	G	H	I
1	n	Trigram	Expec Prob	(E)xpec Freq	(O)bser Freq	(E-O)	(E-O)^2	((E-O)^2)/E	SUM
2	6001	111	0.04167	250.06167	1029	-778.93833	606744.922	2426.38115	
3		112	0.125	750.125	839	-88.875	7898.76563	10.5299325	2436.91106
4		121	0.20833	1250.18833	714	536.18833	287497.925	229.963693	2666.87477
5		122	0.125	750.125	703	47.125	2220.76563	2.96052741	2669.8353
6		211	0.125	750.125	839	-88.875	7898.76563	10.5299325	2680.36523
7		212	0.20833	1250.18833	578	672.18833	451837.151	361.415268	3041.7805
8		221	0.125	750.125	703	47.125	2220.76563	2.96052741	3044.74103
9		222	0.04167	250.06167	596	-345.93833	119673.328	478.575258	
10									
11					6001			Chi-Square	3523.31629

such that a '1' includes all of the Delta-1 price changes that are in the lowest 10% (–6.938 to –0.625) of the Delta-1 price change cumulative probability distribution. A '2' represents the price changes from the

Table 3.11

	A	B	C	D	E	F	G	H	I
1	n	Tetragram	Expected P	(E)xpected P	(O)bser P	(E-O)	(E-O)^2	((E-O)^2)/E	Sum
2	6000	1111	0.00833	49.98	543	-493.02	243068.72	4863.31974	
3		1112	0.03333	199.98	485	-285.02	81236.4004	406.222624	5269.54236
4		1121	0.075	450	427	23	529	1.17555556	5270.71792
5		1122	0.05	300	412	-112	12544	41.8133333	5312.53125
6		1211	0.075	450	428	22	484	1.07555556	5313.6068
7		1212	0.13333	799.98	286	513.98	264175.44	330.227556	5643.83436
8		1221	0.096166	576.996	366	210.996	44519.312	77.1570548	5720.99142
9		1222	0.03333	199.98	337	-137.02	18774.4804	93.8817902	5814.87321
10		2111	0.03333	199.98	485	-285.02	81236.4004	406.222624	6221.09583
11		2112	0.09166	549.98	354	195.96	38400.3216	69.8238446	6290.91967
12		2121	0.13333	799.98	287	512.98	263148.48	328.943824	6619.8635
13		2122	0.075	450	291	159	25281	56.18	6676.0435
14		2211	0.05	300	411	-111	12321	41.07	6717.1135
15		2212	0.075	450	292	158	24964	55.4755556	6772.58905
16		2221	0.03333	199.98	337	-137.02	18774.4804	93.8817902	6866.47084
17		2222	0.00833	49.98	259	-209.02	43689.3604	874.136863	7740.60771
18									
19					6000				
20								Chi-Square	7740.60771

Figure 3.18

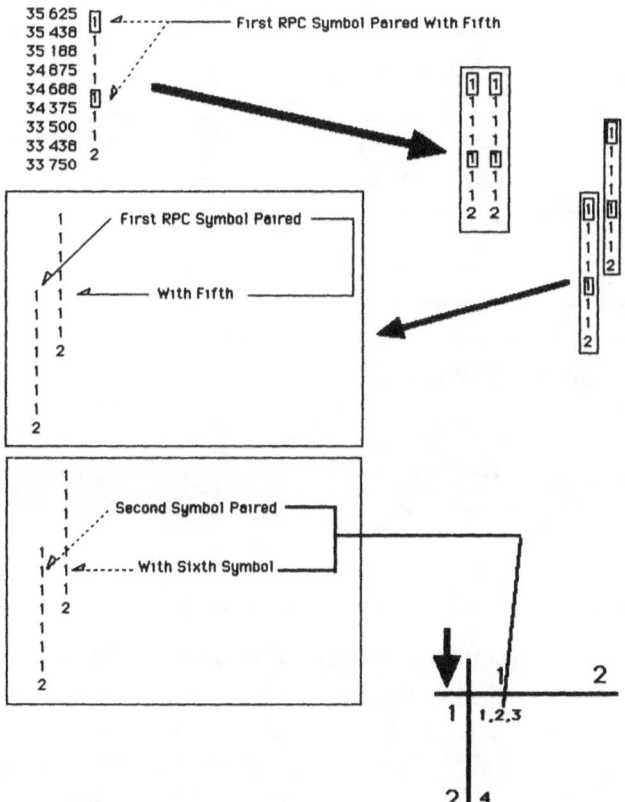

Table 3.12   Temporal Window–5

	A	B	C	D	E	F	G	H	I
1	n	Digram	Expec Prob	(E)xpec Freq	(O)bser Freq	(E-O)	(E-O)^2	((E-O)^2)/E	SUM
2	5999	11	0.1667	1000.0333	1817	-816.9667	667434.589	667.412364	
3		12	0.3333	1999.4667	1491	508.4667	258538.385	129.303671	796.716035
4		21	0.3333	1999.4667	1491	508.4667	258538.385	129.303671	926.019707
5		22	0.1667	1000.0333	1200	-199.9667	39986.6811	39.9853496	1055.32338
6									
7			Sum=	5999	5999				
8								Chi-Square	1095.30873

Table 3.13a   Temporal Window–10

	A	B	C	D	E	F	G	H	I
1	n	Digram	Expec Prob	(E)xpec Freq	(O)bser Freq	(E-O)	(E-O)^2	((E-O)^2)/E	SUM
2	5994	11	0.1667	999.1998	1797	-797.8002	636485.159	636.994882	
3		12	0.3333	1997.8002	1509	488.8002	238925.636	119.59436	756.589242
4		21	0.3333	1997.8002	1507	490.8002	240884.836	120.575039	877.164281
5		22	0.1667	999.1998	1181	-181.8002	33051.3127	33.0777816	997.739319
6			Sum=	5994	5994				
7									
8								Chi-Square	1030.8171

Table 3.13b   Temporal Window–20

	A	B	C	D	E	F	G	H	I
1	n	Digram	Expec Prob	(E)xpec Freq	(O)bser Freq	(E-O)	(E-O)^2	((E-O)^2)/E	SUM
2	5984	11	0.1667	997.5328	1813	-815.4672	664986.754	666.631467	
3		12	0.3333	1994.4672	1487	507.4672	257522.959	129.118673	795.750141
4		21	0.3333	1994.4672	1484	510.4672	260576.762	130.649811	926.399952
5		22	0.1667	997.5328	1200	-202.4672	40992.9671	41.0943551	1057.04976
6			Sum=	5984	5984				
7									
8								Chi-Square	1098.14412

Table 3.13c   Temporal Window–30

	A	B	C	D	E	F	G	H	I
1	n	Digram	Expec Prob	(E)xpec Freq	(O)bser Freq	(E-O)	(E-O)^2	((E-O)^2)/E	SUM
2	5974	11	0.1667	995.8658	1806	-810.1342	656317.422	659.042034	
3		12	0.3333	1991.1342	1488	503.1342	253144.023	127.135591	786.177625
4		21	0.3333	1991.1342	1486	505.1342	255160.56	128.148349	914.325973
5		22	0.1667	995.8658	1194	-198.1342	39257.1612	39.4201319	1042.47432
6			Sum=	5974	5974				
7									
8								Chi-Square	1081.89445

11th percentile to the 25th percentile (–0.624 to –0.312), a '3' from the 26th to the 50th percentile (–0.311 to 0.01), a '4' from the 51st to the 75th percentile (0.011 to 0.312), and a '5' from the 75th to the 90th percentile (0.313 to 0.625), and a '6' all of the price changes above the 91st percentile (0.626 to 5.5). The frequency of occurrence of the six symbols is shown at the top of Table 3.14 (column 2), while the probability of occurrence of the symbols is shown in column 3. The expected frequency of occurrence of each of the digrams under the assumption of independence is shown at the bottom of Table 3.14 (column 2). The observed frequency of occurrence of each of the

Table 3.14

	A	B	C	D	E	F	G	H
1	Symbols	Frequency	Probability					
2	1	624	0.1040					
3	2	894	0.1490					
4	3	1768	0.2946					
5	4	1398	0.2329					
6	5	784	0.1306					
7	6	534	0.0890					
8	Sum	6002						
9								
10								
11								
12	Digrams	(E)xpected	(O)bserved	(E-O)	(E-O)^2	((E-O)^2)/E	Sum	
13	11	64.874	92.000	-27.126	735.800	11.342		
14	12	92.945	88.000	4.945	24.453	0.263	11.605	
15	13	183.811	161.000	22.811	520.329	2.831	14.436	
16	14	145.344	140.000	5.344	28.554	0.196	14.632	
17	15	81.509	76.000	5.509	30.347	0.372	15.005	
18	16	55.517	67.000	-11.483	131.848	2.375	17.379	
19	21	92.945	101.000	-8.055	64.883	0.698	15.134	
20	22	133.162	161.000	-27.838	774.976	5.820	20.954	
21	23	263.344	241.000	22.344	499.264	1.896	22.850	
22	24	208.233	218.000	-9.767	95.402	0.458	23.308	
23	25	116.777	110.000	6.777	45.929	0.393	23.701	
24	26	79.539	63.000	16.539	273.555	3.439	27.140	
25	31	183.811	179.000	4.811	23.143	0.126	22.975	
26	32	263.344	257.000	6.344	40.249	0.153	23.128	
27	33	520.797	589.000	-68.203	4651.640	8.932	32.060	
28	34	411.807	393.000	18.807	353.693	0.859	32.919	
29	35	230.942	234.000	-3.058	9.353	0.041	32.959	
30	36	157.300	116.000	41.300	1705.654	10.843	43.803	
31	41	145.344	116.000	29.344	861.044	5.924	49.727	
32	42	208.233	208.000	0.233	0.054	0.000	49.727	
33	43	411.807	430.000	-18.193	330.995	0.804	50.531	
34	44	325.625	363.000	-37.375	1396.856	4.290	54.821	
35	45	182.611	165.000	17.611	310.152	1.698	56.519	
36	46	124.381	116.000	8.381	70.233	0.565	57.084	
37	51	81.509	71.000	10.509	110.436	1.355	58.439	
38	52	116.777	105.000	11.777	138.699	1.188	59.626	
39	53	230.942	198.000	32.942	1085.155	4.699	64.325	
40	54	182.611	183.000	-0.389	0.151	0.001	64.326	
41	55	102.409	131.000	-28.591	817.472	7.982	72.309	
42	56	69.753	96.000	-26.247	688.918	9.877	82.185	
43	61	55.517	66.000	-10.483	109.883	1.979	84.164	
44	62	79.539	75.000	4.539	20.607	0.259	84.424	
45	63	157.300	148.000	9.300	86.482	0.550	84.973	
46	64	124.381	101.000	23.381	546.650	4.395	89.368	
47	65	69.753	68.000	1.753	3.072	0.044	89.412	
48	66	47.510	76.000	-28.490	811.671	17.084	106.496	
49								
50		n=	6002			Chi-Square	106.496472	
51								
52								
53						14.5942778	8.36660027	6.22767749

digrams is shown in column 3. The Chi-Square values are shown in column 7. The Chi-Square value is 106.496. We can use Equation 2.1 to determine if the Chi-Square is significant.

In this case 'z' is equal to 6.23, which is higher than 1.96. This means that the observed frequency of occurrence of the digrams differs significantly from the frequency of occurrence of these digrams under the assumption of independence. Therefore, the GM Delta-1 time series does contain serial dependencies. If 'z' were less than 1.96, then it would indicate that the observed distribution of digrams was not significantly different from the distribution of digrams under the assumption of independence and thus the underlying GM Delta-1 time series did not contain serial dependencies. Since our 'z' value is higher than 1.96, it would be appropriate to continue our analysis to determine our serial dependencies extended to trigrams, tetragrams, etc. Five digrams make a large contribution to the ultimate Chi-Square; they are 1,1, ,3, 5,6, and 6,6, which all occur more frequently than expected and 3,6, which occurs less frequently.

We can parse our GM Delta-1 time series using the categories (the symbols 1–6) as our parsing agent. The summary statistics for each of the six categories (1–6) are shown in Table 3.15a-f. Table 3.16 a-f shows the value of the first price in the Delta-1 price change for each of the six categories, while Table 3.17 a-f shows the value of the second price in the Delta-1 price change for each of the six categories. If we examine these tables, we will note that the absolute prices that represent the first price in the Delta-1 price changes are not very different over the distribution of price changes. For example, the median price for a '1' (lowest 10% of the Delta-1 price changes) is 37, while the median prices for '2' through '6' are 34.9, 34.2, 34.1, 34.8, and 35.9, respectively, as shown in Table 3.16a-f (50th %). The median price for the second price in the Delta-1 price change is also similar over symbols (36, 34.5, 34, 34.2, 35.3, and 36.9, respectively (Table 3.17a-f 50th %). This suggests that the prices themselves may not show serial dependencies.

We can also determine the temporal distribution of the symbols, as shown in Figure 3.19. In this case, we number the symbols sequentially from 1 to 6,004 (using only trading days, ignoring weekends, holidays, etc.). Then, we parse this series using our categories 1–6. Then, we determine the number of days that occurred between each occurrence of the category (see Figure 3.19). Table 3.18 a-f shows the temporal distribution of the categories. If we examine this table, we

Table 3.15a-b

X_1: 1

Mean	Std. Dev.	Std. Error	Variance	Coef. Var.	Count
−.9	.4	1.7E–2	.2	−47	625

Minimum	Maximum	Range	Sum	Sum of Sqr.	# Missing
−6.9	−.6	6.3	−574.2	644	1143

# < 10th %	10th %	25th %	50th %	75th %	90th %
45	−1.4	−1	−.8	−.6	−.6

# > 90th %	Mode	Geo. Mean	Har. Mean	Kurtosis	Skewness
0	−.6	•	•	63.7	−5.8

X_2: 2

Mean	Std. Dev.	Std. Error	Variance	Coef. Var.	Count
−.4	.1	2.9E–3	7.4E–3	−20.4	894

Minimum	Maximum	Range	Sum		Sum of Sqr.
−.6	−.3	.2	−377.8	166.2	874

# < 10th %	10th %	25th %	50th %	75th %	90th %
0	−.6	−.5	−.4	−.4	−.3

# > 90th %	Mode	Geo. Mean	Har. Mean	Kurtosis	Skewness
0	−.4	•	•	−1.3	−.2

will note that the categories are unevenly distributed in time. For example, category 1 (the lowest 10% of the Delta-1 price changes) had as many as 120 days between sequential occurrences of this level of price change.

We can use other characteristics to parse our time series into component parts. For example, we can generate a Close-Open time

Table 3.15c-d

X_3: 3

Mean	Std. Dev.	Std. Error	Variance	Coef. Var.	Count
−.1	.1	2.2E–3	8.2E–3	−78.4	1768

Minimum	Maximum	Range	Sum	Sum of Sqr.	# Missing
−.2	0	.2	−204.8	38.3	0

# < 10th %	10th %	25th %	50th %	75th %	90th %
0	−.2	−.2	−.1	0	0

# > 90th %	Mode	Geo. Mean	Har. Mean	Kurtosis	Skewness
0	0	•	•	−1.3	−.1

X_4: 4

Mean	Std. Dev.	Std. Error	Variance	Coef. Var.	Count
.2	.1	2.3E–3	7.4E–3	49.2	1398

Minimum	Maximum	Range	Sum	Sum of Sqr.	# Missing
.1	.3	.2	244.1	52.9	370

# < 10th %	10th %	25th %	50th %	75th %	90th %
0	.1	.1	.2	.2	.3

# > 90th %	Mode	Geo. Mean	Har. Mean	Kurtosis	Skewness
0	.1	.2	.1	−1.2	.2

series by determining if the opening price is larger or smaller than the closing price and encode this relationship as a 1 or a 2, respectively. We can then collect these symbols into a digram transition matrix (Table 3.19, column 3), which can be compared to the digram matrix generated under the assumption of independence (Table 3.19, column 2). The Chi-Square value is not statistically significant, which

Table 3.15e–f

X_5: 5

Mean	Std. Dev.	Std. Error	Variance	Coef. Var.	Count
.5	.1	3.2E–3	8.2E–3	19	784

Minimum	Maximum	Range	Sum	Sum of Sqr.	# Missing
.4	.6	.2	373.1	184	984

# < 10th %	10th %	25th %	50th %	75th %	90th %
0	.4	.4	.4	.6	.6

# > 90th %	Mode	Geo. Mean	Har. Mean	Kurtosis	Skewness
0	.4	.5	.5	–1.2	.4

X_6: 6

Mean	Std. Dev.	Std. Error	Variance	Coef. Var.	Count
1	.4	1.8E–2	.2	39.9	534

Minimum	Maximum	Range	Sum	Sum of Sqr.	# Missing
.7	5.5	4.8	545.6	645.9	1234

# < 10th %	10th %	25th %	50th %	75th %	90th %
0	.7	.8	.9	1.1	1.5

# > 90th %	Mode	Geo. Mean	Har. Mean	Kurtosis	Skewness
49	.8	1	.9	31.3	4

means that the Close-Open time series does not contain serial dependencies.

As indicated above, it is important to test your time series over a shorter time span. We will use the closing prices for GM for 09/30/88 to 09/30/93 for these tests. The cumulative probability density functions for the Half1 and Half2 are shown in Figure 3.20,

Table 3.16a-b

X_1: 1

Mean	Std. Dev.	Std. Error	Variance	Coef. Var.	Count
36.1	6.9	.3	47	19	625

Minimum	Maximum	Range	Sum	Sum of Sqr.	# Missing
15.6	50.2	34.6	22553.3	843187.4	1143

# < 10th %	10th %	25th %	50th %	75th %	90th %
62	26.1	32.2	37	40.6	44.1

# > 90th %	Mode	Geo. Mean	Har. Mean	Kurtosis	Skewness
62	•	35.4	34.5	4.7E–4	–.5

X_2: 2

Mean	Std. Dev.	Std. Error	Variance	Coef. Var.	Count
33.9	7.1	.2	50.3	20.9	894

Minimum	Maximum	Range	Sum	Sum of Sqr.	# Missing
15.4	50.4	34.9	30281.3	1070599.6	874

# < 10th %	10th %	25th %	50th %	75th %	90th %
89	23.4	28.8	34.9	38.9	42.2

# > 90th %	Mode	Geo. Mean	Har. Mean	Kurtosis	Skewness
87	35.6	33.1	32.1	–.5	–.4

while the summary statistics are shown in Table 3.20 a-b. These two probability density functions are similar (they are essentially parallel), so it is likely that this time series is stationary.

The cumulative probability density functions for the Delta-1 price changes for this same time period are shown in Figure 3.21 and the summary statistics are shown in Table 3.21 a-b. These two prob-

Table 3.16c-d

X_3: 3

Mean	Std. Dev.	Std. Error	Variance	Coef. Var.	Count
33.2	7.2	.2	52	21.7	1768

Minimum	Maximum	Range	Sum	Sum of Sqr.	# Missing
14.7	50.2	35.6	58674	2039098.5	0

# < 10th %	10th %	25th %	50th %	75th %	90th %
174	22.6	28.2	34.2	38.4	41.8

# > 90th %	Mode	Geo. Mean	Har. Mean	Kurtosis	Skewness
173	34	32.3	31.4	−.5	−.3

X_4: 4

Mean	Std. Dev.	Std. Error	Variance	Coef. Var.	Count
33	7	.2	48.5	21.1	1398

Minimum	Maximum	Range	Sum	Sum of Sqr.	# Missing
14.6	50	35.4	46201.8	1594594.6	370

# < 10th %	10th %	25th %	50th %	75th %	90th %
138	22.8	27.9	34.1	38.1	41.5

# > 90th %	Mode	Geo. Mean	Har. Mean	Kurtosis	Skewness
140	35.2	32.2	31.4	−.5	−.3

ability density functions are identical, thus indicating that the Delta-1 prices changes are stationary.

The Odd and Even cumulative probability density functions for GM closing prices for this time period are shown in Figures 3.22 and 3.23, respectively. These cumulative probability density functions are essentially identical, indicating that the underlying GM closing prices

Table 3.16e-f

X_5: 5

Mean	Std. Dev.	Std. Error	Variance	Coef. Var.	Count
34	6.9	.2	48.2	20.4	784

Minimum	Maximum	Range	Sum	Sum of Sqr.	# Missing
14.7	49.6	34.9	26622.6	941738.6	984

# < 10th %	10th %	25th %	50th %	75th %	90th %
78	23.9	29.5	34.8	39	42

# > 90th %	Mode	Geo. Mean	Har. Mean	Kurtosis	Skewness
76	36	33.2	32.3	–.3	–.4

X_6: 6

Mean	Std. Dev.	Std. Error	Variance	Coef. Var.	Count
34.9	7.1	.3	49.9	20.3	534

Minimum	Maximum	Range	Sum	Sum of Sqr.	# Missing
15.9	49.6	33.8	18610.1	675187.3	1234

# < 10th %	10th %	25th %	50th %	75th %	90th %
53	24.2	30.8	35.9	39.6	43.1

# > 90th %	Mode	Geo. Mean	Har. Mean	Kurtosis	Skewness
52	•	34	33.1	–.1	–.5

are random. This is confirmed by the summary statistics shown in Table 3.22 a-b. The Delta-1 price changes are also randomly distributed as indicated by the fact that the Odd and Even cumulative probability density functions are essentially identical, as shown in Figure 3.24 and Table 3.23a-b.

Table 3.17a-b

$X_1: 1$

Mean	Std. Dev.	Std. Error	Variance	Coef. Var.	Count
35.2	6.8	.3	46.5	19.4	625

Minimum	Maximum	Range	Sum	Sum of Sqr.	# Missing
15	49.2	34.2	21979.1	801965.1	1143

# < 10th %	10th %	25th %	50th %	75th %	90th %
61	25.1	31.2	36	39.8	43.2

# > 90th %	Mode	Geo. Mean	Har. Mean	Kurtosis	Skewness
62	34	34.4	33.6	−9.6E−3	−.5

$X_2: 2$

Mean	Std. Dev.	Std. Error	Variance	Coef. Var.	Count
33.4	7.1	.2	50.2	21.2	894

Minimum	Maximum	Range	Sum	Sum of Sqr.	# Missing
15.1	49.9	34.8	29903.6	1045108.8	874

# < 10th %	10th %	25th %	50th %	75th %	90th %
87	23	28.4	34.5	38.5	41.9

# > 90th %	Mode	Geo. Mean	Har. Mean	Kurtosis	Skewness
87	38.4	32.6	31.7	−.5	−.4

The differential spectrum is shown in Figure 3.25 with the summary statistics shown in Table 3.24. The differential spectrum is not symmetrical as shown by the box and whiskers plot shown in Figure 3.26 and the summary statistics shown in Table 3.25 a-b. This is confirmed by the fact that the digram, trigram, and tetragram relative

Table 3.17c-d

X_3: 3

Mean	Std. Dev.	Std. Error	Variance	Coef. Var.	Count
33.1	7.2	.2	52	21.8	1768

Minimum	Maximum	Range	Sum	Sum of Sqr.	# Missing
14.6	50	35.4	58469.2	2025515.5	0

# < 10th %	10th %	25th %	50th %	75th %	90th %
176	22.5	28.2	34	38.3	41.6

# > 90th %	Mode	Geo. Mean	Har. Mean	Kurtosis	Skewness
176	34	32.2	31.2	–.5	–.3

X_4: 4

Mean	Std. Dev.	Std. Error	Variance	Coef. Var.	Count
33.2	7	.2	48.5	21	1398

Minimum	Maximum	Range	Sum	Sum of Sqr.	# Missing
14.7	50.1	35.4	46445.9	1610800.3	370

# < 10th %	10th %	25th %	50th %	75th %	90th %
140	22.9	28.1	34.2	38.2	41.6

# > 90th %	Mode	Geo. Mean	Har. Mean	Kurtosis	Skewness
136	•	32.4	31.5	–.5	–.3

price change matrices all diverge from independence, as shown in Tables 3.26 to 3.28, respectively.

The cumulative probability density functions for Half1 and Half2 for GM closing prices for 05/04/92 to 05/04/94 are shown in Figure 3.27 and for the Delta-1 price changes are shown in Figure 3.28 and the summary statistics are shown in Table 3.29 a-b and 3.30

Table 3.17e-f

X_5: 5

Mean	Std. Dev.	Std. Error	Variance	Coef. Var.	Count
34.4	6.9	.2	48.2	20.2	784

Minimum	Maximum	Range	Sum	Sum of Sqr.	# Missing
15.2	50.2	35	26995.7	967306.8	984

# < 10th %	10th %	25th %	50th %	75th %	90th %
78	24.4	29.9	35.3	39.5	42.4

# > 90th %	Mode	Geo. Mean	Har. Mean	Kurtosis	Skewness
76	34.1	33.7	32.8	−.3	−.4

X_6: 6

Mean	Std. Dev.	Std. Error	Variance	Coef. Var.	Count
35.9	7.1	.3	50.7	19.9	534

Minimum	Maximum	Range	Sum	Sum of Sqr.	# Missing
16.7	50.4	33.7	19155.7	714183.4	1234

# < 10th %	10th %	25th %	50th %	75th %	90th %
53	25	31.9	36.9	40.6	44.4

# > 90th %	Mode	Geo. Mean	Har. Mean	Kurtosis	Skewness
52	37.9	35.1	34.2	−.1	−.5

a-b, respectively. The cumulative probability density functions for GM closing prices for Odd and Even are shown in Figure 3.29, while the Delta-1 price changes are shown in Figure 3.30 and the summary statistics are shown in Table 3.31 a-b and 3.32 a-b. The differential spectrum with $0.10 bin size is shown in Figure 3.31. It is not symmetrical, which suggests that GM prices for 92–94 contain serial de-

Figure 3.19

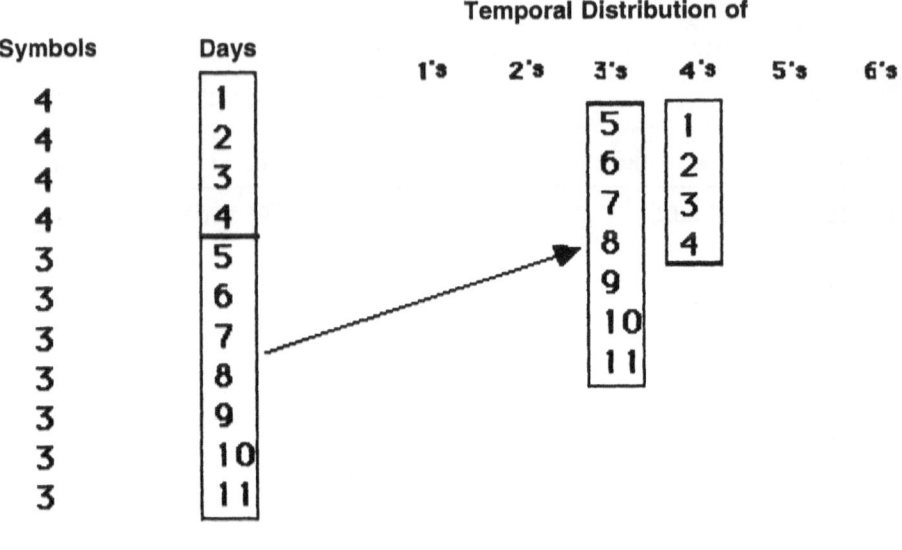

pendencies. This is confirmed by the digram relative price change transition matrix which is shown in Table 3.33.

Chrysler is one of GM's major American competitors. A graph of Chrysler's closing prices for 1977 to 1993 is shown in Figure 3.32. The cumulative probability density function of closing prices is shown in Figure 3.33, while the frequency distribution of closing prices is shown in Figure 3.34. The summary statistics are shown in Table 3.34. The frequency histogram and the cumulative probability density function for Delta-1 price changes are shown in Figures 3.35 and 3.36, while the Delta-1 summary statistics are shown in Table 3.35.

The cumulative probability density functions for Half1 and Half2 for Chrysler closing prices and Delta-1 price changes for 1977 to 1993 are shown in Figure 3.37 a and b, respectively. The summary statistics are shown in Tables 3.36a-b and 3.37a-b, respectively. These cumulative probability density functions suggest that Chrysler closing prices and Delta-1 price changes are probably stationary.

The closing prices and Delta-1 price changes are also random as shown by the fact that the Odd and Even cumulative probability

Table 3.18a-b

X_1

Mean	Std. Dev.	Std. Error	Variance	Coef. Var.	Count
9.6	11.9	.5	142.4	124.1	624

Minimum	Maximum	Range	Sum	Sum of Sqr.	# Missing
1	120	119	5997	146321	1144

# < 10th %	10th %	25th %	50th %	75th %	90th %
0	1	2.5	6	12	23

# > 90th %	Mode	Geo. Mean	Har. Mean	Kurtosis	Skewness
59	1	5.5	3.3	16.6	3.3

X_2

Mean	Std. Dev.	Std. Error	Variance	Coef. Var.	Count
6.7	6.4	.2	40.9	95.3	893

Minimum	Maximum	Range	Sum	Sum of Sqr.	# Missing
1	43	42	5990	76652	875

# < 10th %	10th %	25th %	50th %	75th %	90th %
0	1	2	5	10	15

# > 90th %	Mode	Geo. Mean	Har. Mean	Kurtosis	Skewness
78	1	4.4	2.8	5.5	2

density functions overlap (see Figures 3.38 and 3.39 and Tables 3.38 a-b and 3.39 a-b, respectively).

The digram, trigram, tetragram, and pentagram relative price change transition matrices are all highly significantly different from those generated under the assumption of independence, as shown in Tables 3.40 to 3.43.

Table 3.18c-d

X_3

Mean	Std. Dev.	Std. Error	Variance	Coef. Var.	Count
3.4	3.1	.1	9.7	91.7	1767

Minimum	Maximum	Range	Sum	Sum of Sqr.	# Missing
1	21	20	6000	37500	1

# < 10th %	10th %	25th %	50th %	75th %	90th %
0	1	1	2	4	7

# > 90th %	Mode	Geo. Mean	Har. Mean	Kurtosis	Skewness
165	1	2.4	1.9	5.5	2.1

X_4

Mean	Std. Dev.	Std. Error	Variance	Coef. Var.	Count
4.3	4.2	.1	17.8	98.5	1397

Minimum	Maximum	Range	Sum	Sum of Sqr.	# Missing
1	46	45	5987	50551	371

# < 10th %	10th %	25th %	50th %	75th %	90th %
0	1	1	3	6	9

# > 90th %	Mode	Geo. Mean	Har. Mean	Kurtosis	Skewness
129	1	3	2.2	15.7	3

The category price change for Close-Open distribution, where a '1' codes a decrease in price and a '2' an increase in price, is shown in Table 3.44. The Chi-Square is not significant. So, the Close-Open distribution does not contain significant serial dependencies. However, while the within-day price increases and price decreases are similar, they are not identical, as shown in Table 3.45 a-b. The box and whisk-

Table 3.18e-f

X_5

Mean	Std. Dev.	Std. Error	Variance	Coef. Var.	Count
7.6	7.2	.3	52.2	94.7	783

Minimum	Maximum	Range	Sum	Sum of Sqr.	# Missing
1	45	44	5977	86459	985

# < 10th %	10th %	25th %	50th %	75th %	90th %
0	1	2	5	10	17

# > 90th %	Mode	Geo. Mean	Har. Mean	Kurtosis	Skewness
75	1	4.9	3	4.2	1.8

X_6

Mean	Std. Dev.	Std. Error	Variance	Coef. Var.	Count
11.2	15.2	.7	230	135.5	533

Minimum	Maximum	Range	Sum	Sum of Sqr.	# Missing
1	169	168	5964	189094	1235

# < 10th %	10th %	25th %	50th %	75th %	90th %
0	1	3	6	15	26

# > 90th %	Mode	Geo. Mean	Har. Mean	Kurtosis	Skewness
51	1	6.1	3.4	31.8	4.4

ers diagram in Figure 3.40 also highlights this relationship. In this figure, the absolute value of the negative price changes was used so that the two sets of price changes could be plotted on the same axis. The within-day negative price changes tended to be smaller and less variable than the within-day positive price changes. While not statistically significant, the closing price associated with a within-day price

Table 3.19

	A	B	C	D	E	F	G
1	Symbols	Frequency	Probability				
2	1	3311	0.5516				
3	2	2692	0.4484				
4							
5							
6	Sum	6003					
7							
8							
9							
10	Digrams	(E)xpected	(O)bserved	(E-O)	(E-O)^2	((E-O)^2)/E	Sum
11	11	1826.21	1871	-44.79	2006.407	1.099	
12	12	1484.79	1440	44.79	2006.407	1.351	2.450
13	21	1484.79	1441	43.79	1917.821	1.292	3.742
14	22	1207.21	1251	-43.79	1917.821	1.589	
15						Chi-Square	5.330

Figure 3.20

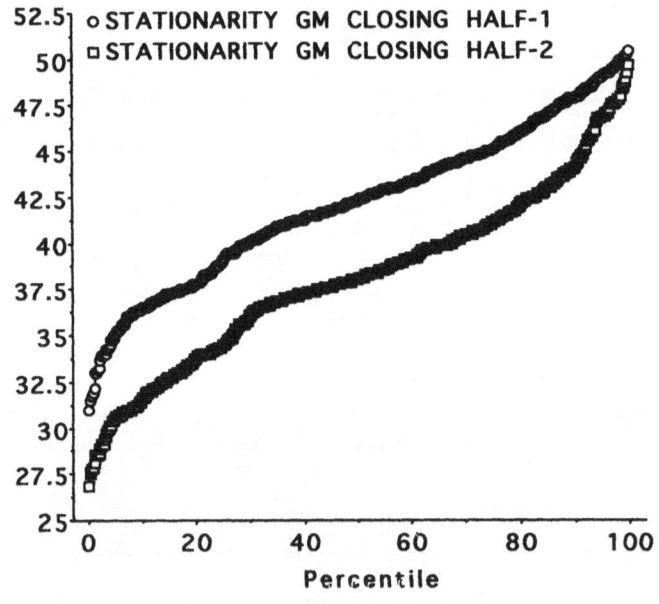

decrease tended to be smaller that the closing price associated with a within-day price increase, as shown in Table 3.46a-b (the standard deviation is 10.8 and 11.5, respectively, while the 10–90% range was 24.1 and 30.2). The distribution of days between sequential within price increases and within price decreases is also different. The

Table 3.20a-b

X_1: *Stationarity GM Closing Half1*

Mean	Std. Dev.	Std. Error	Variance	Coef. Var.	Count
42.2	4.3	.2	18.1	10.1	632

Minimum	Maximum	Range	Sum	Sum of Sqr.	# Missing
31	50.4	19.4	26640.9	1134425.8	633

# < 10th %	10th %	25th %	50th %	75th %	90th %
62	36.5	39.2	42.2	45.1	47.9

# > 90th %	Mode	Geo. Mean	Har. Mean	Kurtosis	Skewness
62	•	41.9	41.7	−.6	−.2

X_2: *Stationarity GM Closing Half2*

Mean	Std. Dev.	Std. Error	Variance	Coef. Var.	Count
38.1	4.7	.2	22.3	12.4	632

Minimum	Maximum	Range	Sum	Sum of Sqr.	# Missing
26.9	49.6	22.8	24085.6	931993	633

# < 10th %	10th %	25th %	50th %	75th %	90th %
62	31.6	34.5	38	41.1	44.1

# > 90th %	Mode	Geo. Mean	Har. Mean	Kurtosis	Skewness
62	39.9	37.8	37.5	−.4	.1

within-day price decreases tended to occur closer together than within-day price increases, shown in the box and whiskers plot in Figure 3.41 and Table 3.47 a-b.

Interestingly, when the daily High-Low price change distribution is parsed into four approximately equal quarters, where the lowest quarter is coded as a '1' (0 to 0.22), from the 26th to the 50th

Figure 3.21

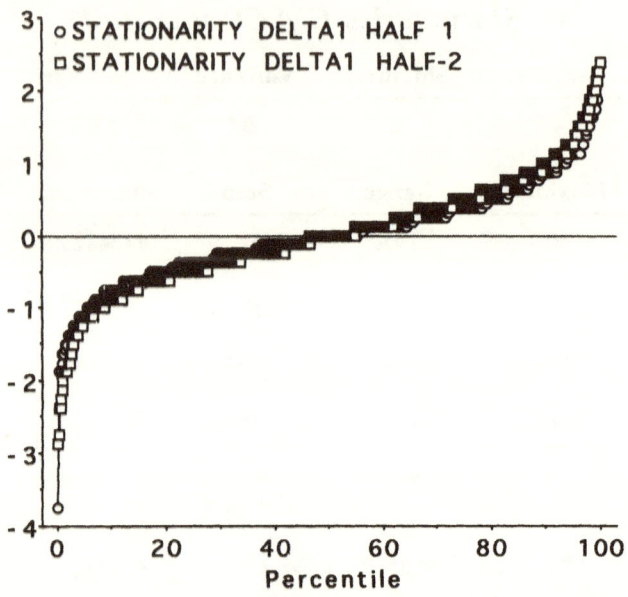

○ STATIONARITY DELTA1 HALF 1
□ STATIONARITY DELTA1 HALF-2

percentile as a '2' (0.221 to 0.424), the 51st to 75th percentile as a '3' (0.424 to 0.660), and the highest quarter as a '4' (0.661 to 12.8), the digram category price changes are significantly different from the independence case, as shown in Table 3.48. Five of the digrams made a large contribution to the ultimate Chi-Square value. They are 1,1, 2,2 and 4,4, which occurred more frequently than would be expected and 3,1 and 4,1, which occurred less frequently.

The trigram transition matrix is also significantly different from the independence case as shown in Table 3.49. The summary statistics for each of the four quarters are shown in Table 3.50 a-d. Seven trigrams contain the digram 1,1 and all but 1, the trigram 1,1,1, occurred less frequently than expected. Seven trigrams contain the digram 2,2 and three 2,2,2, 2,2,3, and 3,2,2 made relatively large contributions to the trigram Chi-Square. Seven trigrams contain the digram 4,4 and one, 4,4,4 made a large contribution to the trigram Chi-Square.

The cumulative probability density functions for Half1 and Half2 for Chrysler's closing prices and Delta-1 price changes are shown in Figures 3.42 and 3.43, respectively, while the summary

Table 3.21a-b

X_1: *Stationarity Delta-1 Half1*

Mean	Std. Dev.	Std. Error	Variance	Coef. Var.	Count
1.6E–3	.7	2.7E–2	.5	42352	631

Minimum	Maximum	Range	Sum	Sum of Sqr.	# Missing
–3.8	2.4	6.1	1	283.8	634

# < 10th %	10th %	25th %	50th %	75th %	90th %
54	–.8	–.4	0	.4	.9

# > 90th %	Mode	Geo. Mean	Har. Mean	Kurtosis	Skewness
51	0	•	•	2.1	–.1

X_2: *Stationarity Delta-1 Half2*

Mean	Std. Dev.	Std. Error	Variance	Coef. Var.	Count
8.7E–3	.8	3.1E–2	.6	8965.9	631

Minimum	Maximum	Range	Sum	Sum of Sqr.	# Missing
–2.9	2.4	5.2	5.5	384.8	634

# < 10th %	10th %	25th %	50th %	75th %	90th %
55	–.9	–.5	0	.5	1

# > 90th %	Mode	Geo. Mean	Har. Mean	Kurtosis	Skewness
53	0	•	•	.8	6.0E–3

statistics are shown in Tables 3.51 a-b and 3.52 a-b. If we examine Figure 3.42, we will note that the two cumulative probability density functions overlap up to the 60th percentile, but they do not overlap above the 61st percentile. This suggests that Chrysler's closing prices for 1988–1992 may not be stationary. Figures 3.44 and 3.45 show the frequency histograms for Half1 and Half2 and it is clear that closing

Figure 3.22

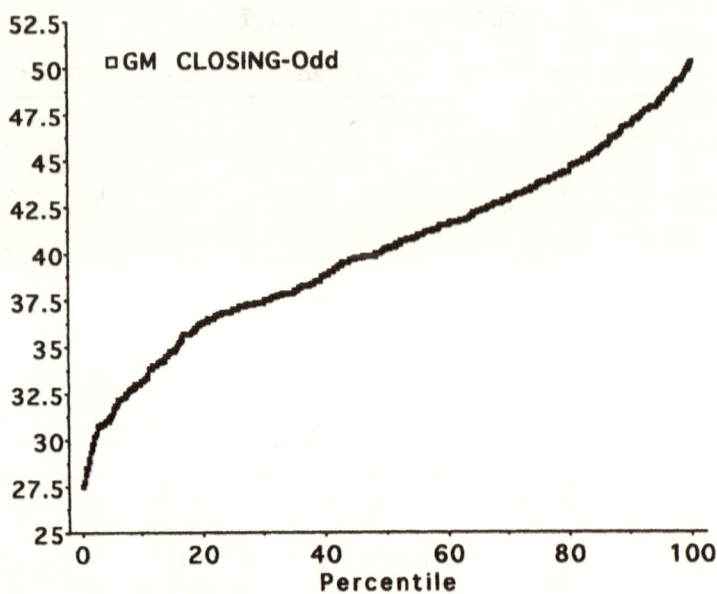

Figure 3.23

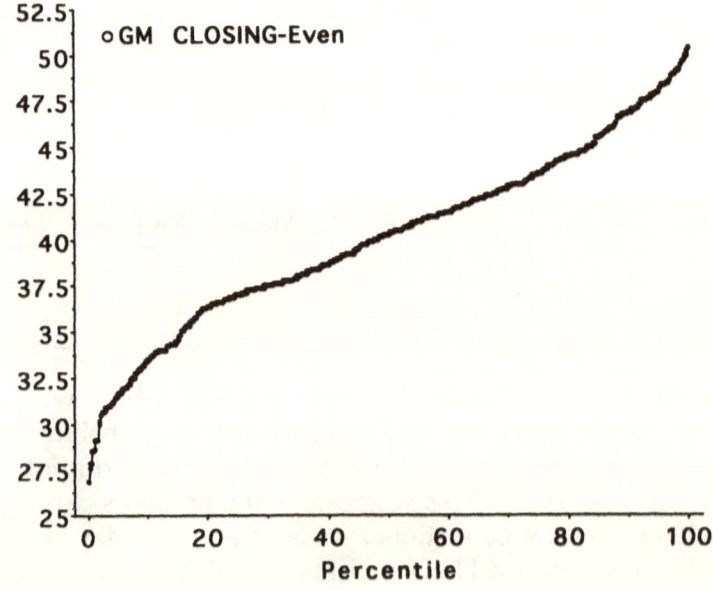

Table 3.22a-b

X_1: GM Closing–Even

Mean	Std. Dev.	Std. Error	Variance	Coef. Var.	Count
40.106	4.906	.195	24.068	12.232	631

Minimum	Maximum	Range	Sum	Sum of Sqr.	# Missing
26.875	50.375	23.5	25307.188	1030144.996	0

# < 10th %	10th %	25th %	50th %	75th %	90th %
63	33.45	37	40.25	43.625	46.875

# > 90th %	Mode	Geo. Mean	Har. Mean	Kurtosis	Skewness
60	43	39.798	39.481	−.486	−.155

X_2: GM Closing–Odd

Mean	Std. Dev.	Std. Error	Variance	Coef. Var.	Count
40.132	4.931	.196	24.319	12.288	631

Minimum	Maximum	Range	Sum	Sum of Sqr.	# Missing
27.5	50.25	22.75	25323	1031572.086	0

# < 10th %	10th %	25th %	50th %	75th %	90th %
62	33.125	37	40.25	43.625	46.875

# > 90th %	Mode	Geo. Mean	Har. Mean	Kurtosis	Skewness
62	39.875	39.821	39.501	−.508	−.144

prices in Half1 and Half2 do not have the same distribution. But the Delta-1 price changes do appear stationary.

The cumulative probability density functions for Odd and Even for closing prices and Delta-1 closing prices are shown in Figures 3.46 and 3.47 and the summary statistics are shown in Tables 3.53 a-b and

Figure 3.24

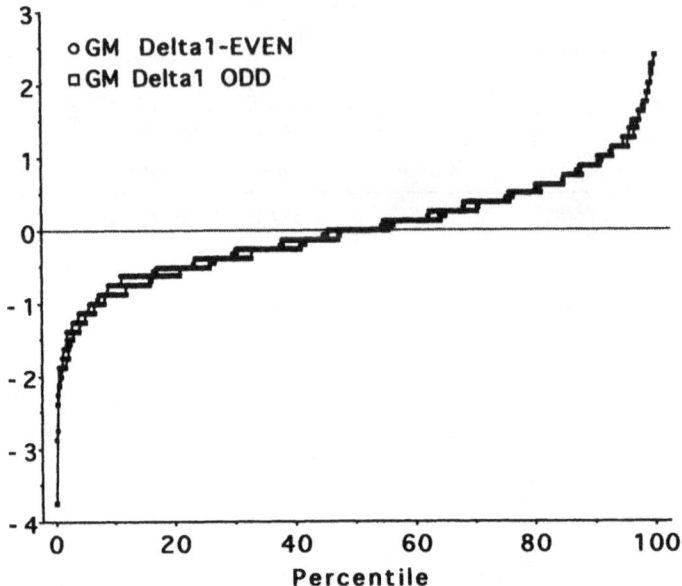

Figure 3.25

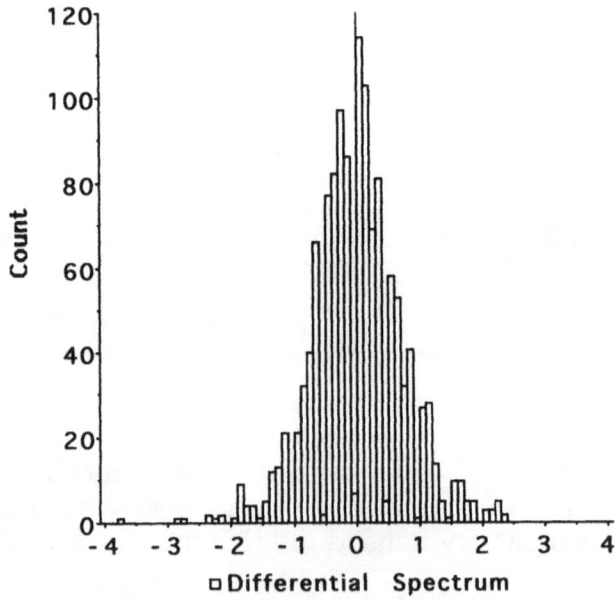

Table 3.23a-b

X_1: GM Delta-1–Even

Mean	Std. Dev.	Std. Error	Variance	Coef. Var.	Count
.032	.713	.028	.509	2219.059	630

Minimum	Maximum	Range	Sum	Sum of Sqr.	# Missing
–2.875	2.375	5.25	20.25	320.656	1

# < 10th %	10th %	25th %	50th %	75th %	90th %
56	–.75	–.375	0	.375	.875

# > 90th %	Mode	Geo. Mean	Har. Mean	Kurtosis	Skewness
61	0	•	•	1.196	.061

X_2: GM Delta-1–Odd

Mean	Std. Dev.	Std. Error	Variance	Coef. Var.	Count
–.025	.747	.03	.558	–2982.071	631

Minimum	Maximum	Range	Sum	Sum of Sqr.	# Missing
–3.75	2.375	6.125	–15.812	352.215	0

# < 10th %	10th %	25th %	50th %	75th %	90th %
53	–.875	–.5	0	.375	.875

# > 90th %	Mode	Geo. Mean	Har. Mean	Kurtosis	Skewness
58	•	•	•	1.462	–.095

3.54 a-b, respectively, which clearly indicate that both the closing prices and Delta-1 price changes are random.

The differential spectrum with a bin size of $1.00 is shown in Figure 3.48. The differential spectrum is not symmetrical and this is confirmed by the box and whiskers plot shown in Figure 3.49 (top) and the compare percentile plot (bottom). The Delta-25, Delta-50, and

Table 3.24

X_1: *Differential Spectrum*

Mean	Std. Dev.	Std. Error	Variance	Coef. Var.	Count
3.3E–3	.7	2.1E–2	.5	22027.8	1263

Minimum	Maximum	Range	Sum	Sum of Sqr.	# Missing
–3.8	2.4	6.1	4.2	673.2	1

# < 10th %	10th %	25th %	50th %	75th %	90th %
99	–.9	–.4	0	.4	.9

# > 90th %	Mode	Geo. Mean	Har. Mean	Kurtosis	Skewness
119	0	•	•	1.4	–2.7E–2

Figure 3.26

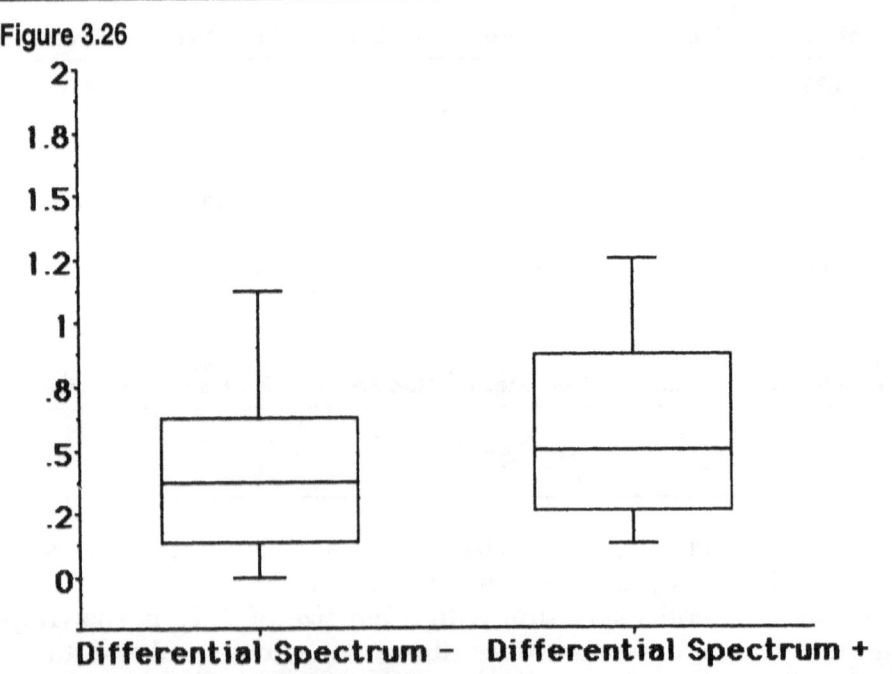

Table 3.25a-b

X_1: *Differential Spectrum* –

Mean	Std. Dev.	Std. Error	Variance	Coef. Var.	Count
–.5	.5	1.8E–2	.2	–96.6	696

Minimum	Maximum	Range	Sum	Sum of Sqr.	# Missing
–3.8	0	3.8	–344.3	329.2	0

# < 10th %	10th %	25th %	50th %	75th %	90th %
57	–1.1	–.6	–.4	–.1	0

# > 90th %	Mode	Geo. Mean	Har. Mean	Kurtosis	Skewness
0	0	•	•	5.4	–1.8

X_2: *Differential Spectrum* +

Mean	Std. Dev.	Std. Error	Variance	Coef. Var.	Count
.6	.5	2.0E–2	.2	77.9	567

Minimum	Maximum	Range	Sum	Sum of Sqr.	# Missing
.1	2.4	2.3	348.5	344	129

# < 10th %	10th %	25th %	50th %	75th %	90th %
6	.1	.2	.5	.9	1.2

# > 90th %	Mode	Geo. Mean	Har. Mean	Kurtosis	Skewness
49	.1	.4	.3	1.6	1.3

Delta-75 differential spectrums are shown in Figures 3.50 to 3.52, respectively, while the summary statistics are shown in Table 3.55 a-c. It is clear that as the length of the comparison increases, the differential spectrum becomes more asymmetrical.

The Half1 and Half2 cumulative probability density functions for Chrysler closing prices and Delta-1 price changes for 05/04/92 to

Table 3.26

	A	B	C	D	E	F	G	H	I
1	n	Digram	Expec Prob	(E)xpec Freq	(O)bser Freq	(E-O)	(E-O)^2	((E-O)^2)/E	SUM
2	1262	11	0.1667	210.3754	391	-180.6246	32625.2461	155.081089	
3		12	0.3333	420.6246	304	116.6246	13601.2973	32.3359531	187.417042
4		21	0.3333	420.6246	304	116.6246	13601.2973	32.3359531	219.752995
5		22	0.1667	210.3754	263	-52.6246	2769.34853	13.163842	252.088948
6									
7									
8								Chi-Square	265.25279

Tables 3.27

	A	B	C	D	E	F	G	H	I
1	n	Trigram	Expec Prob	(E)xpec Freq	(O)bser Freq	(E-O)	(E-O)^2	((E-O)^2)/E	SUM
2	1261	111	0.042	52.546	200.000	-147.454	21742.720	413.786	
3		112	0.125	157.625	190.000	-32.375	1048.141	6.650	420.435
4		121	0.208	262.704	150.000	112.704	12702.221	48.352	468.787
5		122	0.125	157.625	154.000	3.625	13.141	0.083	468.870
6		211	0.125	157.625	190.000	-32.375	1048.141	6.650	475.520
7		212	0.208	262.704	114.000	148.704	22112.918	84.174	559.694
8		221	0.125	157.625	154.000	3.625	13.141	0.083	559.777
9		222	0.042	52.546	109.000	-56.454	3187.069	60.653	
10									
11								Chi-Square	620.431

Tables 3.28

	A	B	C	D	E	F	G	H	I
1	n	Tetragram	Expected P	(E)xpected f	(O)bser f	(E-O)	(E-O)^2	((E-O)^2)/E	Sum
2	1260	1111	0 008	10 496	94 000	-83 504	6972 951	664 356	
3		1112	0 033	41 996	105 000	-63 004	3969 529	94 522	758 878
4		1121	0 075	94 500	100 000	-5 500	30 250	0 320	759 199
5		1122	0 050	63 000	90 000	-27 000	729 000	11 571	770 770
6		1211	0 075	94 500	98 000	-3 500	12 250	0 130	770 900
7		1212	0 133	167 996	52 000	115 996	13455 026	80 091	850 991
8		1221	0 096	121 169	91 000	30 169	910 176	7 512	858 503
9		1222	0 033	41 996	63 000	-21 004	441 176	10 505	869 008
10		2111	0 033	41 996	105 000	-63 004	3969 529	94 522	963 530
11		2112	0 092	115 496	85 000	30 492	929 738	8 050	971 580
12		2121	0 133	167 996	50 000	117 996	13923 009	82 877	1054 457
13		2122	0 075	94 500	64 000	30 500	930 250	9 844	1064 301
14		2211	0 050	63 000	92 000	-29 000	841 000	13 349	1077 650
15		2212	0 075	94 500	62 000	32 500	1056 250	11 177	1088 828
16		2221	0 033	41 996	63 000	-21 004	441 176	10 505	1099 333
17		2222	0 008	10 496	46 000	-35 504	1260 548	120 100	1219 433
18									
19									
20								Chi-Square	1219 433

05/04/94 are shown in Figures 3.53 and 3.54 and the summary statistics are shown in Tables 3.56 a-b and 3.57 a-b, respectively. The cumulative probability density functions in Figure 3.53 are essentially parallel and the cumulative probability density functions in Figure 3.54 overlap, so Chrysler price and Delta-1 price changes are stationary.

Figure 3.27

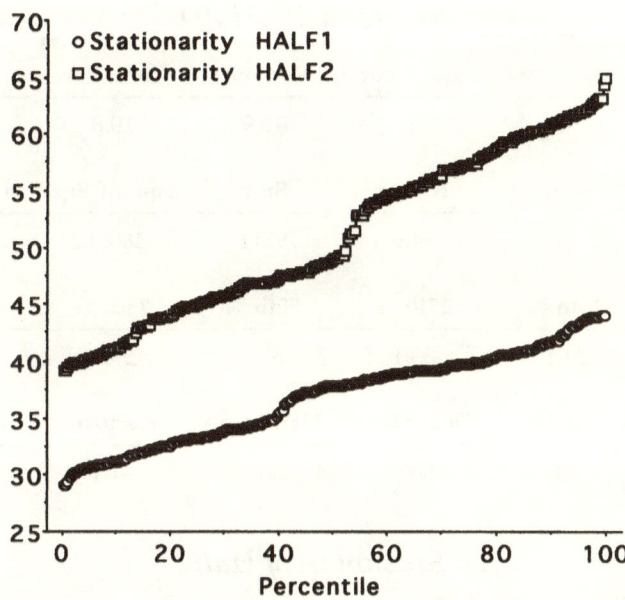

Figure 3.28

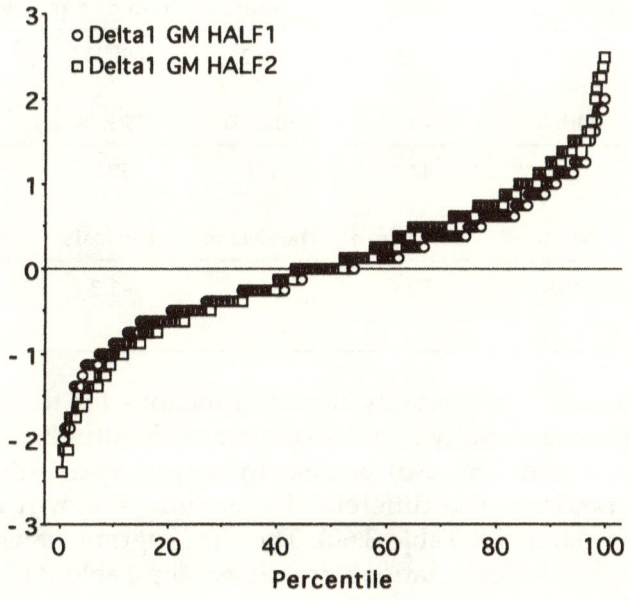

Table 3.29a-b

X_1: Stationarity Half1

Mean	Std. Dev.	Std. Error	Variance	Coef. Var.	Count
36.8	4	.3	15.9	10.8	253

Minimum	Maximum	Range	Sum	Sum of Sqr.	# Missing
29.1	44	14.9	9311	346661.3	0

# < 10th %	10th %	25th %	50th %	75th %	90th %
23	31.1	33.1	37.8	39.8	41.6

# > 90th %	Mode	Geo. Mean	Har. Mean	Kurtosis	Skewness
25	34	36.6	36.4	−1.1	−.1

X_2: Stationarity Half2

Mean	Std. Dev.	Std. Error	Variance	Coef. Var.	Count
50.9	7.2	.5	52.1	14.2	253

Minimum	Maximum	Range	Sum	Sum of Sqr.	# Missing
39.2	64.8	25.5	12882.4	669079.7	0

# < 10th %	10th %	25th %	50th %	75th %	90th %
23	41.1	44.9	48.9	57.3	60.9

# > 90th %	Mode	Geo. Mean	Har. Mean	Kurtosis	Skewness
24	46.9	50.4	49.9	−1.3	.1

The cumulative probability density functions for the Even and Odd parsing are essentially identical for prices (Figures 3.55 and 3.56, Tables 3.58 a-b and 3.59 a-b) and so the prices and Delta-1 price changes are random. The differential spectrum is shown in Figure 3.57 (summary statistics Table 3.60). The RPC digram transition matrix and Chi-Square calculations are shown in Table 3.61 and the

Table 3.30a-b

X_1: Delta-1 GM Half1

Mean	Std. Dev.	Std. Error	Variance	Coef. Var.	Count
6.9E–3	.8	4.7E–2	.6	10830.6	252

Minimum	Maximum	Range	Sum	Sum of Sqr.	# Missing
–2.4	2	4.4	1.8	142	1

# < 10th %	10th %	25th %	50th %	75th %	90th %
24	–.9	–.5	0	.5	1

# > 90th %	Mode	Geo. Mean	Har. Mean	Kurtosis	Skewness
23	0	•	•	.3	–1.4E–2

X_2: Delta-1 GM Half2

Mean	Std. Dev.	Std. Error	Variance	Coef. Var.	Count
.1	.9	.1	.8	1502.8	253

Minimum	Maximum	Range	Sum	Sum of Sqr.	# Missing
–2.4	2.5	4.9	15	200.9	0

# < 10th %	10th %	25th %	50th %	75th %	90th %
24	–1	–.5	0	.6	1.2

# > 90th %	Mode	Geo. Mean	Har. Mean	Kurtosis	Skewness
21	0	•	•	4.2E–2	–2.7E–3

Chi-Square is statistically significant, which strongly suggests that this time series contains serial dependencies. The summary statistics for the portions of the time series parsed into price increases or decreases is shown in Table 3.62 a-b. The category price change transition matrix, where the time series is divided into three unequal parts, the lowest 10% encoded as a '1', the middle 80% as a '2', and the

Figure 3.29

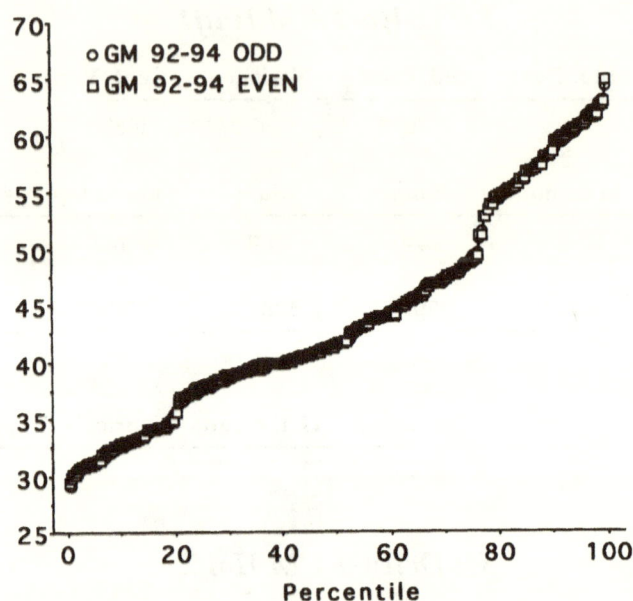

Figure 3.30

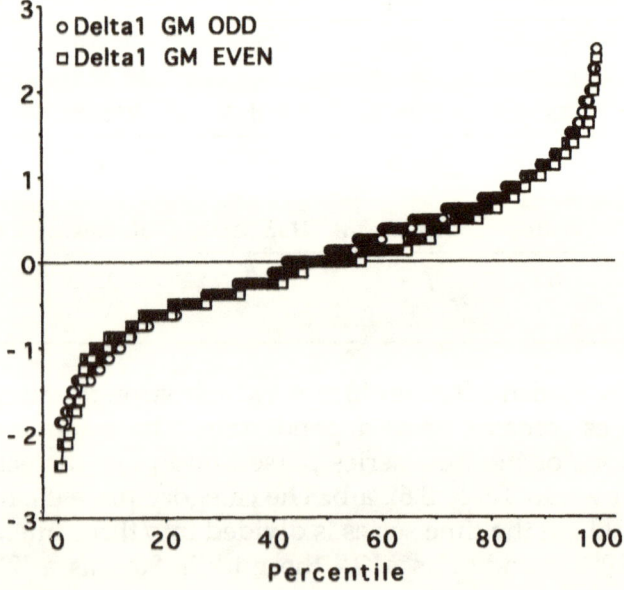

Table 3.31a-b

X_1: GM 92–94 Odd

Mean	Std. Dev.	Std. Error	Variance	Coef. Var.	Count
43.9	9.2	.6	84.1	20.9	253

Minimum	Maximum	Range	Sum	Sum of Sqr.	# Missing
29.1	64.2	35.1	11096	507836.4	0

# < 10th %	10th %	25th %	50th %	75th %	90th %
24	32.5	37.9	41.5	49.1	58.6

# > 90th %	Mode	Geo. Mean	Har. Mean	Kurtosis	Skewness
25	39.2	42.9	42.1	–.8	.5

X_2: GM 92–94 Even

Mean	Std. Dev.	Std. Error	Variance	Coef. Var.	Count
43.9	9.2	.6	83.9	20.9	253

Minimum	Maximum	Range	Sum	Sum of Sqr.	# Missing
29.4	64.8	35.4	11097.4	507904.6	0

# < 10th %	10th %	25th %	50th %	75th %	90th %
25	32.8	37.6	41.5	48.8	58.6

# > 90th %	Mode	Geo. Mean	Har. Mean	Kurtosis	Skewness
25	39.9	42.9	42.1	–.8	.5

upper 10% as a '3' and the Chi-Square calculation are shown in Table 3.63. The summary statistics for each of the three parts of the time series are shown in Table 3.64 a-c.

Ford is GM's other major American competitor. The closing prices for Ford from 1/2/85 to 5/4/94 are shown in Figure 3.58, the summary statistics in Table 3.65, while the frequency histogram of

Table 3.32a-b

X_1: Delta-1 GM Odd

Mean	Std. Dev.	Std. Error	Variance	Coef. Var.	Count
.1	.8	.1	.7	1211.7	253

Minimum	Maximum	Range	Sum	Sum of Sqr.	# Missing
−1.9	2.5	4.4	17.5	178.2	0

# < 10th %	10th %	25th %	50th %	75th %	90th %
25	−1	−.5	0	.6	1.1

# > 90th %	Mode	Geo. Mean	Har. Mean	Kurtosis	Skewness
21	0	•	•	−.1	.1

X_2: Delta-1 GM Even

Mean	Std. Dev.	Std. Error	Variance	Coef. Var.	Count
−5.4E−3	.8	.1	.7	−14893.4	253

Minimum	Maximum	Range	Sum	Sum of Sqr.	# Missing
−2.4	2.4	4.8	−1.4	165.1	0

# < 10th %	10th %	25th %	50th %	75th %	90th %
24	−.9	−.5	0	.5	1.1

# > 90th %	Mode	Geo. Mean	Har. Mean	Kurtosis	Skewness
20	0	•	•	.6	−.1

prices is shown in Figure 3.59 and the cumulative probability of prices is shown in Figure 3.60. The cumulative probability density function for Delta-1 price changes is shown in Figure 3.61.

The cumulative probability density functions for Half1 and Half2 are shown in Figure 3.62. The two cumulative probability density functions are approximately parallel and the box and whiskers

Figure 3.31

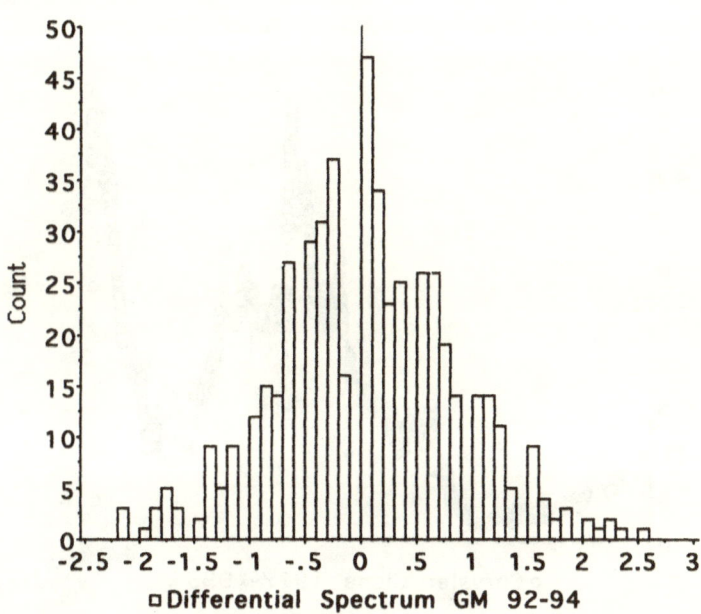

□ Differential Spectrum GM 92-94

Table 3.33

	A	B	C	D	E	F	G	H	I
1	n	Digram	Expec Prob	(E)xpec Freq	(O)bser Freq	(E-O)	(E-O)^2	((E-O)^2)/E	SUM
2	505	11	0.1667	84.1835	141	-56.8165	3228.11467	38.3461685	
3		12	0.3333	168.3165	128	40.3165	1625.42017	9.65692711	48.0030956
4		21	0.3333	168.3165	128	40.3165	1625.42017	9.65692711	57.6600227
5		22	0.1667	84.1835	108	-23.8165	567.225672	6.73796732	67.3169498
6			Sum=	505	505				
7									
8								Chi-Square	74.0549171

plots shown in Figure 3.63 confirm that these two cumulative probability density functions are not significantly different from each other and therefore Ford's closing prices are probably stationary. The summary statistics are shown in Table 3.66 a-b. The Delta-1 price changes are stationary (see Figure 3.64 and Table 3.67 a-b, as well as Table 3.68 a-b, which shows the Delta-1 price changes for each of the two halves). The Ford prices and Delta-1 price changes are also random (see Figures 3.65 and 3.66, as well as Tables 3.69a-b and 3.70a-b.).

The differential spectrum with a bin size of $0.25, $0.10, and $0.05 is shown in Figures 3.67–3.69. Each of these differential spec-

Figure 3.32

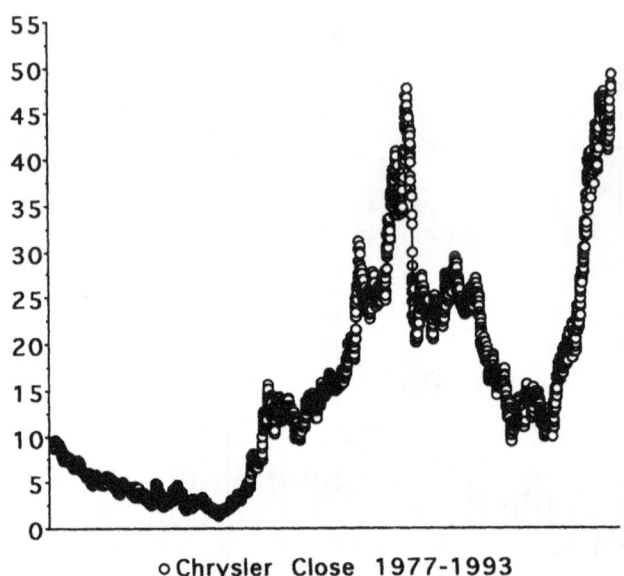

∘ Chrysler Close 1977-1993

Figure 3.33

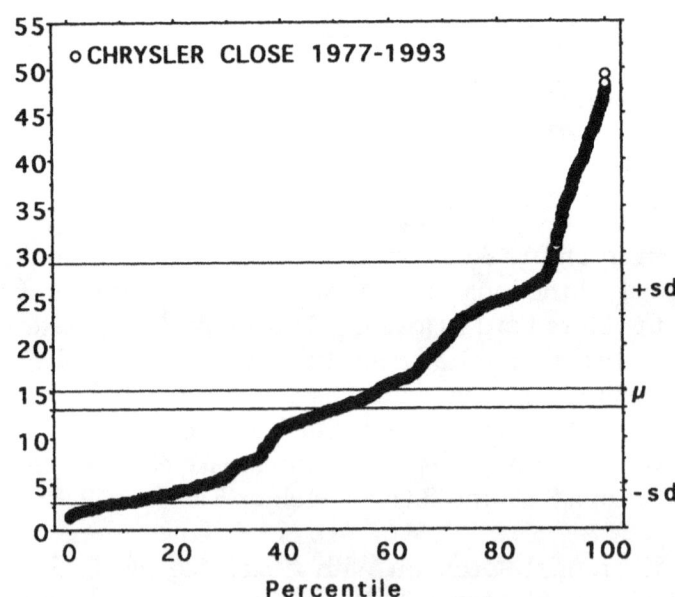

Figure 3.34

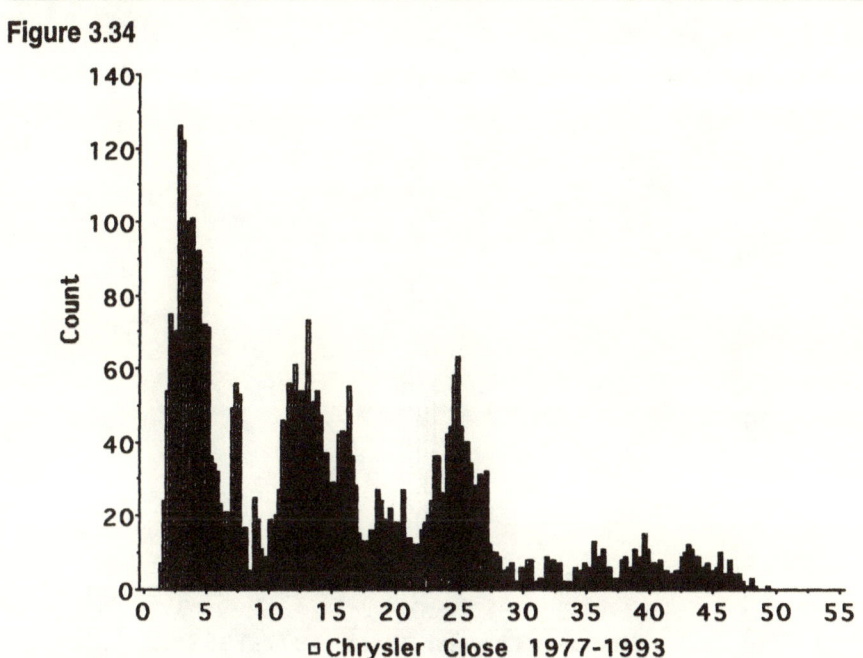

□ Chrysler Close 1977–1993

Table 3.34

X_1: *Chrysler Close 1977–1993*

Mean	Std. Dev.	Std. Error	Variance	Coef. Var.	Count
15.1	11.1	.2	124.2	73.8	4234

Minimum	Maximum	Range	Sum	Sum of Sqr.	# Missing
1.4	49.2	47.9	63931	1491082.5	0

# < 10th %	10th %	25th %	50th %	75th %	90th %
422	3	4.9	13	23.2	28.9

# > 90th %	Mode	Geo. Mean	Har. Mean	Kurtosis	Skewness
423	2.9	10.9	7.3	.1	.9

Figure 3.35

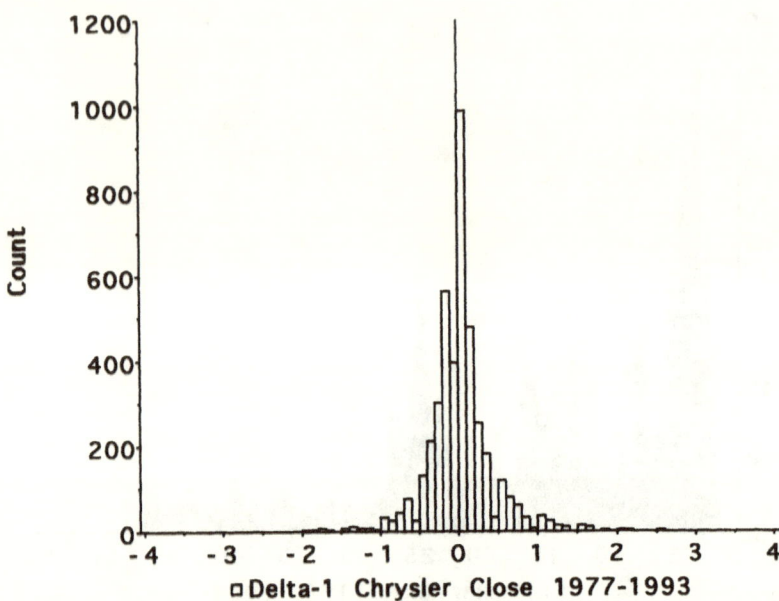

□Delta-1 Chrysler Close 1977-1993

Figure 3.36

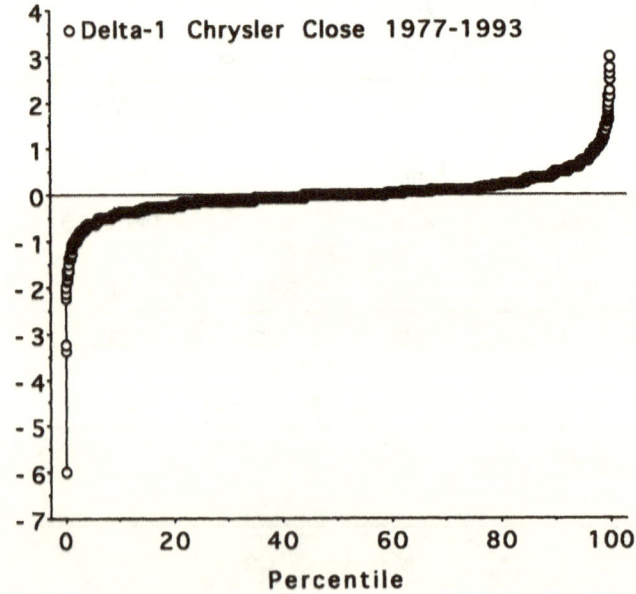

Table 3.35

X_1: Delta-1 Chrysler Close 1977–1993

Mean	Std. Dev.	Std. Error	Variance	Coef. Var.	Count
9.1E–3	.4	6.7E–3	.2	4797	4233

Minimum	Maximum	Range	Sum	Sum of Sqr.	# Missing
–6	3	9	38.6	811.2	1

# < 10th %	10th %	25th %	50th %	75th %	90th %
411	–.4	–.1	0	.1	.5

# > 90th %	Mode	Geo. Mean	Har. Mean	Kurtosis	Skewness
335	0	•	•	14.5	–.4

trums' overall shape suggests symmetry, but if you examine the distribution of counts on either side of the '0', you will note that the counts are not symmetrical, which strongly suggests that Ford prices contain serial dependencies. The summary statistics are shown in Table 3.71.

This is confirmed by the highly significant Chi-Square values for the digram, trigram, and tetragram RPC transition matrices as shown in Tables 3.72–3.74. Three trigrams contribute heavily to the trigram Chi-Square. They are 1,1,1 which occur about three times as often as would be expected if the prices were independent and 2,2,2 which occurs about two times as often. The other trigram, 2,1,2 occurs about half as often as would be expected. Three tetragrams contain the trigram 1,1,1. Each of the tetragrams, 1,1,1,1, 1,1,1,2, and 2,1,1,1 all occur more frequently than would be expected. Each of the three tetragrams that contain the trigram 2,2,2 (i.e., 2,2,2,2, 2,2,2,1, and 1,2,2,2) all occur more often than would be expected. The four tetragrams that contain the trigram 2,1,2 (i.e., 1,2,1,2, 2,1,2,1, 2,1,2,2, and 2,2,1,2) all occur less often than would be expected.

We can also determine if very high and very low Delta-1 price changes diverge from independence. For example, we can parse the Delta-1 price changes into uneven thirds, so that the lowest third contains the lowest 10% of the price changes (i.e., –$7.60 to –$0.90)

Figure 3.37a

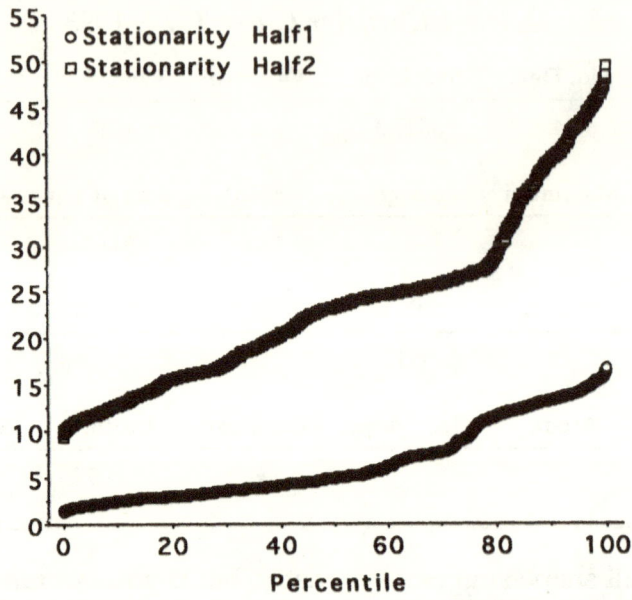

Figure 3.37b

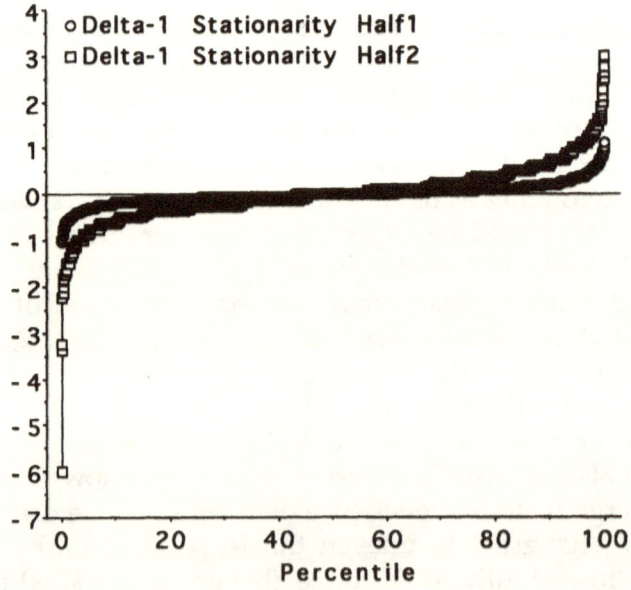

Table 3.36a-b

X_1: *Stationarity Half1*

Mean	Std. Dev.	Std. Error	Variance	Coef. Var.	Count
6.5	4.1	.1	16.8	62.9	2117

Minimum	Maximum	Range	Sum	Sum of Sqr.	# Missing
1.4	16.8	15.4	13820.7	125878.4	0

# < 10th %	10th %	25th %	50th %	75th %	90th %
205	2.4	3.2	4.9	9.4	13.2

# > 90th %	Mode	Geo. Mean	Har. Mean	Kurtosis	Skewness
210	2.9	5.4	4.5	–.7	.8

X_2: *Stationarity Half2*

Mean	Std. Dev.	Std. Error	Variance	Coef. Var.	Count
23.7	9.2	.2	84.6	38.9	2117

Minimum	Maximum	Range	Sum	Sum of Sqr.	# Missing
9.4	49.2	39.9	50110.3	1365204.1	0

# < 10th %	10th %	25th %	50th %	75th %	90th %
204	12.8	16.3	23.2	26.8	39.2

# > 90th %	Mode	Geo. Mean	Har. Mean	Kurtosis	Skewness
209	24.8	22	20.5	–.1	.8

and encode them as a '1'. The middle third contains approximately 80% of the price changes (–$0.89 to $0.90) and we encode them as a '2'. The highest Delta-1 price changes (i.e., $0.91 to $3.50) are encoded as a '3'. The summary statistics are shown in Table 3.75a-c. The digram (Table 3.76) and trigram (Table 3.77) category price change transition matrices diverge from independence as indicated by the

Table 3.37a-b

X_1: Delta-1 Stationarity Half1

Mean	Std. Dev.	Std. Error	Variance	Coef. Var.	Count
3.0E–3	.2	4.5E–3	4.2E–2	6749.8	2116

Minimum	Maximum	Range		Sum	Sum of Sqr.
–1	1.1	2.2	6.5	89.6	1

# < 10th %	10th %	25th %	50th %	75th %	90th %
212	–.2	–.1	0	.1	.2

# > 90th %	Mode	Geo. Mean	Har. Mean	Kurtosis	Skewness
204	0	•	•	4.7	.4

X_2: Delta-1 Stationarity Half2

Mean	Std. Dev.	Std. Error	Variance	Coef. Var.	Count
1.5E–2	.6	1.3E–2	.3	3841.2	2117

Minimum	Maximum	Range	Sum	Sum of Sqr.	# Missing
–6	3	9	32.2	721.5	0

# < 10th %	10th %	25th %	50th %	75th %	90th %
154	–.6	–.2	0	.2	.7

# > 90th %	Mode	Geo. Mean	Har. Mean	Kurtosis	Skewness
210	0	•	•	8	–.4

significant Chi-Square values. This means that when we parse the Delta-1 price changes in this manner, they contain significant serial dependencies.

We can also perform the same tests on higher order Delta price changes. For example, the differential spectrum for Delta-50 price changes are shown in Figure 3.70 and the summary statistics are

Figure 3.38

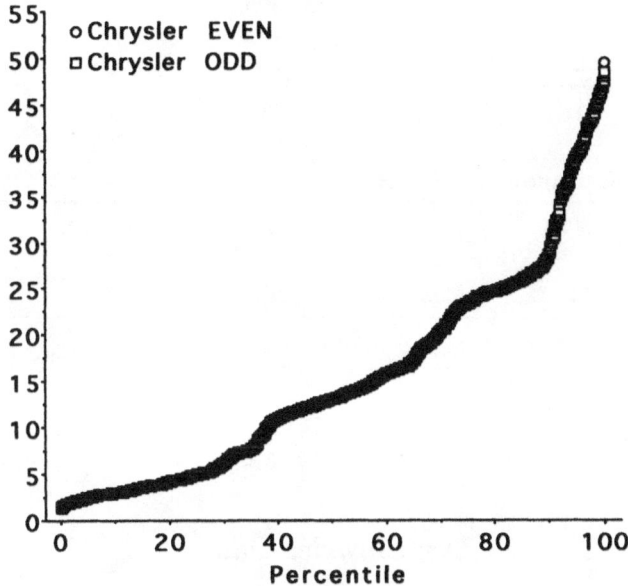

Figure 3.39

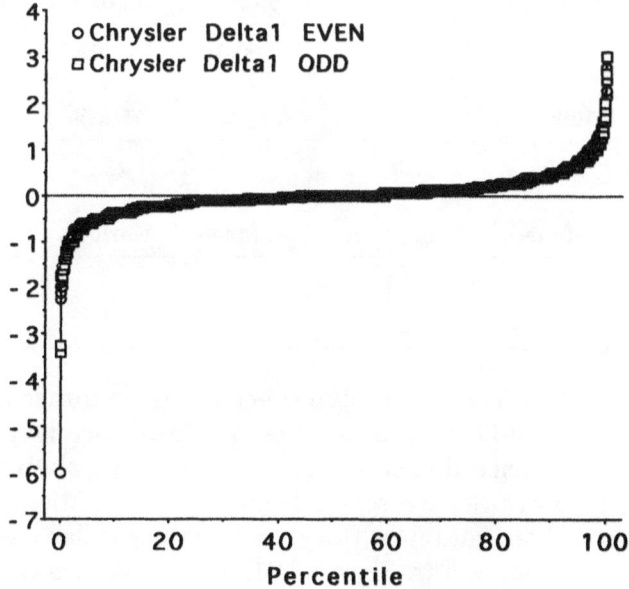

Table 3.38a-b

X_1: *Chrysler Even*

Mean	Std. Dev.	Std. Error	Variance	Coef. Var.	Count
15.1	11.1	.2	124.2	73.8	2117

Minimum	Maximum	Range	Sum	Sum of Sqr.	# Missing
1.4	49.2	47.9	31970.2	745679.7	0

# < 10th %	10th %	25th %	50th %	75th %	90th %
210	3	4.9	13.1	23.2	29

# > 90th %	Mode	Geo. Mean	Har. Mean	Kurtosis	Skewness
212	•	10.9	7.3	.1	.9

X_2: *Chrysler Odd*

Mean	Std. Dev.	Std. Error	Variance	Coef. Var.	Count
15.1	11.1	.2	124.2	73.8	2117

Minimum	Maximum	Range	Sum	Sum of Sqr.	# Missing
1.4	48.2	46.9	31960.8	745402.7	0

# < 10th %	10th %	25th %	50th %	75th %	90th %
212	2.9	4.9	13	23.2	28.9

# > 90th %	Mode	Geo. Mean	Har. Mean	Kurtosis	Skewness
212	2.9	10.9	7.3	.1	.9

shown in Table 3.78. The differential spectrum is asymmetrical. However, if we parse the Delta-50 price changes into price increases (encoded as a '2') or price decreases (encoded as a '1'), there are more price increases than price decreases (see summary statistics in Table 3.79a-b) but the distribution of price changes is similar (see the box and whiskers plot shown in Figure 3.71, where we use the absolute

Table 3.39a-b

X_1: *Chrysler Delta-1 Even*

Mean	Std. Dev.	Std. Error	Variance	Coef. Var.	Count
1.4E–2	.4	9.7E–3	.2	3237.2	2116

Minimum	Maximum	Range	Sum	Sum of Sqr.	# Missing
–6	2.8	8.8	29.3	425.3	1

# < 10th %	10th %	25th %	50th %	75th %	90th %
206	–.4	–.1	0	.1	.5

# > 90th %	Mode	Geo. Mean	Har. Mean	Kurtosis	Skewness
171	0	•	•	19.4	–.7

X_2: *Chrysler Delta-1 Odd*

Mean	Std. Dev.	Std. Error	Variance	Coef. Var.	Count
4.4E–3	.4	9.3E–3	.2	9690.4	2117

Minimum	Maximum	Range	Sum	Sum of Sqr.	# Missing
–3.4	3	6.4	9.3	385.9	0

# < 10th %	10th %	25th %	50th %	75th %	90th %
205	–.4	–.1	0	.1	.4

# > 90th %	Mode	Geo. Mean	Har. Mean	Kurtosis	Skewness
212	0	•	•	8.6	4.3E–2

value of the negative price changes ('1s') so we can plot the distributions on the same axis). The temporal distribution of Delta-50 price changes shows significant serial dependencies as indicated by the significant Chi-Square value for the digram RPC transition matrix as shown in Table 3.80.

Table 3.40

	A	B	C	D	E	F	G	H	I
1	n	Digram	Expec Prob	(E)xpec Freq	(O)bser Freq	(E-O)	(E-O)^2	((E-O)^2/E	Sum
2	4252	11	0 1667	708 808	1528 000	-819 192	671074 878	946 765	
3		12	0 3333	1417 192	993 000	424 192	179938 514	126 968	1073 733
4		21	0 3333	1417 192	993 000	424 192	179938 514	126 968	1200 702
5		22	0 1667	708 808	738 000	-29 192	852 150	1 202	1327 670
6									
7									
8								Chi-Square	1328 872

Table 3.41

	A	B	C	D	E	F	G	H	I
1	n	Tngram	Expec Prob	(E)xpec Freq	(O)bser Freq	(E-O)	(E-O)^2	((E-O)^2/E	Sum
2	4231	111	0 042	176 306	904 000	-727 694	529538 892	3003 526	
3		112	0 125	528 875	603 000	-74 125	5494 516	10 389	3013 915
4		121	0 208	881 444	552 000	329 444	108533 501	123 131	3137 046
5		122	0 125	528 875	441 000	87 875	7722 016	14 601	3151 647
6		211	0 125	528 875	603 000	-74 125	5494 516	10 389	3162 036
7		212	0 208	881 44	390 000	491 444	241517 431	274 002	3436 038
8		221	0 125	528 875	441 000	87 875	7722 016	14 601	3450 639
9		222	0 042	176 306	297 000	-120 694	14567 097	82 624	
10									
11								Chi-Square	3533 263

Table 3.42

	A	B	C	D	E	F	G	H	I
1	n	Tetragram	Expected P	(E)xpected f	(O)bser f	(E-O)	(E-O)^2	((E-O)^2)/E	Sum
2	4230	1111	0.008	35.236	557.000	-521.764	272237.776	7726.148	
3		1112	0.033	140.986	347.000	-206.014	42441.809	301.036	8027.184
4		1121	0.075	317.250	334.000	-16.750	280.563	0.884	8028.068
5		1122	0.050	211.500	269.000	-57.500	3306.250	15.632	8043.701
6		1211	0.075	317.250	339.000	-21.750	473.063	1.491	8045.192
7		1212	0.133	563.986	213.000	350.986	123191.102	218.429	8263.621
8		1221	0.096	406.782	271.000	135.782	18436.800	45.324	8308.945
9		1222	0.033	140.986	170.000	-29.014	841.818	5.971	8314.916
10		2111	0.033	140.986	347.000	-206.014	42441.809	301.036	8615.951
11		2112	0.092	387.722	255.000	132.722	17615.076	45.432	8661.384
12		2121	0.133	563.986	218.000	345.986	119706.243	212.250	8873.634
13		2122	0.075	317.250	172.000	145.250	21097.563	66.501	8940.135
14		2211	0.050	211.500	264.000	-52.500	2756.250	13.032	8953.167
15		2212	0.075	317.250	177.000	140.250	19670.063	62.002	9015.169
16		2221	0.033	140.986	170.000	-29.014	841.818	5.971	9021.140
17		2222	0.008	35.236	127.000	-91.764	8420.650	238.979	9260.119
18									
19									
20								Chi-Square	9260.119

The cumulative probability density functions for Half1 and Half2 for prices (Figure 3.72, Table 3.81 a-b) and Delta-1 price changes (Figure 3.73, Table 3.82 a-b) show that the prices and Delta-1 price changes are stationary. The cumulative probability density functions for Even and Odd prices (Figure 3.74, Table 3.83 a-b) and

Table 3.43

	A	B	C	D	E	F	G	H	I
1	N	PENTAGRAM	(E)XPECTED P	(E)XPECTED f	(O)BSERVED f	(O-E)	(O-E)^2	((O-E)^2)/E	SUM
2	4229	11111	0.00139	5.878	334	328.12	107663.84	18315.44	
3		11112	0.00694	29.349	223	193.65	37500.61	1277.74	19593.18
4		11121	0.01944	82.212	198	115.79	13406.92	163.08	19756.26
5		11122	0.01389	58.741	149	90.26	8146.72	138.69	19894.94
6		11211	0.02639	111.603	204	92.40	8537.15	76.50	19971.44
7		11212	0.04861	205.572	130	-75.57	5711.08	27.78	19999.22
8		11221	0.03611	152.709	156	3.29	10.83	0.07	19999.29
9		11222	0.01389	58.741	113	54.26	2944.06	50.12	20049.41
10		12111	0.01944	82.212	196	113.79	12947.76	157.49	20206.91
11		12112	0.0555	234.710	143	-91.71	8410.63	35.83	20242.74
12		12121	0.08472	358.281	124	-234.28	54887.53	153.20	20395.94
13		12122	0.04861	205.572	89	-116.57	13588.96	66.10	20462.04
14		12211	0.03611	152.709	170	17.29	298.97	1.96	20464.00
15		12212	0.0555	234.710	101	-133.71	17878.23	76.17	20540.17
16		12221	0.02639	111.603	90	-21.60	466.70	4.18	20544.35
17		12222	0.00694	29.349	80	50.65	2565.50	87.41	20631.76
18		21111	0.00694	29.349	223	193.65	37500.61	1277.74	21909.50
19		21112	0.02639	111.603	124	12.40	153.68	1.38	21910.88
20		21121	0.0555	234.710	136	-98.71	9743.57	41.51	21952.39
21		21122	0.03611	152.709	119	-33.71	1136.31	7.44	21959.83
22		21211	0.04861	205.572	135	-70.57	4980.36	24.23	21984.06
23		21212	0.08472	358.281	83	-275.28	75779.56	211.51	22195.57
24		21221	0.0555	234.710	115	-119.71	14330.36	61.06	22256.62
25		21222	0.01944	82.212	57	-25.21	635.63	7.73	22264.35
26		22111	0.01389	58.741	151	92.26	8511.76	144.90	22409.26
27		22112	0.03611	152.709	112	-40.71	1657.24	10.85	22420.11
28		22121	0.04861	205.572	94	-111.57	12448.24	60.55	22480.66
29		22122	0.02639	111.603	83	-28.60	818.15	7.33	22487.99
30		22211	0.01389	58.741	94	35.26	1243.21	21.16	22509.16
31		22212	0.01944	82.212	76	-6.21	38.59	0.47	22509.63
32		22221	0.00694	29.349	80	50.65	2565.50	87.41	22597.04
33		22222	0.00139	5.878	47	41.12	1690.99	287.67	
34									
35				n	4229		Chi-Square		22884.7079

Table 3.44

	A	B	C	D	E	F	G
1	Symbols	Frequency	Probability				
2	1	2531	0.5979				
3	2	1702	0.4021				
4							
5							
6	Sum	4233					
7							
8							
9							
10	Digrams	(E)xpected	(O)bserved	(E-O)	(E-O)^2	((E-O)^2)/E	Sum
11	11	1513.34	1551	-37.66	1418.404	0.937	
12	12	1017.66	980	37.66	1418.404	1.394	2.331
13	21	1017.66	980	37.66	1418.404	1.394	3.725
14	22	684.34	722	-37.66	1418.404	2.073	
15						Chi-Square	5.798

Delta-1 price changes (Figure 3.75, Table 3.84 a-b) show that the prices and Delta-1 price changes are random.

Table 3.45a-b

X_1: 1–Chrysler Close–Open

Mean	Std. Dev.	Std. Error	Variance	Coef. Var.	Count
–.2	.3	5.6E–3	.1	–145.6	2531

Minimum	Maximum	Range	Sum	Sum of Sqr.	# Missing
–5.2	0	5.2	–488.3	293.8	0

# < 10th %	10th %	25th %	50th %	75th %	90th %
198	–.5	–.2	–.1	0	0

# > 90th %	Mode	Geo. Mean	Har. Mean	Kurtosis	Skewness
0	0	•	•	50.1	–4.7

X_2: 2–Chrysler Close–Open

Mean	Std. Dev.	Std. Error	Variance	Coef. Var.	Count
.3	.3	7.8E–3	.1	105.4	1703

Minimum	Maximum	Range	Sum	Sum of Sqr.	# Missing
4.7E–2	2.8	2.7	519.3	334.2	828

# < 10th %	10th %	25th %	50th %	75th %	90th %
150	.1	.1	.2	.4	.8

# > 90th %	Mode	Geo. Mean	Har. Mean	Kurtosis	Skewness
136	.1	.2	.1	8.1	2.4

The frequency histogram of prices for 1992–1994 is shown in Figure 3.76. The differential spectrum (Figure 3.77) is asymmetrical and thus suggests that the Delta-1 price changes for this time period contain serial dependencies and this is confirmed by the significant Chi-Square for the RPC digram transition matrix (Table 3.85). The

Figure 3.40

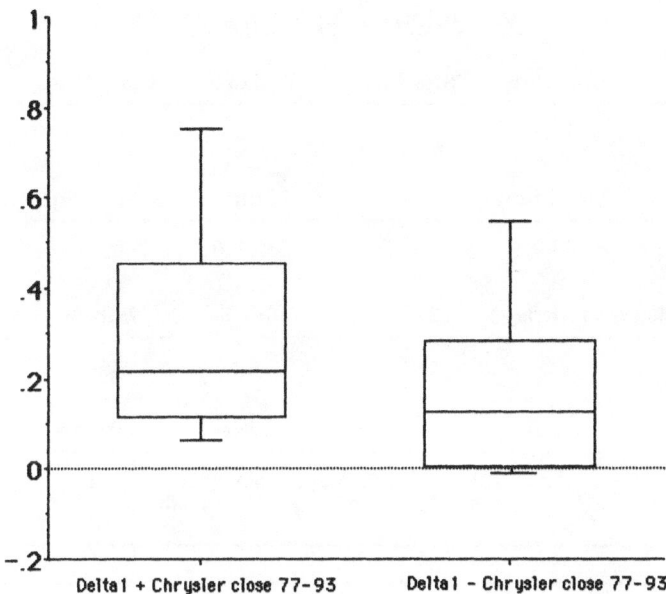

Table 3.46a

X_1: 1–Chrysler Close 77–93

Mean	Std. Dev.	Std. Error	Variance	Coef. Var.	Count
14.1	10.8	.2	116.6	76.4	2502

Minimum	Maximum	Range	Sum	Sum of Sqr.	# Missing
1.4	48.2	46.9	35346	790913.5	0

# < 10th %	10th %	25th %	50th %	75th %	90th %
228	2.8	4.4	12.2	22.2	26.9

# > 90th %	Mode	Geo. Mean	Har. Mean	Kurtosis	Skewness
246	•	10	6.8	.3	.9

Table 3.46b

X_2: 2–Chrysler Close 77–93

Mean	Std. Dev.	Std. Error	Variance	Coef. Var.	Count
16.5	11.5	.3	132	69.6	1731

Minimum	Maximum	Range	Sum	Sum of Sqr.	# Missing
1.4	49.2	47.8	28575.6	700081.1	771

# < 10th %	10th %	25th %	50th %	75th %	90th %
170	3.3	6	14.2	24.5	33.5

# > 90th %	Mode	Geo. Mean	Har. Mean	Kurtosis	Skewness
173	3.7	12.2	8.3	–.1	.8

Figure 3.41

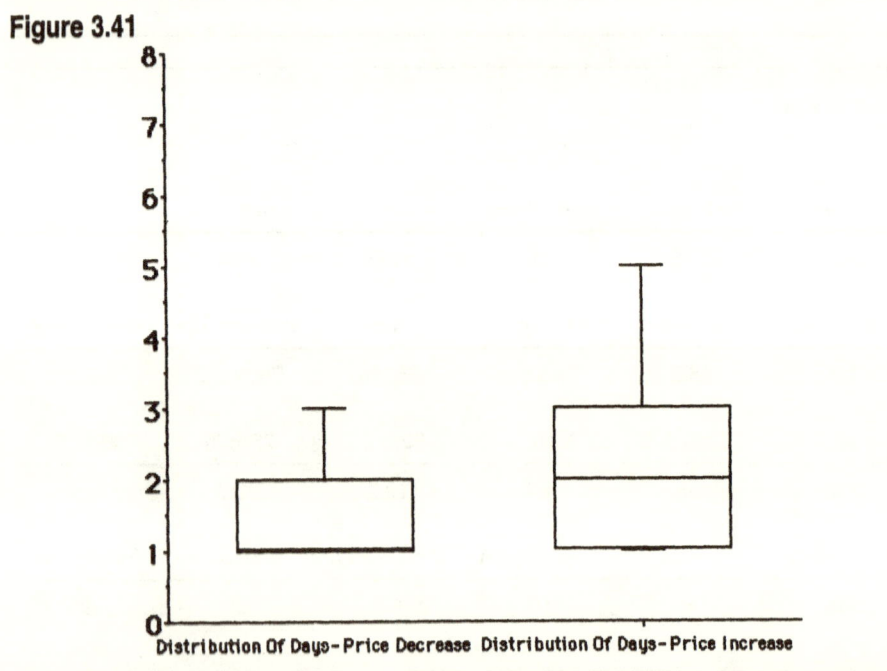

Table 3.47a-b

X_1: *Difference of 1–Date*

Mean	Std. Dev.	Std. Error	Variance	Coef. Var.	Count
1.7	1.1	2.2E–2	1.2	64.6	2501

Minimum	Maximum	Range	Sum	Sum of Sqr.	# Missing
1	9	8	4232	10150	1

# < 10th %	10th %	25th %	50th %	75th %	90th %
0	1	1	1	2	3

# > 90th %	Mode	Geo. Mean	Har. Mean	Kurtosis	Skewness
170	1	1.5	1.3	5.1	2.1

X_2: *Difference of 2–Date*

Mean	Std. Dev.	Std. Error	Variance	Coef. Var.	Count
2.4	2	4.7E–2	3.8	79.8	1730

Minimum	Maximum	Range	Sum	Sum of Sqr.	# Missing
1	18	17	4228	16916	772

# < 10th %	10th %	25th %	50th %	75th %	90th %
0	1	1	2	3	5

# > 90th %	Mode	Geo. Mean	Har. Mean	Kurtosis	Skewness
129	1	1.9	1.6	6.8	2.2

summary statistics for Delta-1 price increases and decreases are shown in Table 3.86.

The cumulative probability density function for Half1 and Half2 for Ford closing prices for 05/04/93 to 04/05/94 are shown in Figure 3.78 (summary statistics Table 3.87 a-b). The middle portion of these cumulative probability density functions appear to be parallel, but

Table 3.48

	A	B	C	D	E	F	G	H
1	Symbols	Frequency	Probability					
2	1	1448	0.3420					
3	2	1072	0.2532					
4	3	802	0.1894					
5	4	912	0.2154					
6	Sum	4234						
7								
8								
9								
10	Digrams	(E)xpected	(O)bserved	(E-O)	(E-O)^2	((E-O)^2)/E	Sum	
11	11	495.206	1221.000	-725.794	526776.315	1063.751		
12	12	366.617	169.000	197.617	39052.443	106.521	1170.272	
13	13	274.279	48.000	226.279	51202.048	186.679	1356.951	
14	14	311.898	10.000	301.898	91142.384	292.219	1649.170	
15	21	366.617	188.000	178.617	31904.001	87.023	1736.192	
16	22	271.418	470.000	-198.582	39434.793	145.292	1881.484	
17	23	203.057	257.000	-53.943	2909.830	14.330	1895.814	
18	24	230.908	157.000	73.908	5462.376	23.656	1919.470	
19	31	274.279	34.000	240.279	57733.852	210.493	2129.964	
20	32	203.057	294.000	-90.943	8270.601	40.730	2170.694	
21	33	151.914	259.000	-107.086	11467.405	75.486	2246.180	
22	34	172.750	215.000	-42.250	1785.053	10.333	2256.513	
23	41	311.898	5.000	306.898	94186.363	301.978	2558.491	
24	42	230.908	139.000	91.908	8447.060	36.582	2595.073	
25	43	172.750	237.000	-64.250	4128.047	23.896	2618.969	
26	44	196.444	530.000	-333.556	111259.589	566.368	3185.337	
27								
28		n=	4233			Chi-Square	3185.33733	
29								
30								
31								
32								
33								
34								
35								
36								
37								
38								
39								
40								
41								
42								
43								
44								
45								
46								
47								
48								
49								
50								
51								0

the tails are not. This is confirmed by box and whiskers plots shown in Figure 3.79. The plots in the bottom half of the figure represent the first 24 weeks of price data, where each box and whiskers plot represents ten days of price data. The leftmost plot represents the first two weeks of the year, the second, the second two weeks, etc. The box and whiskers plots in the upper portion of the figure represent the

Table 3.49

Symbols	Frequency	Probability					
1	1448	0.3422					
2	1072	0.2533					
3	802	0.1895					
4	910	0.2150					
Sum	4232						
Digrams	(E)xpected	(O)bserved	(E-O)	(E-O)^2	((E-O)^2)/E	Sum	
111	169.517	1103.000	-933.483	871389.703	5140.414		
112	125.499	102.000	23.499	552.207	4.400	5144.814	
113	93.890	15.000	78.890	6223.661	66.287	5211.100	
114	106.534	1.000	105.534	11137.372	104.543	5315.643	
121	125.499	95.000	30.499	930.195	7.412	5323.055	
122	92.911	57.000	35.911	1289.595	13.880	5336.935	
123	69.510	11.000	58.510	3423.403	49.251	5386.186	
124	78.870	6.000	72.870	5310.079	67.327	5453.513	
131	274.408	9.000	265.408	70441.575	256.703	5710.216	
132	69.510	24.000	45.510	2071.147	29.796	5740.012	
133	52.003	11.000	41.003	1681.222	32.330	5772.342	
134	59.006	4.000	55.006	3025.613	51.277	5823.619	
141	311.361	2.000	309.361	95704.265	307.374	6130.993	
142	78.870	0.000	78.870	6220.522	78.870	6209.863	
143	59.006	5.000	54.006	2916.602	49.429	6259.292	
144	66.951	3.000	63.951	4089.789	61.086	6320.378	
211	366.790	106.000	260.790	68011.513	185.423	6505.802	
212	92.911	52.000	40.911	1673.704	18.014	6523.816	
213	93.890	24.000	69.890	4884.637	52.025	6575.841	
214	78.870	6.000	72.870	5310.079	67.327	6643.167	
221	271.546	72.000	199.546	39818.731	146.637	6789.804	
222	68.785	231.000	-162.215	26313.742	382.551	7172.356	
223	51.460	105.000	-53.540	2866.496	55.703	7228.059	
224	58.390	62.000	-3.610	13.031	0.223	7228.282	
231	203.153	11.000	192.153	36922.821	181.749	7410.030	
232	51.460	109.000	-57.540	3310.813	64.337	7474.368	
233	52.003	87.000	-34.997	1224.810	23.553	7497.920	
234	43.684	50.000	-6.316	39.896	0.913	7498.834	
241	230.510	2.000	228.510	52217.002	226.528	7725.362	
242	58.390	57.000	1.390	1.933	0.033	7725.395	
243	43.684	46.000	-2.316	5.365	0.123	7725.517	
244	49.566	52.000	-2.434	5.923	0.119	7725.637	
311	274.408	10.000	264.408	69911.758	254.773	7980.410	
312	69.510	14.000	55.510	3081.344	44.330	8024.739	
313	52.003	7.000	45.003	2025.244	38.945	8063.684	
314	59.006	3.000	56.006	3136.624	53.158	8116.842	
321	203.153	16.000	187.153	35026.290	172.413	8289.256	
322	51.460	128.000	-76.540	5858.320	113.841	8403.097	
323	38.499	95.000	-56.501	3192.335	82.919	8486.016	
324	43.684	55.000	-11.316	128.059	2.932	8488.948	
331	151.986	12.000	139.986	19596.030	128.933	8617.881	
332	38.499	95.000	-56.501	3192.335	82.919	8700.801	
333	28.803	93.000	-64.197	4121.305	143.088	8843.889	
334	32.681	59.000	-26.319	692.676	21.195	8865.084	
341	172.453	0.000	172.453	29739.948	172.453	9037.536	

Table 3.49 continues

Table 3.49 (Continued)

342	43.684	44.000	-0.316	0.100	0.002	9037.539	
343	32.681	73.000	-40.319	1625.601	49.741	9087.280	
344	37.082	98.000	-60.918	3710.975	100.074	9187.354	
411	311.361	2.000	309.361	95704.265	307.374	9494.728	
412	78.870	1.000	77.870	6063.782	76.883	9571.611	
413	59.006	2.000	57.006	3249.635	55.073	9626.684	
414	66.951	0.000	66.951	4482.498	66.951	9693.636	
421	230.510	5.000	225.510	50854.939	220.619	9914.254	
422	58.390	54.000	4.390	19.273	0.330	9914.584	
423	43.684	46.000	-2.316	5.365	0.123	9914.707	
424	49.566	34.000	15.566	242.309	4.889	9919.596	
431	172.453	2.000	170.453	29054.137	168.476	10088.072	
432	43.684	65.000	-21.316	454.385	10.402	10098.474	
433	32.681	68.000	-35.319	1247.413	38.169	10136.643	
434	37.082	102.000	-64.918	4214.317	113.648	10250.290	
441	195.676	1.000	194.676	37898.668	193.681	10443.971	
442	49.566	38.000	11.566	133.779	2.699	10446.670	
443	37.082	113.000	-75.918	5763.508	155.425	10602.095	
444	42.076	377.000	-334.924	112174.189	2666.000	13268.095	
	n=	4232			Chi-Square	13268.095	

-13268.095

second 24 weeks. The 13th box and whiskers plots that represent the 13th and 26th two-week price data overlapped and are not shown. Clearly, the prices for 93–94 are not stationary. The cumulative probability density functions for Delta-1 price changes for Half1 and Half2 (Figure 3.80, Table 3.88 a-b) do overlap and suggest that the Delta-1 price changes are stationary.

The differential spectrum with a bin size of $0.10 is shown in Figure 3.81 and it is asymmetrical, which suggests that the Delta-1 price changes contain serial dependencies. This is confirmed by the

Table 3.50a-b

X_1: 1–Column 1

Mean	Std. Dev.	Std. Error	Variance	Coef. Var.	Count
.1	.1	1.4E–3	2.7E–3	40.7	1448

Minimum	Maximum	Range	Sum	Sum of Sqr.	# Missing
–.3	.2	.5	185.8	27.8	0

# < 10th %	10th %	25th %	50th %	75th %	90th %
88	.1	.1	.1	.2	.2

# > 90th %	Mode	Geo. Mean	Har. Mean	Kurtosis	Skewness
137	.1	•	•	6.1	–.7

X_2: 2–Column 1

Mean	Std. Dev.	Std. Error	Variance	Coef. Var.	Count
.3	.1	1.8E–3	3.3E–3	17.7	1072

Minimum	Maximum	Range	Sum	Sum of Sqr.	# Missing
.2	.4	.2	349.7	117.6	376

# < 10th %	10th %	25th %	50th %	75th %	90th %
63	.2	.3	.3	.4	.4

# > 90th %	Mode	Geo. Mean	Har. Mean	Kurtosis	Skewness
73	.4	.3	.3	–1.5	–.2

significant Chi-Square for the RPC digram and trigram transition matrices shown in Tables 3.89 and 3.90, respectively. The summary statistics for price increases and decreases are shown in Table 3.91 a-b.

The closing prices for the three automakers are shown in Figure 3.82. We can generate standard scores for each time series (price-

Table 3.50c-d

X_3: 3 – Column 1

Mean	Std. Dev.	Std. Error	Variance	Coef. Var.	Count
.5	.1	2.3E–3	4.2E–3	12	802

Minimum	Maximum	Range	Sum	Sum of Sqr.	# Missing
.4	.7	.2	435	239.3	646

# < 10th %	10th %	25th %	50th %	75th %	90th %
78	.5	.5	.5	.6	.6

# > 90th %	Mode	Geo. Mean	Har. Mean	Kurtosis	Skewness
17	.5	.5	.5	–1.4	.2

X_4: 4 – Column 1

Mean	Std. Dev.	Std. Error	Variance	Coef. Var.	Count
1.1	.6	2.1E–2	.4	57	912

Minimum	Maximum	Range	Sum	Sum of Sqr.	# Missing
.7	12.8	12.1	1018.3	1506.3	536

# < 10th %	10th %	25th %	50th %	75th %	90th %
64	.8	.8	1	1.2	1.8

# > 90th %	Mode	Geo. Mean	Har. Mean	Kurtosis	Skewness
77	.8	1	1	133	8.6

mean/standard deviation) and collect the standard scores into individual cumulative probability density functions, which are shown in Figure 3.83. In contrast to the raw prices (Figure 3.82), the standard score plot suggests that the underlying probability density functions for the three auto makers are quite similar. We can determine the impact of GM price on the prices of the other auto makers by sub-

Figure 3.42

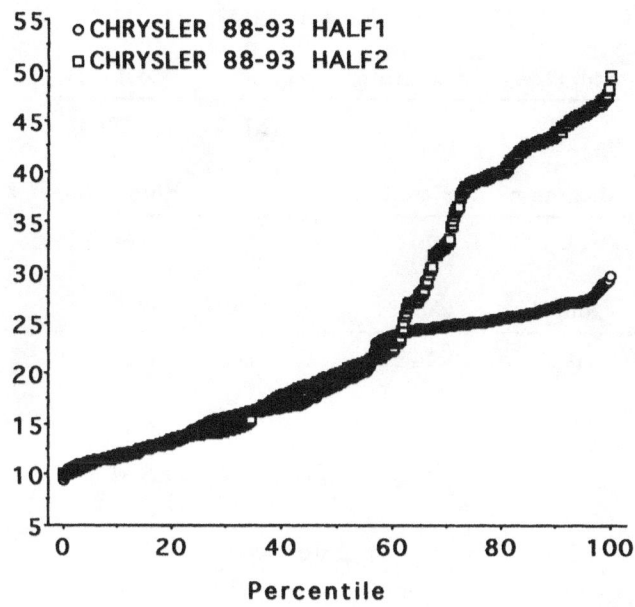

Figure 3.43

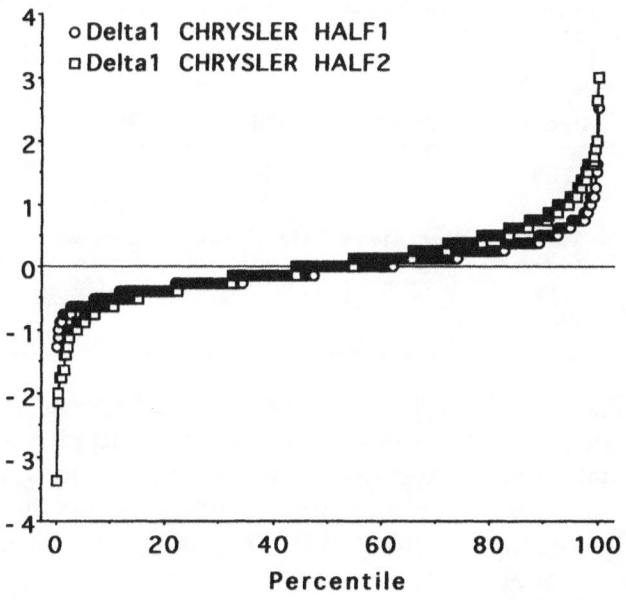

Table 3.51a-b

X_1: Chrysler 88–93 Half1

Mean	Std. Dev.	Std. Error	Variance	Coef. Var.	Count
19.5	5.7	.2	32.1	29	632

Minimum	Maximum	Range	Sum	Sum of Sqr.	# Missing
9.4	29.5	20.1	12329.9	260812.5	0

# < 10th %	10th %	25th %	50th %	75th %	90th %
59	11.9	14.5	18.8	24.9	26.5

# > 90th %	Mode	Geo. Mean	Har. Mean	Kurtosis	Skewness
60	24.8	18.6	17.8	–1.5	–4.2E–2

X_2: Chrysler 88–93 Half2

Mean	Std. Dev.	Std. Error	Variance	Coef. Var.	Count
24.7	12.4	.5	153.3	50.2	632

Minimum	Maximum	Range	Sum	Sum of Sqr.	# Missing
10	49.2	39.2	15590.4	481303.6	0

# < 10th %	10th %	25th %	50th %	75th %	90th %
60	11.8	13.9	19.9	38.9	43.7

# > 90th %	Mode	Geo. Mean	Har. Mean	Kurtosis	Skewness
63	13.9	21.8	19.4	–1.3	.6

tracting GM's standard score from the standard scores of the other two auto makers and then collecting the resulting standard scores into individual cumulative probability density functions. As shown in Figure 3.84, the new cumulative probability density functions for Ford and Chrysler are very similar. This suggests that GM prices affect the prices of Ford and Chrysler.

Table 3.52a-b

X_1: Delta-1 Chrysler Half1

Mean	Std. Dev.	Std. Error	Variance	Coef. Var.	Count
$-1.5E-2$	.4	$1.6E-2$	.2	-2615.4	632

Minimum	Maximum	Range	Sum	Sum of Sqr.	# Missing
-1.2	2.5	3.8	-9.8	102.9	0

# < 10th %	10th %	25th %	50th %	75th %	90th %
43	$-.5$	$-.2$	0	.2	.5

# > 90th %	Mode	Geo. Mean	Har. Mean	Kurtosis	Skewness
48	0	•	•	3	.9

X_2: Delta-1 Chrysler Half2

Mean	Std. Dev.	Std. Error	Variance	Coef. Var.	Count
.1	.6	$2.4E-2$	.4	1148.7	632

Minimum	Maximum	Range	Sum	Sum of Sqr.	# Missing
-3.4	3	6.4	33.9	241	0

# < 10th %	10th %	25th %	50th %	75th %	90th %
45	$-.6$	$-.2$	0	.4	.8

# > 90th %	Mode	Geo. Mean	Har. Mean	Kurtosis	Skewness
60	$-.1$	•	•	3.3	.1

Figure 3.44

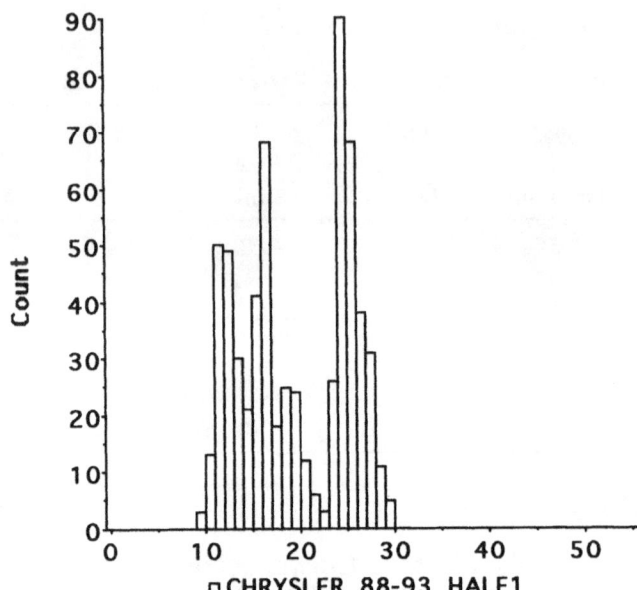

Figure 3.45

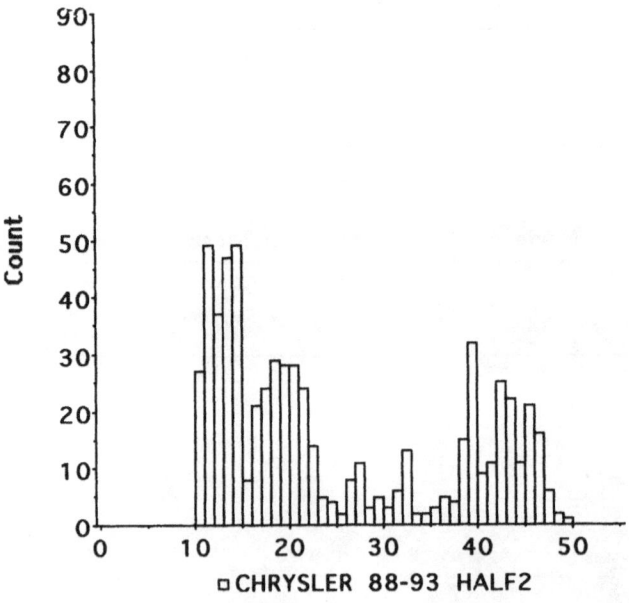

Figure 3.46

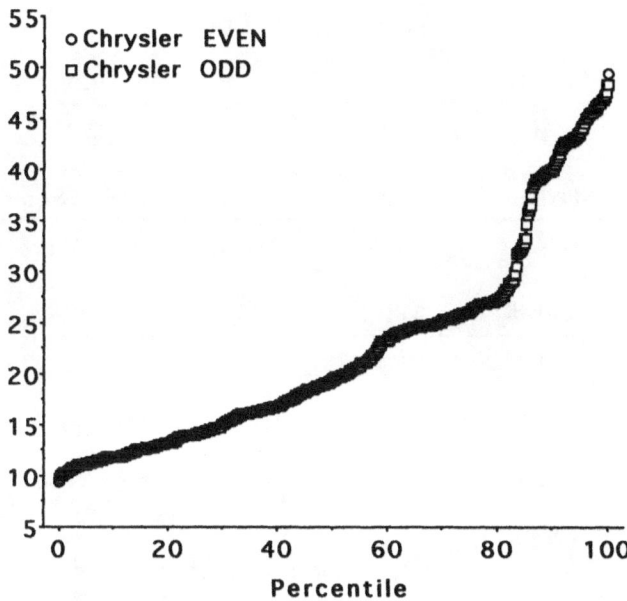

Figure 3.47

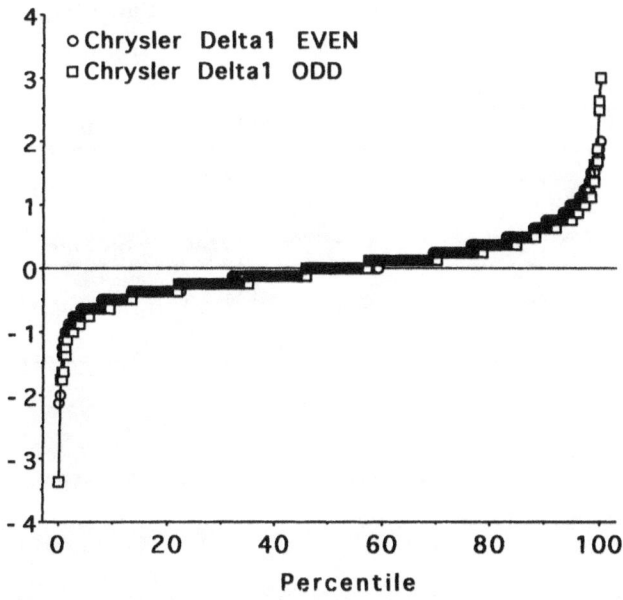

Figure 3.53a-b

X_1: *Chrysler Even*

Mean	Std. Dev.	Std. Error	Variance	Coef. Var.	Count
22.1	10	.4	100.5	45.3	632

Minimum	Maximum	Range	Sum	Sum of Sqr.	# Missing
9.4	49.2	39.9	13991.1	373133.5	0

# < 10th %	10th %	25th %	50th %	75th %	90th %
59	11.8	14	19.5	26.1	40

# > 90th %	Mode	Geo. Mean	Har. Mean	Kurtosis	Skewness
63	12.5	20.2	18.5	.1	1

X_2: *Chrysler Odd*

Mean	Std. Dev.	Std. Error	Variance	Coef. Var.	Count
22.1	10	.4	100.3	45.2	632

Minimum	Maximum	Range	Sum	Sum of Sqr.	# Missing
9.8	48.2	38.5	13987.1	372832	0

# < 10th %	10th %	25th %	50th %	75th %	90th %
55	11.8	14.1	19.4	26.1	39.9

# > 90th %	Mode	Geo. Mean	Har. Mean	Kurtosis	Skewness
62	12.8	20.2	18.6	.1	1

Table 3.54a-b

X_1: *Chrysler Delta-1 Even*

Mean	Std. Dev.	Std. Error	Variance	Coef. Var.	Count
3.1E–2	.5	2.0E–2	.3	1619.9	631

Minimum	Maximum	Range	Sum	Sum of Sqr.	# Missing
–2.1	2	4.1	19.8	163.6	1

# < 10th %	10th %	25th %	50th %	75th %	90th %
49	–.5	–.2	0	.2	.7

# > 90th %	Mode	Geo. Mean	Har. Mean	Kurtosis	Skewness
63	–.1	•	•	2	.5

X_2: *Chrysler Delta-1 Odd*

Mean	Std. Dev.	Std. Error	Variance	Coef. Var.	Count
6.4E–3	.5	2.1E–2	.3	8316	632

Minimum	Maximum	Range	Sum	Sum of Sqr.	# Missing
–3.4	3	6.4	4.1	180.3	0

# < 10th %	10th %	25th %	50th %	75th %	90th %
60	–.5	–.2	0	.2	.6

# > 90th %	Mode	Geo. Mean	Har. Mean	Kurtosis	Skewness
51	–.2	•	•	5.9	.3

Figure 3.48

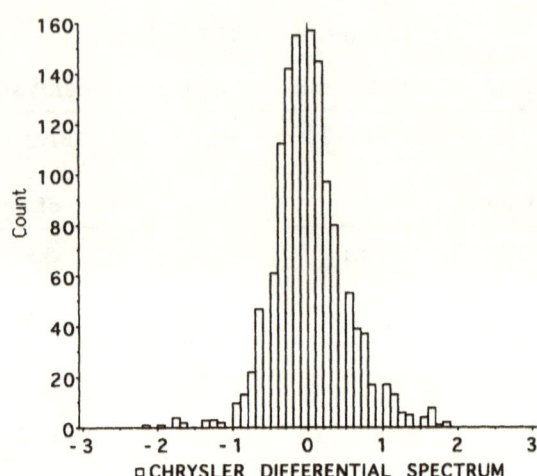

Figure 3.49

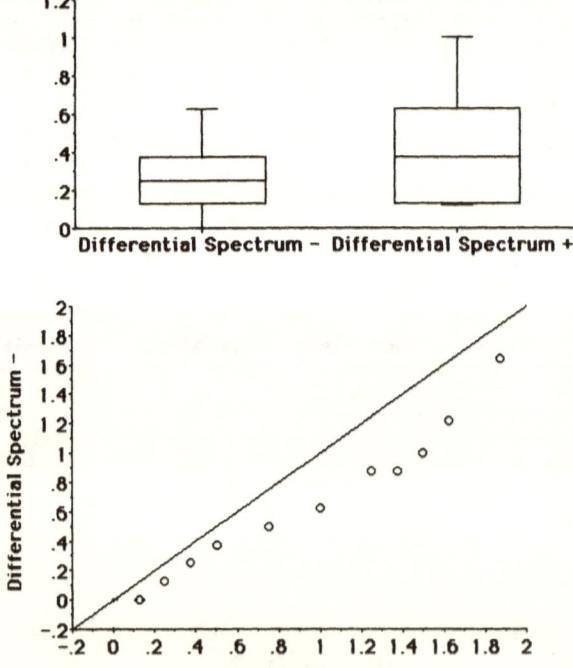

Figure 3.50

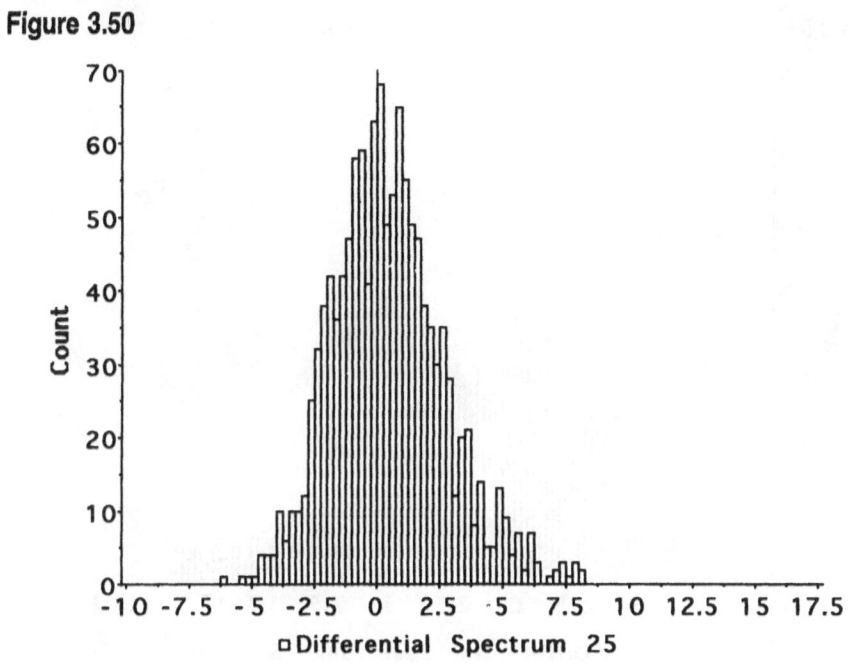

Figure 3.51

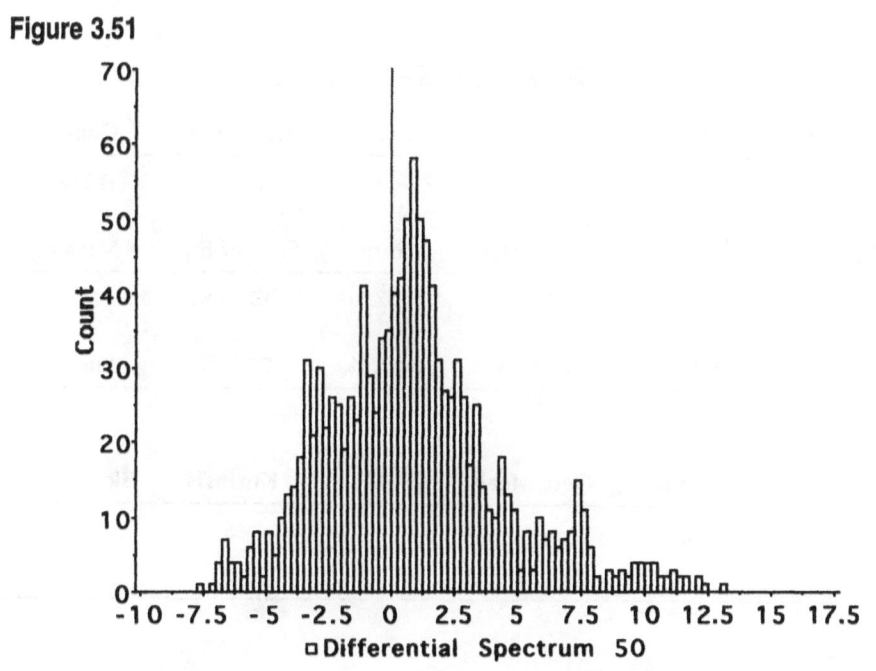

Figure 3.52

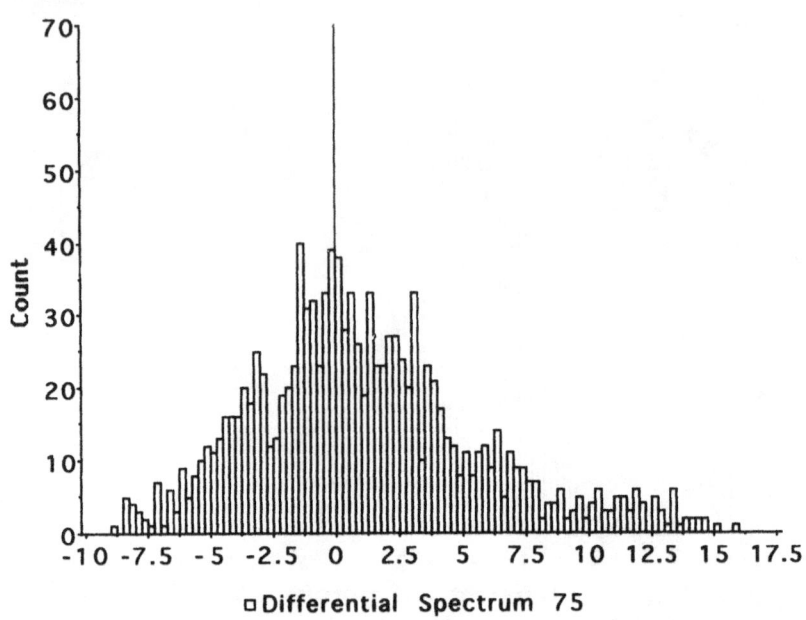

□ Differential Spectrum 75

Table 3.55a

X_1: *Differential Spectrum 25*

Mean	Std. Dev.	Std. Error	Variance	Coef. Var.	Count
.4	2.2	.1	4.9	589.7	1241

Minimum	Maximum	Range	Sum	Sum of Sqr.	# Missing
–6.1	8.1	14.2	467.5	6295.6	24

# < 10th %	10th %	25th %	50th %	75th %	90th %
121	–2.2	–1.1	.2	1.6	3.2

# > 90th %	Mode	Geo. Mean	Har. Mean	Kurtosis	Skewness
120	.1	•	•	.5	.5

Table 3.55b-c

X_2: *Differential Spectrum 50*

Mean	Std. Dev.	Std. Error	Variance	Coef. Var.	Count
.7	3.5	.1	12.3	472.1	1216

Minimum	Maximum	Range	Sum	Sum of Sqr.	# Missing
–7.6	13.1	20.8	904.2	15650.1	49

# < 10th %	10th %	25th %	50th %	75th %	90th %
121	–3.4	–1.6	.6	2.5	5.6

# > 90th %	Mode	Geo. Mean	Har. Mean	Kurtosis	Skewness
121	.8	•	•	.6	.6

X_3: *Differential Spectrum 75*

Mean	Std. Dev.	Std. Error	Variance	Coef. Var.	Count
1.2	4.7	.1	21.9	401.2	1191

Minimum	Maximum	Range	Sum	Sum of Sqr.	# Missing
–8.9	17.9	26.8	1389.2	27676.3	74

# < 10th %	10th %	25th %	50th %	75th %	90th %
117	–4.2	–1.8	.5	3.5	7.2

# > 90th %	Mode	Geo. Mean	Har. Mean	Kurtosis	Skewness
117	–1.5	•	•	.7	.7

Figure 3.53

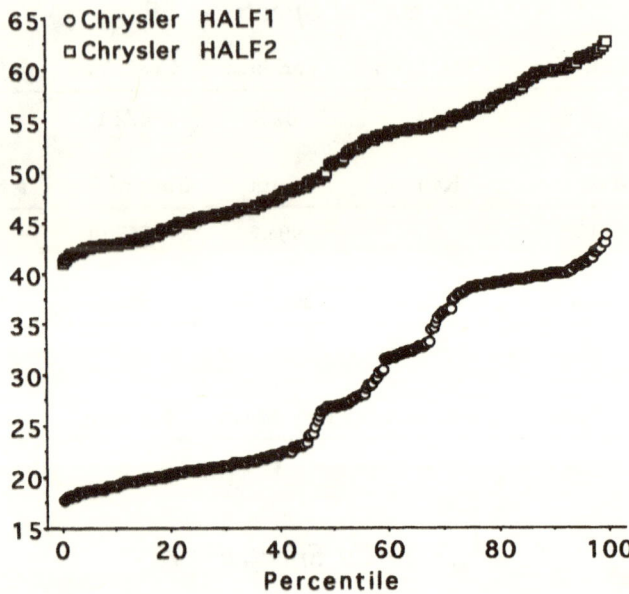

Figure 3.54

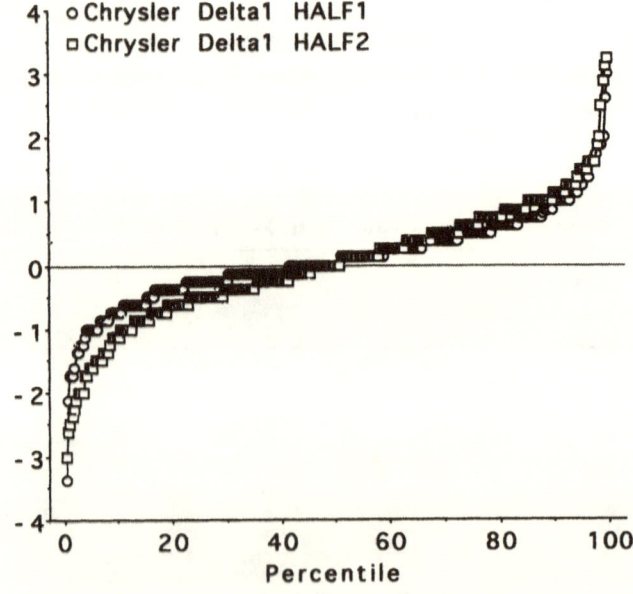

Table 3.56a-b

X_1: *Chrysler Delta-1 Half1*

Mean	Std. Dev.	Std. Error	Variance	Coef. Var.	Count
.1	.7	4.6E–2	.5	786.2	252

Minimum	Maximum	Range	Sum	Sum of Sqr.	# Missing
–3.4	3	6.4	23.5	137.1	1

# < 10th %	10th %	25th %	50th %	75th %	90th %
21	–.8	–.2	0	.5	1

# > 90th %	Mode	Geo. Mean	Har. Mean	Kurtosis	Skewness
18	–.1	•	•	3	–3.5E–2

X_2: *Chrysler Delta-1 Half2*

Mean	Std. Dev.	Std. Error	Variance	Coef. Var.	Count
2.0E–2	1	.1	.9	4810	253

Minimum	Maximum	Range	Sum	Sum of Sqr.	# Missing
–3	3.2	6.2	5	227.8	0

# < 10th %	10th %	25th %	50th %	75th %	90th %
22	–1.1	–.5	0	.6	1.1

# > 90th %	Mode	Geo. Mean	Har. Mean	Kurtosis	Skewness
20	.1	•	•	1	–2.9E–2

Table 3.57a-b

X_1: *Chrysler Half1*

Mean	Std. Dev.	Std. Error	Variance	Coef. Var.	Count
28.5	8.4	.5	70.4	29.4	253

Minimum	Maximum	Range	Sum	Sum of Sqr.	# Missing
17.8	43.8	26	7213.8	223436	0

# < 10th %	10th %	25th %	50th %	75th %	90th %
21	19.1	20.7	26.9	38.5	39.9

# > 90th %	Mode	Geo. Mean	Har. Mean	Kurtosis	Skewness
21	21.4	27.3	26.2	–1.5	.3

X_2: *Chrysler Half2*

Mean	Std. Dev.	Std. Error	Variance	Coef. Var.	Count
50.8	6.2	.4	38.5	12.2	253

Minimum	Maximum	Range	Sum	Sum of Sqr.	# Missing
41	62.5	21.5	12852.8	662631.3	0

# < 10th %	10th %	25th %	50th %	75th %	90th %
24	42.9	45.3	50.8	55.8	59.8

# > 90th %	Mode	Geo. Mean	Har. Mean	Kurtosis	Skewness
25	•	50.4	50.1	–1.3	.2

Figure 3.55

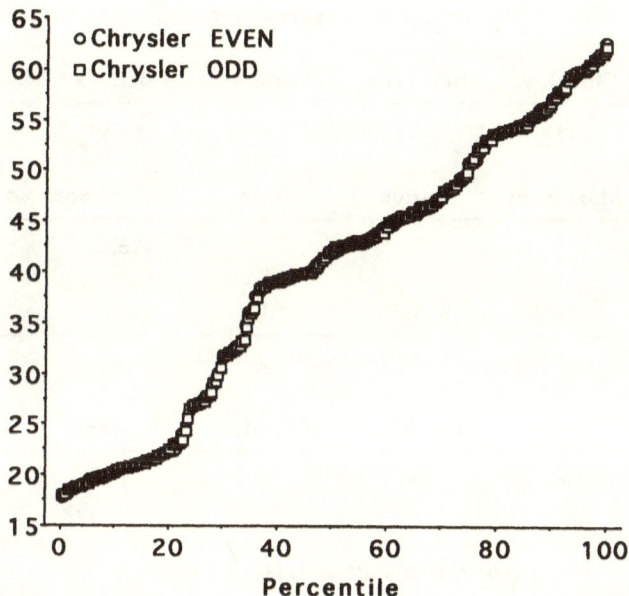

Figure 3.56

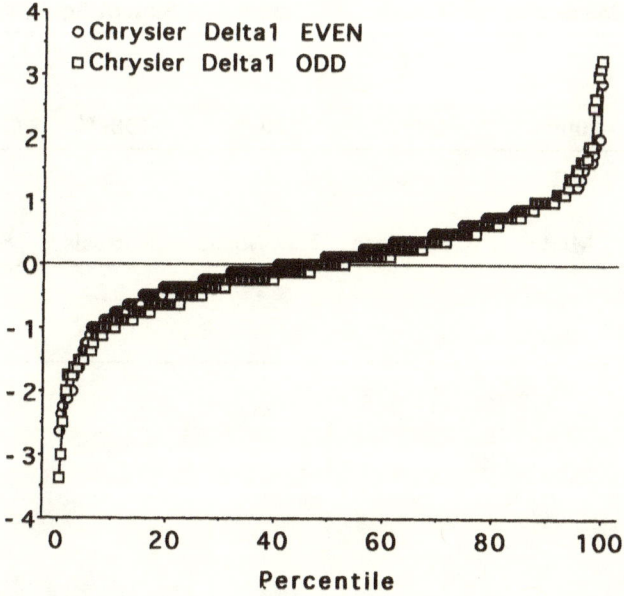

Table 3.58a-b

X_1: *Chrysler Delta-1 Even*

Mean	Std. Dev.	Std. Error	Variance	Coef. Var.	Count
.1	.8	.1	.6	879.4	253

Minimum	Maximum	Range	Sum	Sum of Sqr.	# Missing
−2.6	2.9	5.5	22.9	162.3	0

# < 10th %	10th %	25th %	50th %	75th %	90th %
21	−.9	−.4	.1	.6	1

# > 90th %	Mode	Geo. Mean	Har. Mean	Kurtosis	Skewness
22	•	•	•	1.3	−.3

X_2: *Chrysler Delta-1 Odd*

Mean	Std. Dev.	Std. Error	Variance	Coef. Var.	Count
2.1E−2	.9	.1	.8	4171.7	253

Minimum	Maximum	Range	Sum	Sum of Sqr.	# Missing
−3.4	3.2	6.6	5.4	202.7	0

# < 10th %	10th %	25th %	50th %	75th %	90th %
21	−1	−.5	0	.5	1

# > 90th %	Mode	Geo. Mean	Har. Mean	Kurtosis	Skewness
22	−.2	•	•	2.2	.2

Table 3.59a-b

X_1: Chrysler Even

Mean	Std. Dev.	Std. Error	Variance	Coef. Var.	Count
39.7	13.4	.8	178.9	33.7	253

Minimum	Maximum	Range	Sum	Sum of Sqr.	# Missing
17.8	62.5	44.8	10036	443197.2	0

# < 10th %	10th %	25th %	50th %	75th %	90th %
24	20.1	26.9	41.9	50.8	57.2

# > 90th %	Mode	Geo. Mean	Har. Mean	Kurtosis	Skewness
25	43	37.1	34.4	−1.2	−.2

X_2: Chrysler Odd

Mean	Std. Dev.	Std. Error	Variance	Coef. Var.	Count
39.6	13.4	.8	179.3	33.8	253

Minimum	Maximum	Range	Sum	Sum of Sqr.	# Missing
17.9	62.1	44.2	10030.6	442870.2	0

# < 10th %	10th %	25th %	50th %	75th %	90th %
25	20.3	26.9	42.1	50.7	57.1

# > 90th %	Mode	Geo. Mean	Har. Mean	Kurtosis	Skewness
25	42.8	37.1	34.4	−1.2	−.2

Figure 3.57

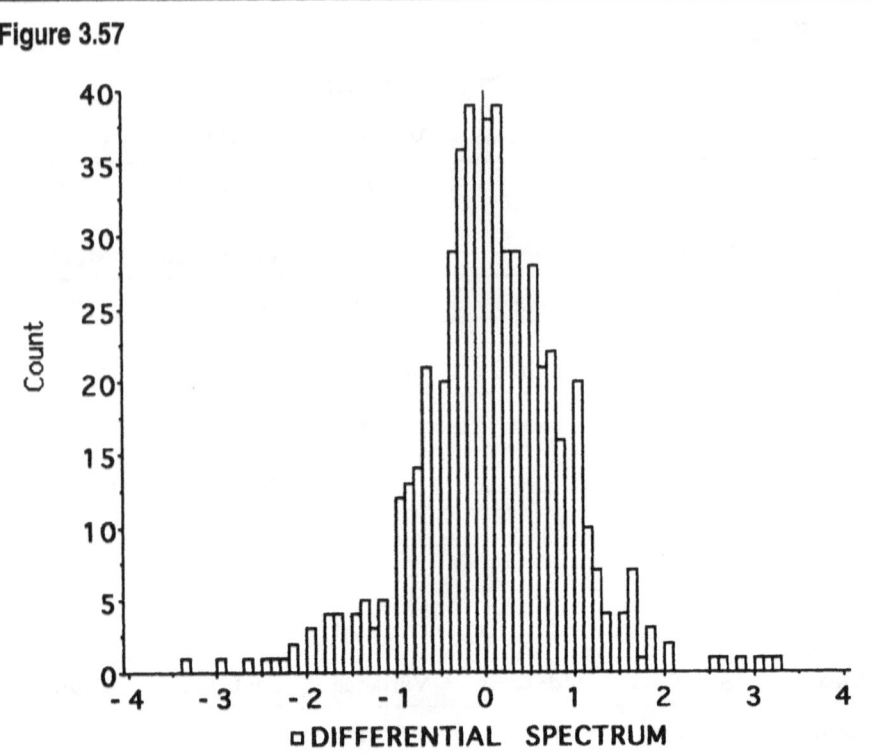

Table 3.60

X₁: *Differential Spectrum*

Mean	Std. Dev.	Std. Error	Variance	Coef. Var.	Count
.1	.8	3.8E–2	.7	1512.6	506

Minimum	Maximum	Range	Sum	Sum of Sqr.	# Missing
–3.4	3.2	6.6	28.4	364.9	1

# < 10th %	10th %	25th %	50th %	75th %	90th %
48	–.9	–.4	0	.5	1

# > 90th %	Mode	Geo. Mean	Har. Mean	Kurtosis	Skewness
44	•	•	•	1.9	–.1

Table 3.61

	A	B	C	D	E	F	G	H	I
1	n	Digram	Expec Prob	(E)xpec Freq	(O)bser Freq	(E-O)	(E-O)^2	((E-O)^2)/E	SUM
2	505	11	0.1667	84.1835	126	-41.8165	1748.61967	20.771525	
3		12	0.3333	168.3165	131	37.3165	1392.52117	8.27323033	29.0447553
4		21	0.3333	168.3165	131	37.3165	1392.52117	8.27323033	37.3179856
5		22	0.1667	84.1835	117	-32.8165	1076.92267	12.7925623	45.5912159
6			SUM=	505	505				
7									
8								Chi-Square	58.3837783

Table 3.62a-b

X_1: 1 – RPC Chrysler 92–94

Mean	Std. Dev.	Std. Error	Variance	Coef. Var.	Count
-.6	.6	3.6E-2	.3	-104.5	258

Minimum	Maximum	Range	Sum	Sum of Sqr.	# Missing
-3.4	0	3.4	-143.9	167.5	0

# < 10th %	10th %	25th %	50th %	75th %	90th %
23	-1.4	-.8	-.4	-.1	0

# > 90th %	Mode	Geo. Mean	Har. Mean	Kurtosis	Skewness
0	-.1	•	•	3.8	-1.8

X_2: 2 – RPC Chrysler 92–94

Mean	Std. Dev.	Std. Error	Variance	Coef. Var.	Count
.7	.6	3.6E-2	.3	80.8	248

Minimum	Maximum	Range	Sum	Sum of Sqr.	# Missing
.1	3.2	3.1	172.2	197.4	10

# < 10th %	10th %	25th %	50th %	75th %	90th %
0	.1	.2	.5	1	1.4

# > 90th %	Mode	Geo. Mean	Har. Mean	Kurtosis	Skewness
23	.1	.5	.4	4.4	1.8

Table 3.63

	A	B	C	D	E	F	G
1	Symbols	Frequency	Probability				
2	1	48	0.0950				
3	2	413	0.8178				
4	3	44	0.0871				
5							
6	Sum	505					
7							
8							
9							
10	Digrams	(E)xpected	(O)bserved	(E-O)	(E-O)^2	((E-O)^2)/E	Sum
11	11	4.56	7	-2.44	5.942	1.302	
12	12	39.26	31	8.26	68.152	1.736	3.039
13	13	4.18	10	-5.82	33.847	8.093	11.132
14	21	39.26	37	2.26	5.087	0.130	11.261
15	22	337.76	345	-7.24	52.412	0.155	11.416
16	23	35.98	31	4.98	24.842	0.690	12.107
17	31	4.18	4	0.18	0.033	0.008	12.115
18	32	35.98	37	-1.02	1.032	0.029	12.143
19	33	3.83	3	0.83	0.695	0.181	
20							
21	SUM=	505	505			Chi-Square	12.325
22					4.96481626		
23					4		
24				Z=	0.96481626		

Table 3.64a

X_1: 1 – RPC Chrysler 92–94

Mean	Std. Dev.	Std. Error	Variance	Coef. Var.	Count
–1.5	.6	.1	.3	–36.7	48

Minimum	Maximum	Range	Sum	Sum of Sqr.	# Missing
–3.4	–1	2.4	–74.1	129.5	366

# < 10th %	10th %	25th %	50th %	75th %	90th %
5	–2.3	–1.8	–1.4	–1.1	–1

# > 90th %	Mode	Geo. Mean	Har. Mean	Kurtosis	Skewness
0	–1	•	•	1.3	–1.3

Table 3.64b-c

X_2: 2 – RPC Chrysler 92–94

Mean	Std. Dev.	Std. Error	Variance	Coef. Var.	Count
.1	.5	2.4E–2	.2	667.4	414

Minimum	Maximum	Range	Sum	Sum of Sqr.	# Missing
–.9	1	1.9	30.8	104.2	0

# < 10th %	10th %	25th %	50th %	75th %	90th %
27	–.6	–.2	0	.5	.8

# > 90th %	Mode	Geo. Mean	Har. Mean	Kurtosis	Skewness
36	•	•	•	–.8	.1

X_3: 3 – RPC Chrysler 92–94

Mean	Std. Dev.	Std. Error	Variance	Coef. Var.	Count
1.6	.6	.1	.3	35.5	44

Minimum	Maximum	Range	Sum	Sum of Sqr.	# Missing
1.1	3.2	2.1	71.7	131.2	370

# < 10th %	10th %	25th %	50th %	75th %	90th %
0	1.1	1.2	1.5	1.8	2.6

# > 90th %	Mode	Geo. Mean	Har. Mean	Kurtosis	Skewness
4	1.1	1.5	1.5	1.3	1.5

Figure 3.58

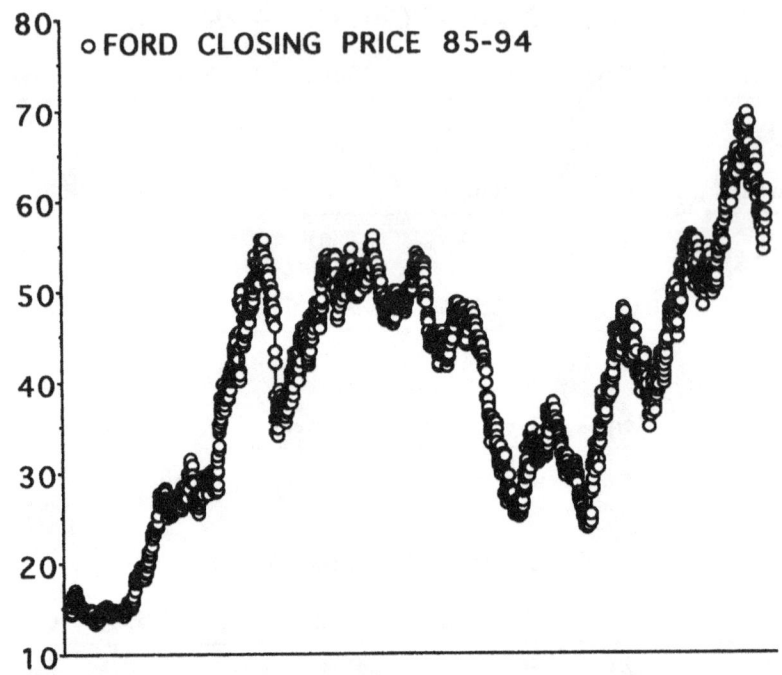

Table 3.65

X₁: Ford Closing Price 85–94

Mean	Std. Dev.	Std. Error	Variance	Coef. Var.	Count
39.5	13.2	.3	173.3	33.3	2361

Minimum	Maximum	Range	Sum	Sum of Sqr.	# Missing
13.4	69.5	56.1	93335	4098592.9	2

# < 10th %	10th %	25th %	50th %	75th %	90th %
234	18.4	29.1	42.2	49.8	53.6

# > 90th %	Mode	Geo. Mean	Har. Mean	Kurtosis	Skewness
232	44	36.9	33.7	−.7	−.3

Figure 3.59

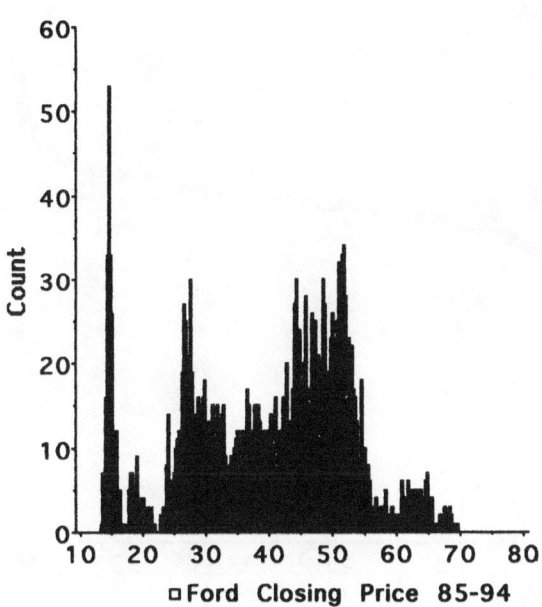

Figure 3.60

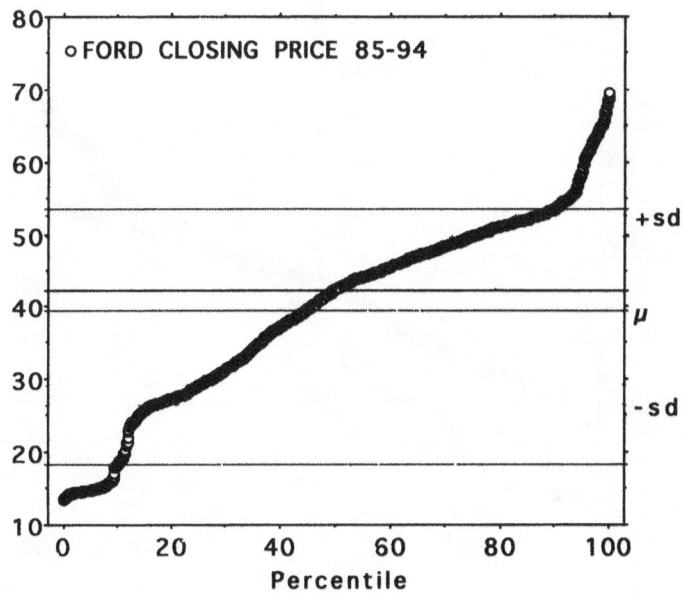

Figure 3.61

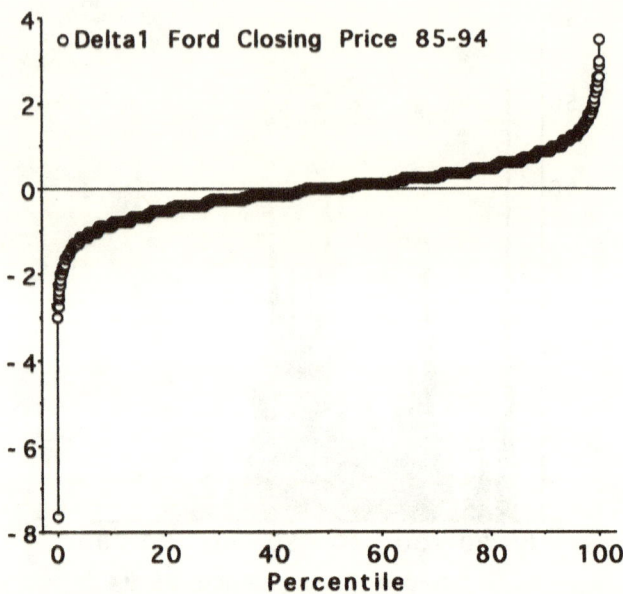

Figure 3.62

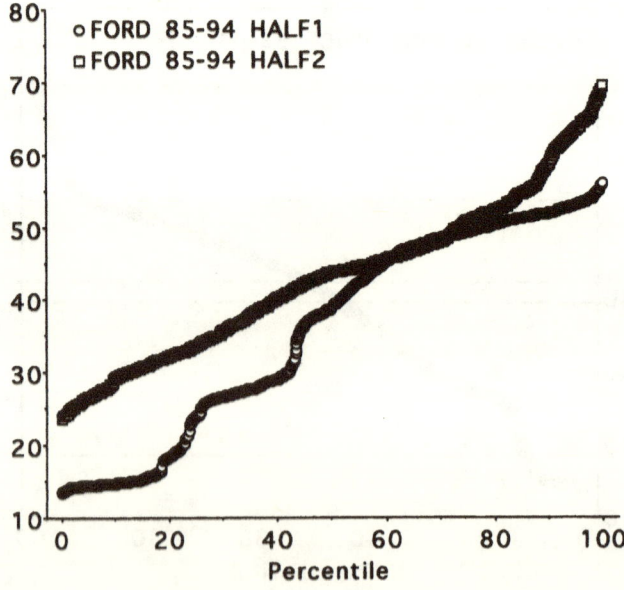

Figure 3.63

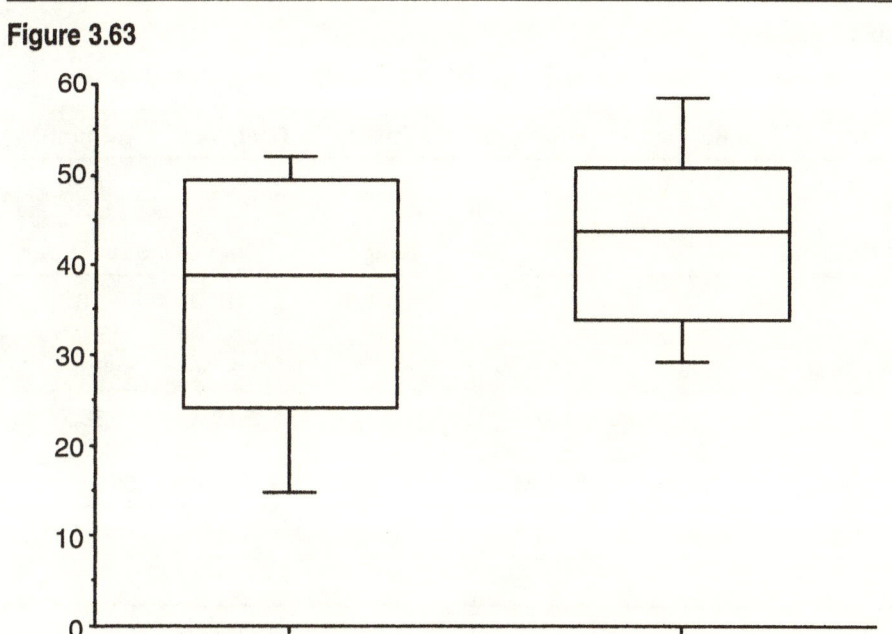

Table 3.66a

X_1: *Ford 85–94 Half1*

Mean	Std. Dev.	Std. Error	Variance	Coef. Var.	Count
35.9	14.2	.4	201.3	39.5	1180

Minimum	Maximum	Range	Sum	Sum of Sqr.	# Missing
13.4	56.1	42.8	42398.5	1760810.6	0

# < 10th %	10th %	25th %	50th %	75th %	90th %
116	14.8	24	38.9	49.4	51.9

# > 90th %	Mode	Geo. Mean	Har. Mean	Kurtosis	Skewness
117	•	32.6	28.9	–1.5	–.3

Table 3.66b

X_2: Ford 85–94 Half2

Mean	Std. Dev.	Std. Error	Variance	Coef. Var.	Count
43.1	10.9	.3	119.3	25.3	1180

Minimum	Maximum		Range	Sum	Sum of Sqr.
23.8	69.5	45.8	50876.5	2334182.3	0

# < 10th %	10th %	25th %	50th %	75th %	90th %
116	29.2	33.8	43.8	50.9	58.7

# > 90th %	Mode	Geo. Mean	Har. Mean	Kurtosis	Skewness
118	44	41.7	40.3	–.7	.3

Figure 3.64

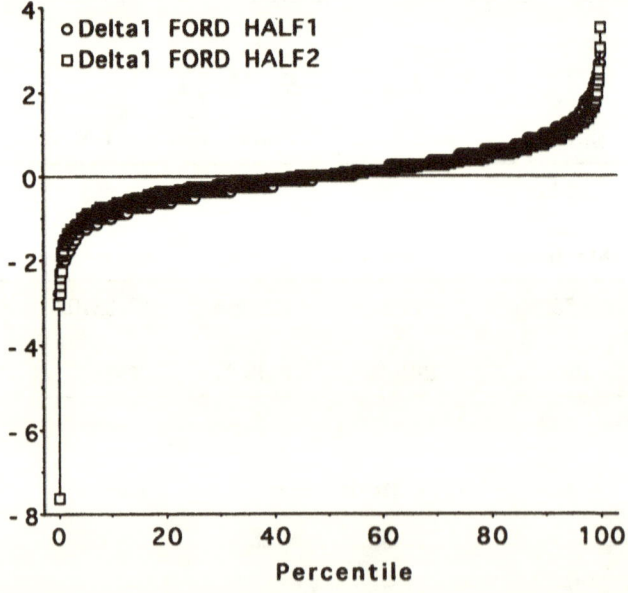

Table 3.67a-b

X_1: Delta-1 Ford Half1

Mean	Std. Dev.	Std. Error	Variance	Coef. Var.	Count
6.0E–3	.8	2.3E–2	.6	12996	1180

Minimum	Maximum	Range	Sum	Sum of Sqr.	# Missing
–3	2.9	5.9	7.1	726	0

# < 10th %	10th %	25th %	50th %	75th %	90th %
117	–.9	–.5	0	.5	1

# > 90th %	Mode	Geo. Mean	Har. Mean	Kurtosis	Skewness
97	0	•	•	1	.1

X_2: Delta-1 Ford Half2

Mean	Std. Dev.	Std. Error	Variance	Coef. Var.	Count
3.3E–2	.7	2.0E–2	.5	2063.7	1179

Minimum	Maximum	Range	Sum	Sum of Sqr.	# Missing
–7.6	3.5	11.1	39	550.3	1

# < 10th %	10th %	25th %	50th %	75th %	90th %
94	–.8	–.3	0	.4	.8

# > 90th %	Mode	Geo. Mean	Har. Mean	Kurtosis	Skewness
116	0	•	•	15.1	–1.1

Table 3.68a-b

X_1: 1–Delta-1 Ford Half1

Mean	Std. Dev.	Std. Error	Variance	Coef. Var.	Count
−.5	.5	1.5E–2	.3	−107	1253

Minimum	Maximum	Range	Sum	Sum of Sqr.	# Missing
−7.6	0	7.6	−602.5	621.2	0

# < 10th %	10th %	25th %	50th %	75th %	90th %
103	−1.1	−.8	−.4	−.1	0

# > 90th %	Mode	Geo. Mean	Har. Mean	Kurtosis	Skewness
0	0	•	•	31.4	−3.4

X_2: 2–Delta-1 Ford Half2

Mean	Std. Dev.	Std. Error	Variance	Coef. Var.	Count
.6	.5	1.5E–2	.3	85.6	1107

Minimum	Maximum	Range	Sum	Sum of Sqr.	# Missing
4.2E–2	3.5	3.5	647.7	656.2	146

# < 10th %	10th %	25th %	50th %	75th %	90th %
77	.1	.2	.4	.8	1.2

# > 90th %	Mode	Geo. Mean	Har. Mean	Kurtosis	Skewness
95	.2	.4	.3	3.5	1.7

Figure 3.65

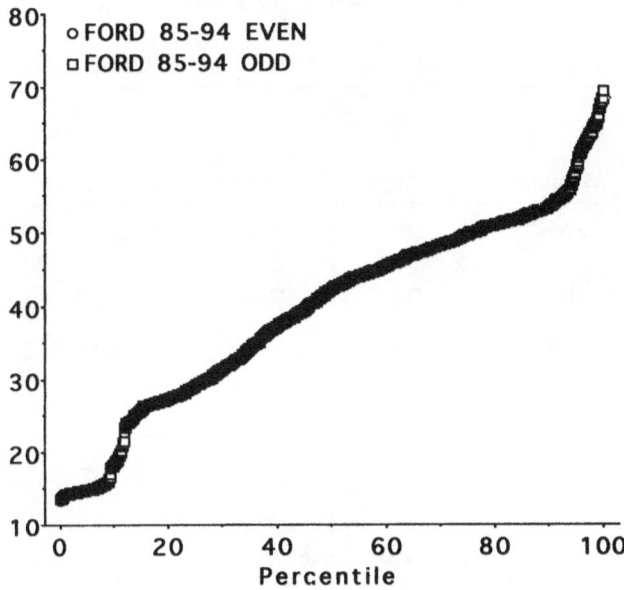

Figure 3.66

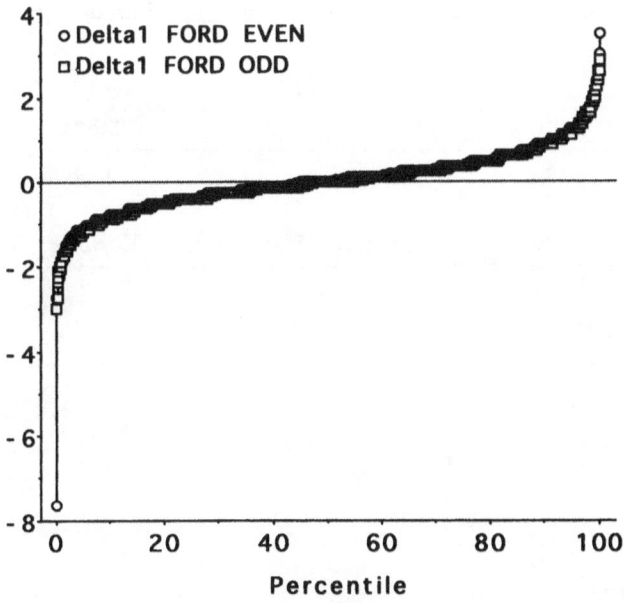

Table 3.69a-b

X_1: Ford 85–94 Even

Mean	Std. Dev.	Std. Error	Variance	Coef. Var.	Count
39.5	13.2	.4	173	33.3	1181

Minimum	Maximum	Range	Sum	Sum of Sqr.	# Missing
13.4	69.4	56	46669.1	2048384.5	0

# < 10th %	10th %	25th %	50th %	75th %	90th %
118	18.4	29	42.2	49.8	53.6

# > 90th %	Mode	Geo. Mean	Har. Mean	Kurtosis	Skewness
115	•	36.9	33.7	–.7	–.3

X_2: Ford 85–94 Odd

Mean	Std. Dev.	Std. Error	Variance	Coef. Var.	Count
39.5	13.2	.4	173.4	33.3	1181

Minimum	Maximum	Range	Sum	Sum of Sqr.	# Missing
13.6	69.5	55.9	46661.2	2048136.7	0

# < 10th %	10th %	25th %	50th %	75th %	90th %
118	18.4	29	42.2	49.8	53.6

# > 90th %	Mode	Geo. Mean	Har. Mean	Kurtosis	Skewness
118	44	36.9	33.7	–.7	–.3

Table 3.70a-b

X_1: Delta-1 Ford Even

Mean	Std. Dev.	Std. Error	Variance	Coef. Var.	Count
3.2E–2	.7	2.2E–2	.6	2372.3	1181

Minimum	Maximum	Range	Sum	Sum of Sqr.	# Missing
–7.6	3.5	11.1	37.2	661.8	0

# < 10th %	10th %	25th %	50th %	75th %	90th %
115	–.8	–.4	0	.4	.9

# > 90th %	Mode	Geo. Mean	Har. Mean	Kurtosis	Skewness
109	0	•	•	10.3	–.7

X_2: Delta-1 Ford Odd

Mean	Std. Dev.	Std. Error	Variance	Coef. Var.	Count
6.7E–3	.7	2.1E–2	.5	10828.2	1181

Minimum	Maximum	Range	Sum	Sum of Sqr.	# Missing
–3	2.9	5.9	7.9	615.6	0

# < 10th %	10th %	25th %	50th %	75th %	90th %
99	–.9	–.4	0	.4	.9

# > 90th %	Mode	Geo. Mean	Har. Mean	Kurtosis	Skewness
103	0	•	•	1.6	.1

Figure 3.67

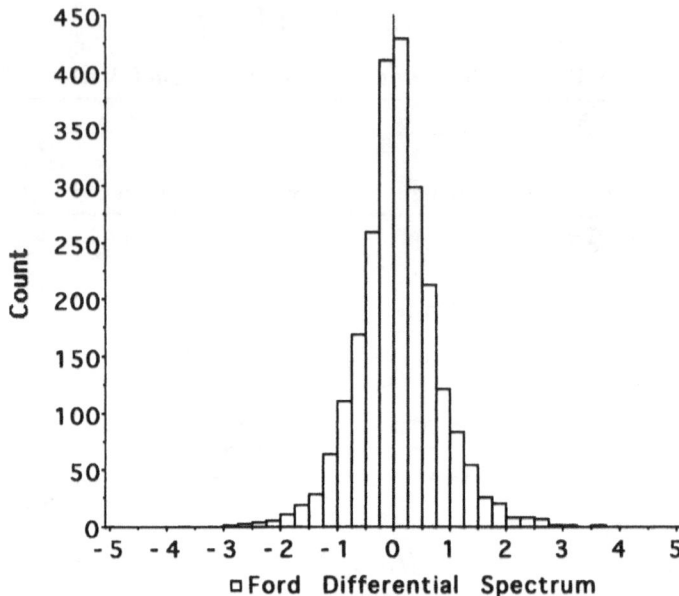

Figure 3.68

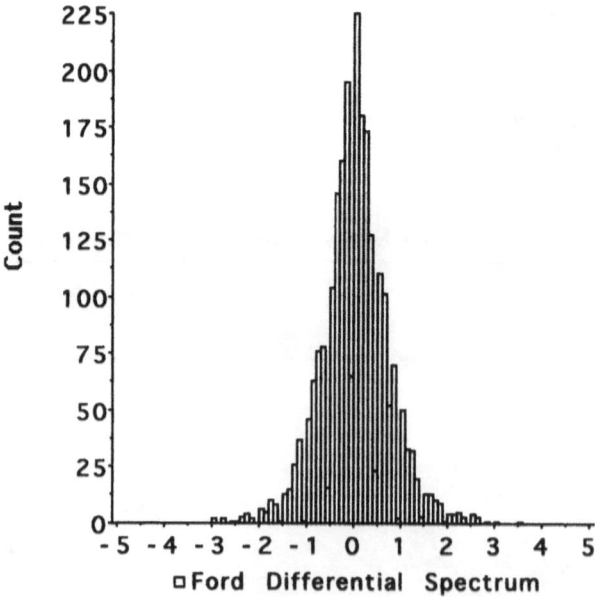

Figure 3.69

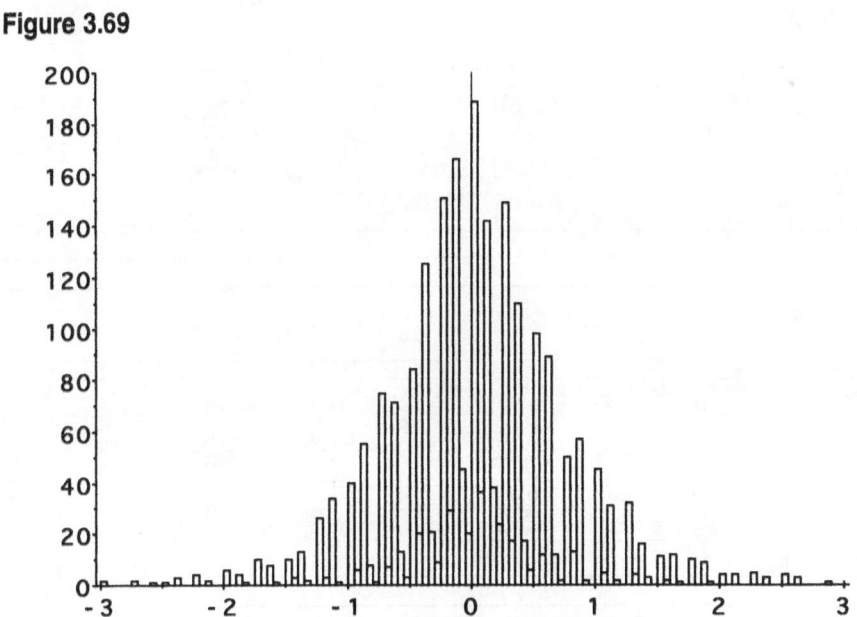

Table 3.71

X_1: *Ford Differential Spectrum*

Mean	Std. Dev.	Std. Error	Variance	Coef. Var.	Count
1.9E–2	.7	1.5E–2	.5	3847	2360

Minimum	Maximum	Range	Sum	Sum of Sqr.	# Missing
–7.6	3.5	11.1	45.1	1277.4	1

# < 10th %	10th %	25th %	50th %	75th %	90th %
184	–.9	–.4	0	.4	.9

# > 90th %	Mode	Geo. Mean	Har. Mean	Kurtosis	Skewness
212	0	•	•	6.2	–.4

Table 3.72

	A	B	C	D	E	· F	G	H	I
1	n	Digram	Expec Prob	(E)xpec Freq	(O)bser Freq	(E-O)	(E-O)^2	((E-O)^2)/E	SUM
2	2359	11	0.1667	393.245	658.000	-264.755	70095.051	178.248	
3		12	0.3333	786.255	594.000	192.255	36961.870	47.010	225.258
4		21	0.3333	786.255	594.000	192.255	36961.870	47.010	272.268
5		22	0.1667	393.245	513.000	-119.755	14341.188	36.469	319.278
6									
7		Sum=		2359.000	2359.000				
8								Chi-Square	355.747

Table 3.73

	A	B	C	D	E	F	G	H	I
1	n	Trigram	Expec Prob	(E)xpec Freq	(O)bser Freq	(E-O)	(E-O)^2	((E-O)^2)/E	SUM
2	2358	111	0.042	98.258	333.000	-234.742	55103.872	560.809	
3		112	0.125	294.750	325.000	-30.250	915.063	3.105	563.913
4		121	0.208	491.242	303.000	188.242	35435.103	72.134	636.047
5		122	0.125	294.750	291.000	3.750	14.063	0.048	636.095
6		211	0.125	294.750	324.000	-29.250	855.563	2.903	638.997
7		212	0.208	491.242	269.000	222.242	49391.569	100.544	739.542
8		221	0.125	294.750	291.000	3.750	14.063	0.048	739.589
9		222	0.042	98.258	222.000	-123.742	15312.117	155.836	
10									
11		Sum=		2358.000	2358.000			Chi-Square	895.425

Table 3.74

	A	B	C	D	E	F	G	H	I
1	n	Tetragram	Expected P	(E)xpected P	(O)bser P	(E-O)	(E-O)^2	((E-O)^2)/E	Sum
2	2357	1111	0.0083	19.634	164.000	-144.366	20841.597	1061.516	
3		1112	0.0333	78.559	169.000	-90.441	8179.609	104.121	1165.636
4		1121	0.0750	176.775	166.000	10.775	116.101	0.657	1166.293
5		1122	0.0500	117.850	159.000	-41.150	1693.322	14.368	1180.662
6		1211	0.0750	176.775	158.000	18.775	352.501	1.994	1182.656
7		1212	0.1333	314.259	145.000	169.259	28648.545	91.162	1273.818
8		1221	0.0917	216.043	161.000	55.043	3029.690	14.024	1287.842
9		1222	0.0333	78.559	130.000	-51.441	2646.196	33.684	1321.526
10		2111	0.0333	78.559	169.000	-90.441	8179.609	104.121	1425.647
11		2112	0.0917	216.043	155.000	61.043	3726.201	17.248	1442.894
12		2121	0.1333	314.259	137.000	177.259	31420.686	99.983	1542.878
13		2122	0.0750	176.775	132.000	44.775	2004.801	11.341	1554.219
14		2211	0.0500	117.850	166.000	-48.150	2318.422	19.673	1573.891
15		2212	0.0750	176.775	124.000	52.775	2785.201	15.756	1589.647
16		2221	0.0333	78.559	130.000	-51.441	2646.196	33.684	1623.331
17		2222	0.0083	19.634	92.000	-72.366	5236.865	266.727	1890.058
18									
19									
20		Sum=		2356.906	2357.000			Chi-Square	1890.058

Table 3.75a-b

X_1: 1–<10%

Mean	Std. Dev.	Std. Error	Variance	Coef. Var.	Count
−1.4	.6	4.7E–2	.4	−44.9	184

Minimum	Maximum	Range	Sum	Sum of Sqr.	# Missing
−7.6	−.9	6.7	−258.8	436.8	1780

# < 10th %	10th %	25th %	50th %	75th %	90th %
16	−2	−1.5	−1.2	−1	−1

# > 90th %	Mode	Geo. Mean	Har. Mean	Kurtosis	Skewness
6	−1	•	•	49.7	−5.7

X_2: 2–11%–90%

Mean	Std. Dev.	Std. Error	Variance	Coef. Var.	Count
1.8E–3	.4	9.8E–3	.2	23543.5	1964

Minimum	Maximum	Range	Sum	Sum of Sqr.	# Missing
−.9	.9	1.8	3.6	371	0

# < 10th %	10th %	25th %	50th %	75th %	90th %
146	−.6	−.3	0	.3	.6

# > 90th %	Mode	Geo. Mean	Har. Mean	Kurtosis	Skewness
134	0	•	•	−.6	−6.7E–3

Table 3.75c

X_3: 3–>90%

Mean	Std. Dev.	Std. Error	Variance	Coef. Var.	Count
1.4	.5	3.1E–2	.2	32.3	212

Minimum	Maximum	Range	Sum	Sum of Sqr.	# Missing
.9	3.5	2.6	300.2	469.5	1752

# < 10th %	10th %	25th %	50th %	75th %	90th %
2	1	1.1	1.2	1.6	2.1

# > 90th %	Mode	Geo. Mean	Har. Mean	Kurtosis	Skewness
18	1	1.4	1.3	2.5	1.5

Table 3.76

	A	B	C	D	E	F	G
1	Symbols	Frequency	Probability				
2	1	183	0.0776				
3	2	1964	0.8326				
4	3	212	0.0899				
5							
6	Sum	2359					
7							
8							
9							
10	Digrams	(E)xpected	(O)bserved	(E-O)	(E-O)^2	((E-O)^2)/E	Sum
11	11	14.20	26	-11.80	139.328	9.814	
12	12	152.36	128	24.36	593.301	3.894	13.709
13	13	16.45	29	-12.55	157.604	9.583	23.292
14	21	152.36	137	15.36	235.861	1.548	24.840
15	22	1635.14	1667	-31.86	1015.040	0.621	25.461
16	23	176.50	160	16.50	272.313	1.543	27.003
17	31	16.45	21	-4.55	20.739	1.261	28.264
18	32	176.50	168	8.50	72.282	0.410	28.674
19	33	19.05	23	-3.95	15.586	0.818	
20							
21	Sum=	2359	2359			Chi-Square	29.492
22							
23							
24							7.68010615
25							4
26						Z=	3.68010615

Table 3.77

	A	B	C	D	E	F	G
1	Symbols	Frequency	Probability				
2	1	183	0.0776				
3	2	1963	0.8325				
4	3	212	0.0899				
5							
6	Sum	2358					
7							
8							
9							
10	Digrams	(E)xpected	(O)bserved	(E-O)	(E-O)^2	((E-O)^2)/E	Sum
11	111	1.10	4	-2.90	8.397	7.618	
12	112	11.82	17	-5.18	26.799	2.267	9.885
13	113	1.28	5	-3.72	13.862	10.856	20.741
14	121	11.82	16	-4.18	17.446	1.476	22.217
15	122	126.82	96	30.82	950.167	7.492	29.708
16	123	13.70	16	-2.30	5.305	0.387	30.096
17	131	1.28	2	-0.72	0.523	0.410	30.505
18	132	13.70	20	-6.30	39.730	2.901	33.406
19	133	1.48	7	-5.52	30.479	20.605	54.011
20	211	11.82	18	-6.18	38.153	3.227	57.238
21	212	126.82	100	26.82	719.568	5.674	62.911
22	213	13.70	18	-4.30	18.517	1.352	64.263
23	221	126.82	108	18.82	354.372	2.794	67.057
24	222	1360.42	1432	-71.58	5123.567	3.766	70.824
25	223	146.92	127	19.92	396.913	2.702	73.525
26	231	13.70	16	-2.30	5.305	0.387	73.912
27	232	146.92	130	16.92	286.377	1.949	75.862
28	233	15.87	14	1.87	3.487	0.220	76.081
29	311	1.28	4	-2.72	7.415	5.807	81.889
30	312	13.70	11	2.70	7.273	0.531	82.420
31	313	1.48	6	-4.52	20.437	13.816	96.236
32	321	13.70	13	0.70	0.486	0.035	96.271
33	322	146.92	138	8.92	79.614	0.542	96.813
34	323	15.87	17	-1.13	1.283	0.081	96.894
35	331	1.48	3	-1.52	2.313	1.563	98.458
36	332	15.87	18	-2.13	4.548	0.287	98.744
37	333	1.71	2	-0.29	0.082	0.048	98.792
38							
39	Sum=	2358	2358			Chi-Square	98.792
40							
41							
42							14.056463
43							7.21110255
44						Z=	6.8453605

Figure 3.70

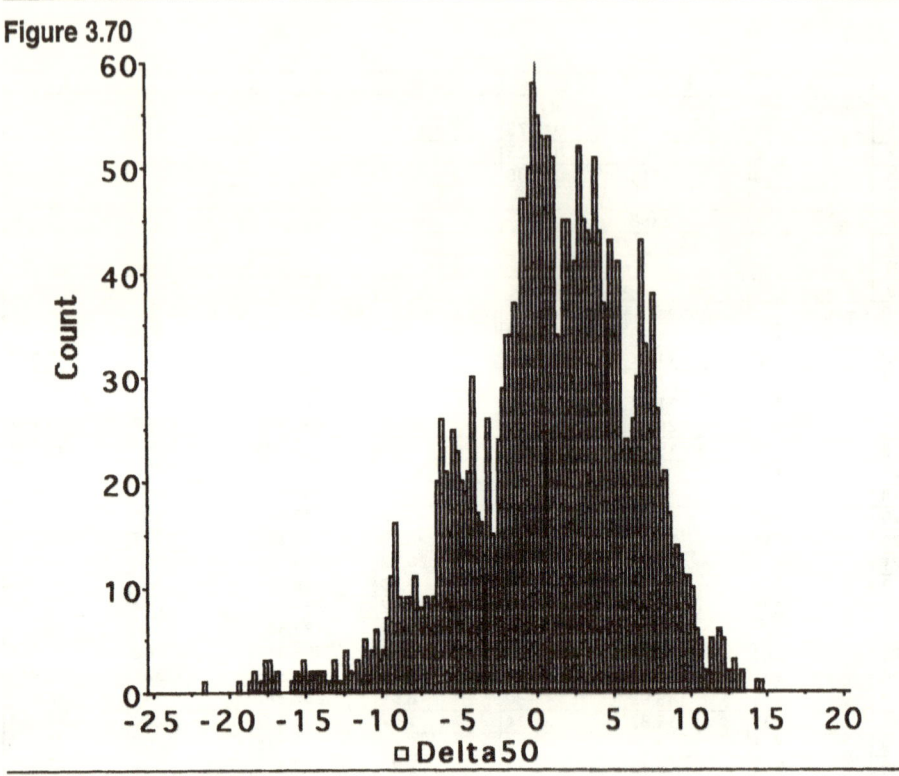

Table 3.78

X_1: *Delta-50*

Mean	Std. Dev.	Std. Error	Variance	Coef. Var.	Count
1	5.4	.1	28.6	557.9	2312

Minimum	Maximum	Range	Sum	Sum of Sqr.	# Missing
−21.7	14.5	36.2	2218	68324.8	49

# < 10th %	10th %	25th %	50th %	75th %	90th %
221	−6.1	−2	1.2	4.8	7.5

# > 90th %	Mode	Geo. Mean	Har. Mean	Kurtosis	Skewness
221	3.8	•	•	.6	−.6

Table 3.79a-b

X_1: 1–Delta-50

Mean	Std. Dev.	Std. Error	Variance	Coef. Var.	Count
–4.2	3.8	.1	14.8	–91.8	922

Minimum	Maximum	Range	Sum	Sum of Sqr.	# Missing
–21.7	0	21.7	–3866.5	29852.2	468

# < 10th %	10th %	25th %	50th %	75th %	90th %
89	–9.2	–6.1	–3.2	–1.1	–.4

# > 90th %	Mode	Geo. Mean	Har. Mean	Kurtosis	Skewness
92	–.1	•	•	2	–1.4

X_2: 2–Delta-50

Mean	Std. Dev.	Std. Error	Variance	Coef. Var.	Count
4.4	2.9	.1	8.5	66.7	1390

Minimum	Maximum	Range	Sum	Sum of Sqr.	# Missing
2.6E–3	14.5	14.5	6084.5	38472.6	0

# < 10th %	10th %	25th %	50th %	75th %	90th %
133	.8	2	4	6.6	8.4

# > 90th %	Mode	Geo. Mean	Har. Mean	Kurtosis	Skewness
134	3.8	3.1	1.1	–.3	.5

Figure 3.71

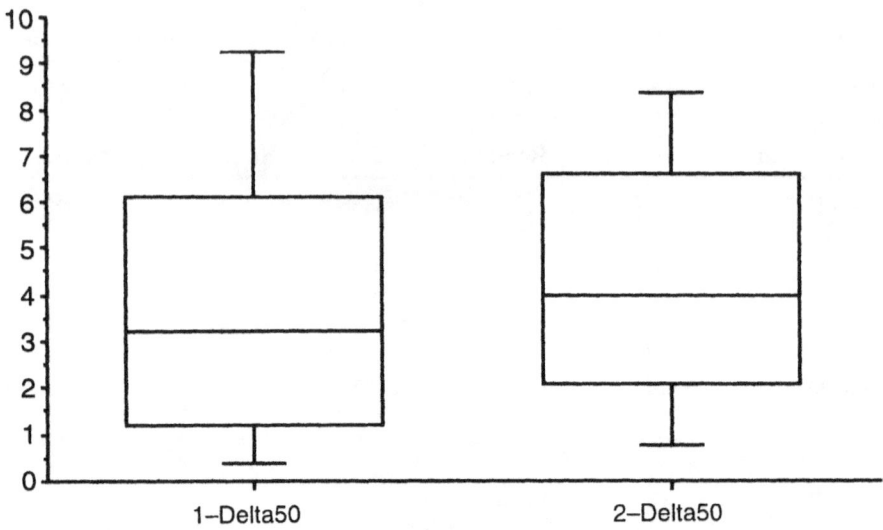

Table 3.80

	A	B	C	D	E	F	G	H	I
1	n	Digram	Expec Prob	(E)xpec Freq	(O)bser Freq	(E-O)	(E-O)^2	((E-O)^2)/E	SUM
2	2311	11	0.1667	385.244	860.000	-474.756	225393.544	585.067	
3		12	0.3333	770.256	61.000	709.256	503044.499	653.087	1238.155
4		21	0.3333	770.256	61.000	709.256	503044.499	653.087	1891.242
5		22	0.1667	385.244	1329.000	-943.756	890675.954	2311 981	2544.329
6									
7		Sum=		2311.000	2311.000				
8								Chi-Square	4856.309

Figure 3.72

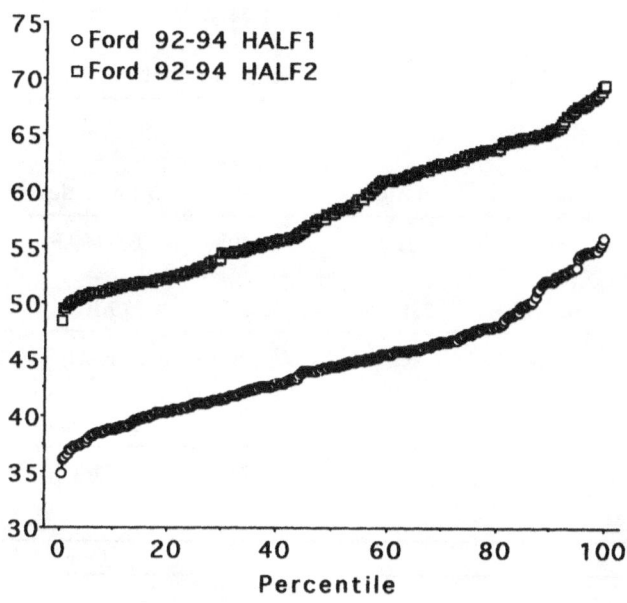

Table 3.81a

X_1: Ford 92–94 Half1

Mean	Std. Dev.	Std. Error	Variance	Coef. Var.	Count
44.6	4.7	.3	22	10.5	253

Minimum	Maximum	Range	Sum	Sum of Sqr.	# Missing
34.9	55.8	20.9	11272.9	507828.3	0

# < 10th %	10th %	25th %	50th %	75th %	90th %
22	38.8	41	44.4	47.2	52

# > 90th %	Mode	Geo. Mean	Har. Mean	Kurtosis	Skewness
24	45.8	44.3	44.1	–.4	.4

Table 3.81b

X$_2$: Ford 92–94 Half2

Mean	Std. Dev.	Std. Error	Variance	Coef. Var.	Count
58.3	5.6	.4	31.1	9.6	254

Minimum	Maximum	Range	Sum	Sum of Sqr.	# Missing
48.4	69.5	21.1	14808	871169.5	0

# < 10th %	10th %	25th %	50th %	75th %	90th %
24	51.2	52.9	58	63.2	65.5

# > 90th %	Mode	Geo. Mean	Har. Mean	Kurtosis	Skewness
24	•	58	57.8	–1.3	.1

Figure 3.73

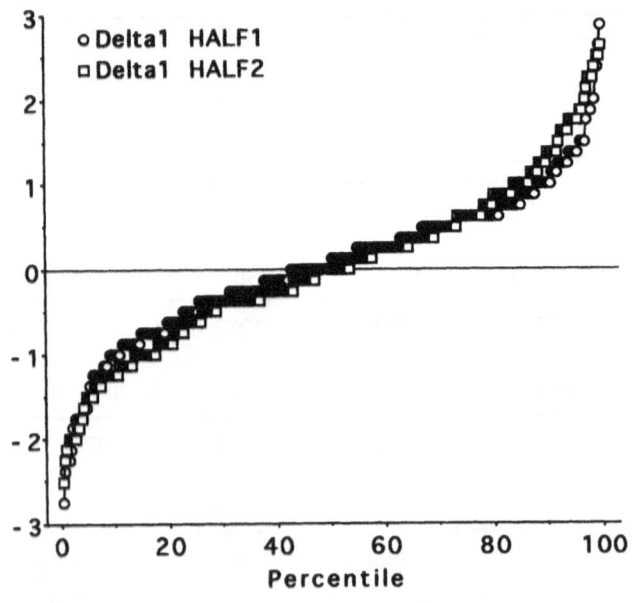

Table 3.82a-b

X_1: Delta-1 Half1

Mean	Std. Dev.	Std. Error	Variance	Coef. Var.	Count
4.4E–2	.9	.1	.8	1960.1	253

Minimum	Maximum	Range	Sum	Sum of Sqr.	# Missing
–2.8	2.9	5.6	11.2	191.9	0

# < 10th %	10th %	25th %	50th %	75th %	90th %
22	–1	–.5	.1	.6	1

# > 90th %	Mode	Geo. Mean	Har. Mean	Kurtosis	Skewness
24	•	•	•	.8	–.1

X_2: Delta-1 Half2

Mean	Std. Dev.	Std. Error	Variance	Coef. Var.	Count
1.5E–2	1	.1	1	6432.6	253

Minimum	Maximum	Range	Sum	Sum of Sqr.	# Missing
–2.5	2.6	5.1	3.9	244.7	1

# < 10th %	10th %	25th %	50th %	75th %	90th %
19	–1.2	–.6	0	.6	1.4

# > 90th %	Mode	Geo. Mean	Har. Mean	Kurtosis	Skewness
21	–.4	•	•	–4.2E–2	.2

Figure 3.74

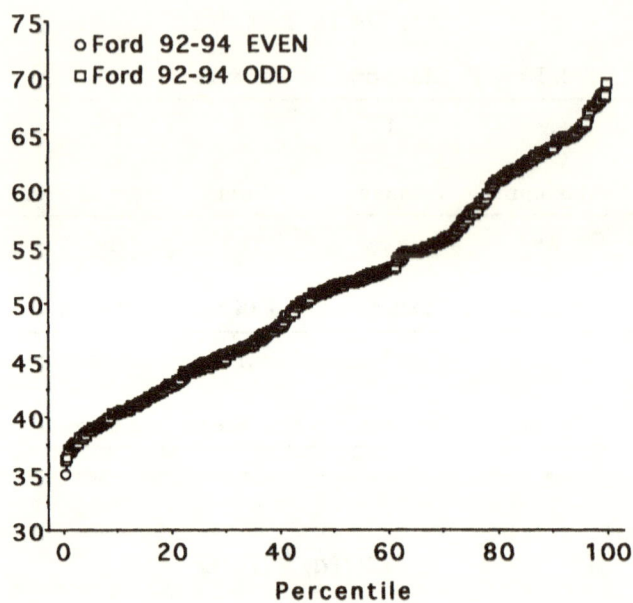

Table 3.83a

X₁: *Ford 92–94 Even*

Mean	Std. Dev.	Std. Error	Variance	Coef. Var.	Count
51.4	8.6	.5	73.9	16.7	253

Minimum	Maximum	Range	Sum	Sum of Sqr.	# Missing
34.9	69.4	34.5	13008.2	687448.8	0

# < 10th %	10th %	25th %	50th %	75th %	90th %
25	40.3	44.5	51.6	58.3	63.9

# > 90th %	Mode	Geo. Mean	Har. Mean	Kurtosis	Skewness
24	•	50.7	50	−1	.2

Table 3.83b

X₂: Ford 92–94 Odd

Mean	Std. Dev.	Std. Error	Variance	Coef. Var.	Count
51.4	8.6	.5	74.1	16.7	253

Minimum	Maximum	Range	Sum	Sum of Sqr.	# Missing
36.1	69.5	33.4	13012.6	687949.1	0

# < 10th %	10th %	25th %	50th %	75th %	90th %
25	40.3	44.2	51.4	57.9	63.8

# > 90th %	Mode	Geo. Mean	Har. Mean	Kurtosis	Skewness
25	50.9	50.7	50	–1	.2

Figure 3.75

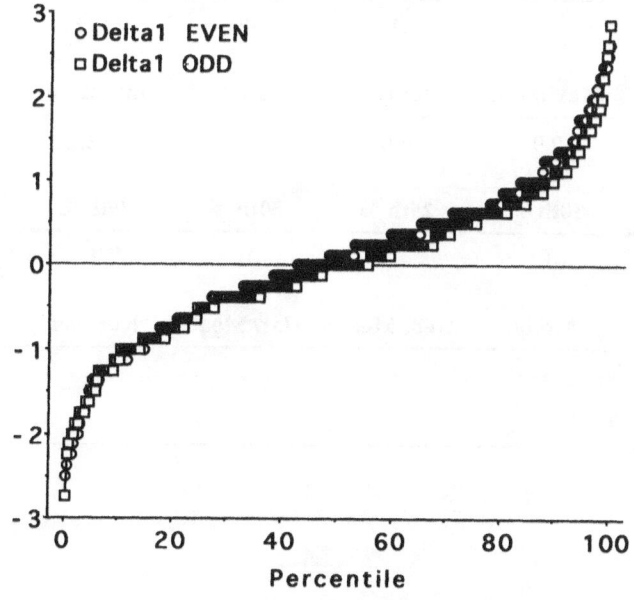

Table 3.84a-b

X_1: *Delta-1 Even*

Mean	Std. Dev.	Std. Error	Variance	Coef. Var.	Count
.1	1	.1	.9	1244.5	253

Minimum	Maximum	Range	Sum	Sum of Sqr.	# Missing
–2.5	2.6	5.1	19.5	233.4	0

# < 10th %	10th %	25th %	50th %	75th %	90th %
23	–1.1	–.5	.1	.6	1.3

# > 90th %	Mode	Geo. Mean	Har. Mean	Kurtosis	Skewness
25	.6	•	•	.1	–1.1E–3

X_2: *Delta-1 Odd*

Mean	Std. Dev.	Std. Error	Variance	Coef. Var.	Count
–1.7E–2	.9	.1	.8	–5192.3	253

Minimum	Maximum	Range	Sum	Sum of Sqr.	# Missing
–2.8	2.9	5.6	–4.4	203.2	0

# < 10th %	10th %	25th %	50th %	75th %	90th %
23	–1.1	–.5	0	.5	1.1

# > 90th %	Mode	Geo. Mean	Har. Mean	Kurtosis	Skewness
20	•	•	•	.6	.1

Figure 3.76

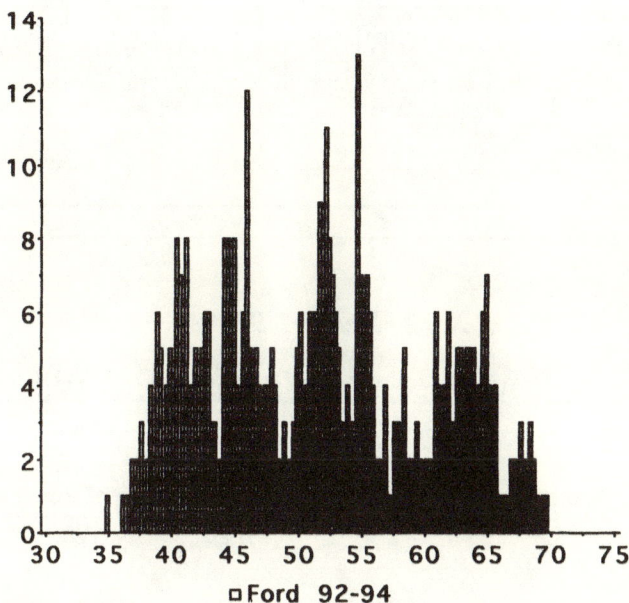

□ Ford 92-94

Figure 3.77

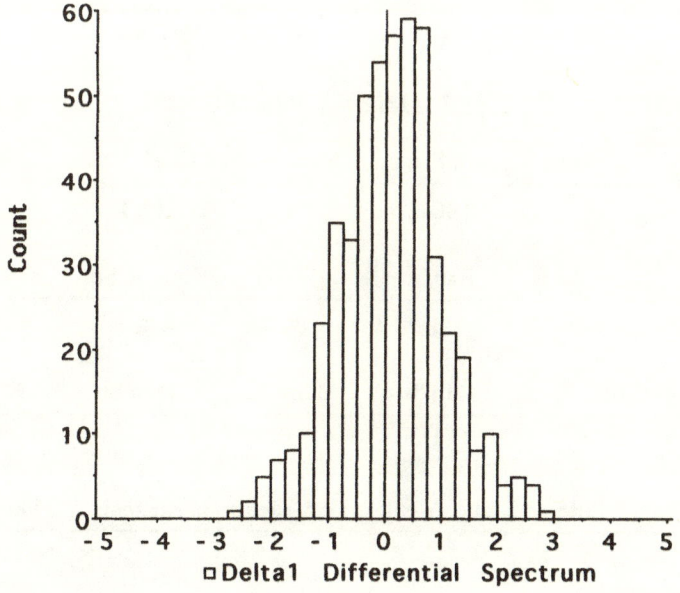

□Delta1 Differential Spectrum

Table 3.85

	A	B	C	D	E	F	G	H	I
1	n	Digram	Expec Prob	(E)xpec Freq	(O)bser Freq	(E-O)	(E-O)^2	((E-O)^2)/E	SUM
2	505	11	0.1667	84.1835	127	-42.8165	1833.25267	21.7768645	
3		12	0.3333	168.3165	135	33.3165	1109.98917	6.59465455	28.371519
4		21	0.3333	168.3165	135	33.3165	1109.98917	6.59465455	34.9661736
5		22	0.1667	84.1835	108	-23.8165	567.225672	6.73796732	41.5608281
6			SUM=	505	505				
7									
8								Chi-Square	48.2987955

Table 3.86a

X_1: 1–Delta-1 92–94

Mean	Std. Dev.	Std. Error	Variance	Coef. Var.	Count
-.7	.6	3.6E-2	.3	-87.9	263

Minimum	Maximum	Range	Sum	Sum of Sqr.	# Missing
-2.8	0	2.8	-174.5	204.8	0

# < 10th %	10th %	25th %	50th %	75th %	90th %
23	-1.5	-1	-.5	-.2	0

# > 90th %	Mode	Geo. Mean	Har. Mean	Kurtosis	Skewness
0	—.4	•	•	.7	-1.1

X_2: 2–Delta-1 92–94

Mean	Std. Dev.	Std. Error	Variance	Coef. Var.	Count
.8	.6	3.8E-2	.3	75.4	243

Minimum	Maximum	Range	Sum	Sum of Sqr.	# Missing
.1	2.9	2.8	189.6	231.8	20

# < 10th %	10th %	25th %	50th %	75th %	90th %
22	.2	.4	.6	1	1.6

# > 90th %	Mode	Geo. Mean	Har. Mean	Kurtosis	Skewness
24	.2	.6	.4	1.2	1.3

Figure 3.78

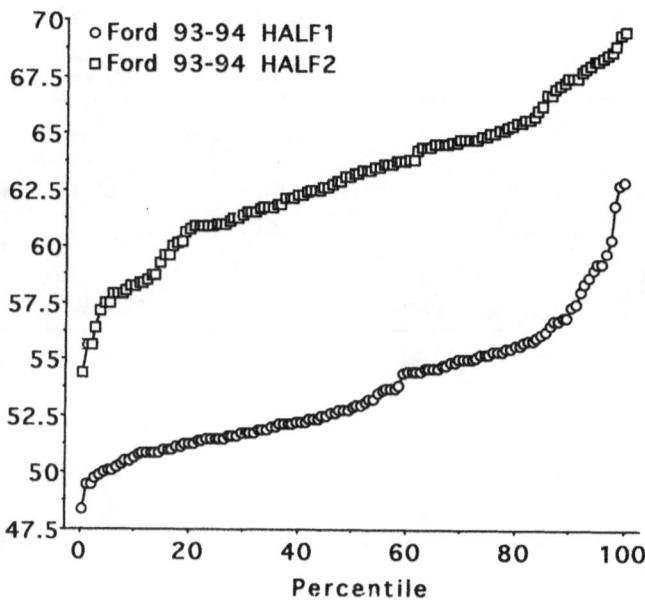

Table 3.87a

X_1: Ford 93–94 Half1

Mean	Std. Dev.	Std. Error	Variance	Coef. Var.	Count
53.6	2.8	.3	8.1	5.3	127

Minimum	Maximum	Range	Sum	Sum of Sqr.	# Missing
48.4	62.9	14.5	6811	366295.3	0

# < 10th %	10th %	25th %	50th %	75th %	90th %
13	50.7	51.5	52.9	55.2	57.3

# > 90th %	Mode	Geo. Mean	Har. Mean	Kurtosis	Skewness
13	•	53.6	53.5	.8	1

Table 3.87b

X_2: Ford 93–94 Half2

Mean	Std. Dev.	Std. Error	Variance	Coef. Var.	Count
63	3.2	.3	10.4	5.1	127

Minimum	Maximum	Range	Sum	Sum of Sqr.	# Missing
54.4	69.5	15.1	7997	504874.2	0

# < 10th %	10th %	25th %	50th %	75th %	90th %
13	58.3	61	63.2	65	67.5

# > 90th %	Mode	Geo. Mean	Har. Mean	Kurtosis	Skewness
11	•	62.9	62.8	–.3	–.2

Figure 3.79

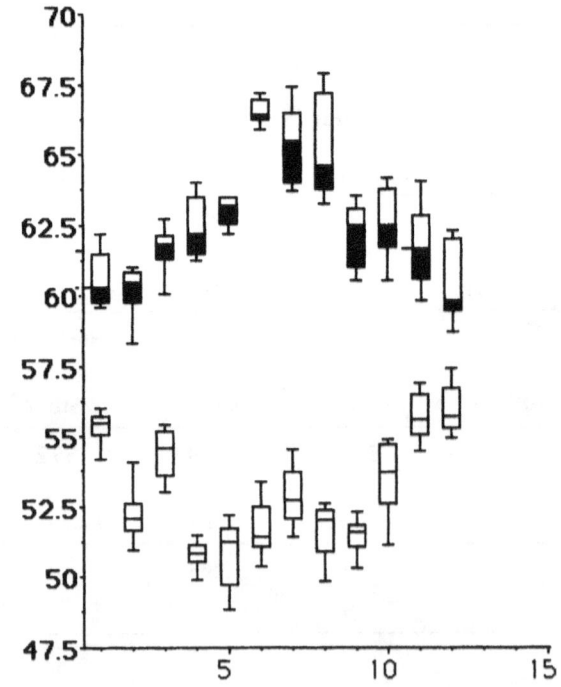

Figure 3.80

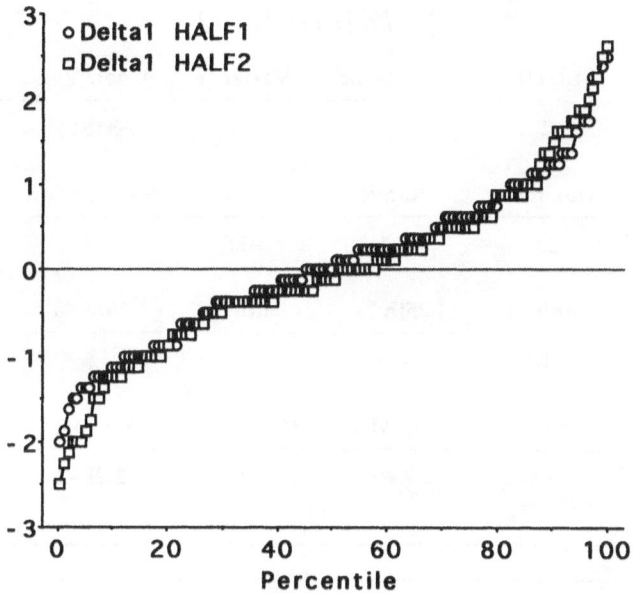

Table 3.88a

X_1: *Delta-1 Half1*

Mean	Std. Dev.	Std. Error	Variance	Coef. Var.	Count
.1	.9	.1	.9	1555.8	127

Minimum	Maximum	Range	Sum	Sum of Sqr.	# Missing
–2	2.5	4.5	7.6	110.4	0

# < 10th %	10th %	25th %	50th %	75th %	90th %
12	–1.1	–.6	0	.6	1.2

# > 90th %	Mode	Geo. Mean	Har. Mean	Kurtosis	Skewness
11	.2	•	•	–.2	.2

Table 3.88b

X_2 Delta-1 Half2

Mean	Std. Dev.	Std. Error	Variance	Coef. Var.	Count
−3.0E−2	1	.1	1.1	−3481.1	126

Minimum	Maximum	Range	Sum	Sum of Sqr.	# Missing
−2.5	2.6	5.1	−3.8	134.3	1

# < 10th %	10th %	25th %	50th %	75th %	90th %
11	−1.2	−.6	−.1	.5	1.5

# > 90th %	Mode	Geo. Mean	Har. Mean	Kurtosis	Skewness
13	−.4	•	•	2.2E−2	.2

Figure 3.81

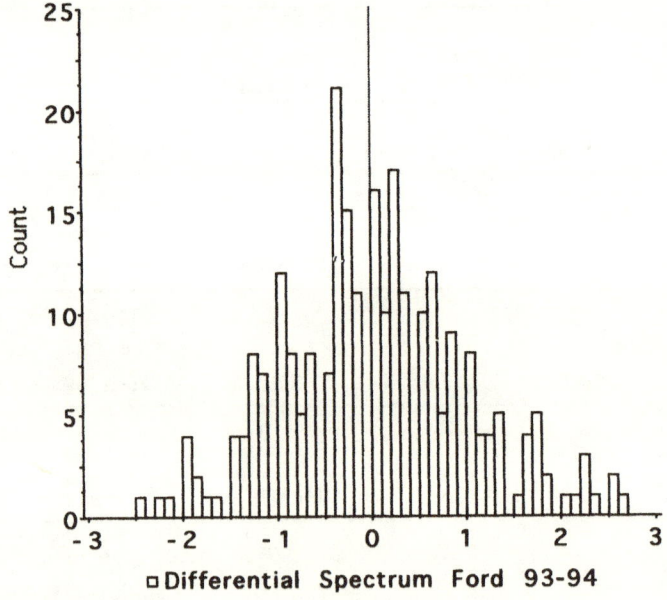

□Differential Spectrum Ford 93-94

Table 3.89

	A	B	C	D	E	F	G	H	I
1	n	Digram	Expec Prob	(E)xpec Freq	(O)bser Freq	(E-O)	(E-O)^2	((E-O)^2)/E	SUM
2	252	11	0.1667	42.008	69.000	-26.992	728.546	17.343	
3		12	0.3333	83.992	67.000	16.992	288.714	3.437	20.780
4		21	0.3333	83.992	67.000	16.992	288.714	3.437	24.218
5		22	0.1667	42 008	49.000	-6.992	48.882	1.164	27.655
6									
7		n=		252.000	252.000				
8								Chi-Square	28.819

Table 3.90

	A	B	C	D	E	F	G	H	I
1	n	Trigram	Expec Prob	(E)xpec Freq	(O)bser Freq	(E-O)	(E-O)^2	((E-O)^2)/E	SUM
2	251	111	0.042	10.459	33.000	-22.541	508.089	48.578	
3		112	0.125	31.375	36.000	-4.625	21.391	0.682	49.260
4		121	0.208	52.291	39.000	13.291	176.646	3.378	52.638
5		122	0.125	31.375	28.000	3.375	11.391	0.363	53.001
6		211	0.125	31.375	35.000	-3.625	13.141	0.419	53.420
7		212	0.208	52.291	31.000	21.291	453.299	8.669	62.089
8		221	0.125	31.375	28.000	3.375	11.391	0.363	62.452
9		222	0.042	10.459	21.000	-10.541	111.109	10.623	
10									
11	n=			251.000	251.000			Chi-Square	73.075

Table 3.91a

X_1: 1–Price Decreases

Mean	Std. Dev.	Std. Error	Variance	Coef. Var.	Count
-.7	.6	4.9E-2	.3	-83	137

Minimum	Maximum	Range	Sum	Sum of Sqr.	# Missing
-2.5	0	2.5	-95.6	112.4	0

# < 10th %	10th %	25th %	50th %	75th %	90th %
11	-1.5	-1	-.5	-.2	0

# > 90th %	Mode	Geo. Mean	Har. Mean	Kurtosis	Skewness
0	-.4	•	•	.1	-.9

Table 3.91b

X₂: 2–Price Increases

Mean	Std. Dev.	Std. Error	Variance	Coef. Var.	Count
.9	.6	.1	.4	74.5	116

Minimum	Maximum	Range	Sum	Sum of Sqr.	# Missing
.1	2.6	2.5	99.5	132.3	21

# < 10th %	10th %	25th %	50th %	75th %	90th %
10	.2	.4	.6	1.2	1.8

# > 90th %	Mode	Geo. Mean	Har. Mean	Kurtosis	Skewness
11	.2	.6	.4	.1	1

Figure 3.82

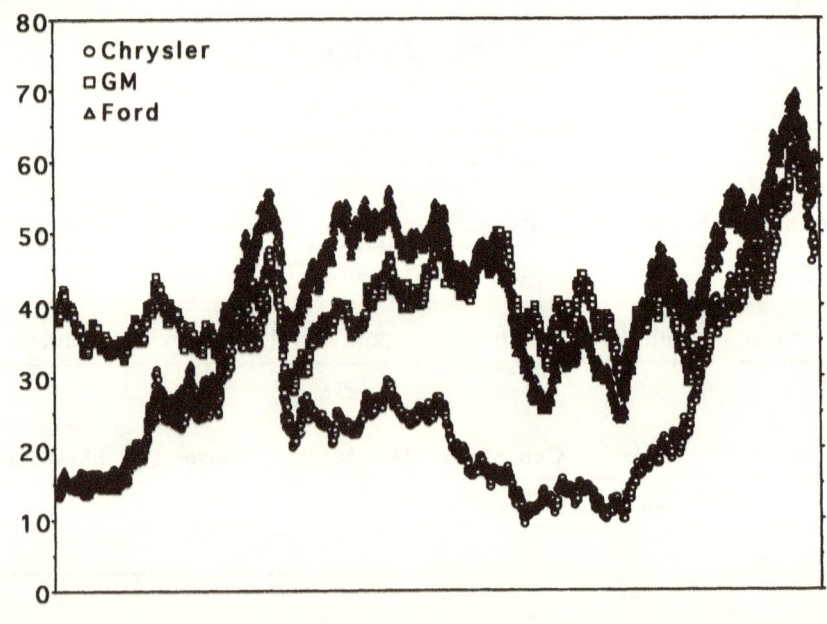

Figure 3.83

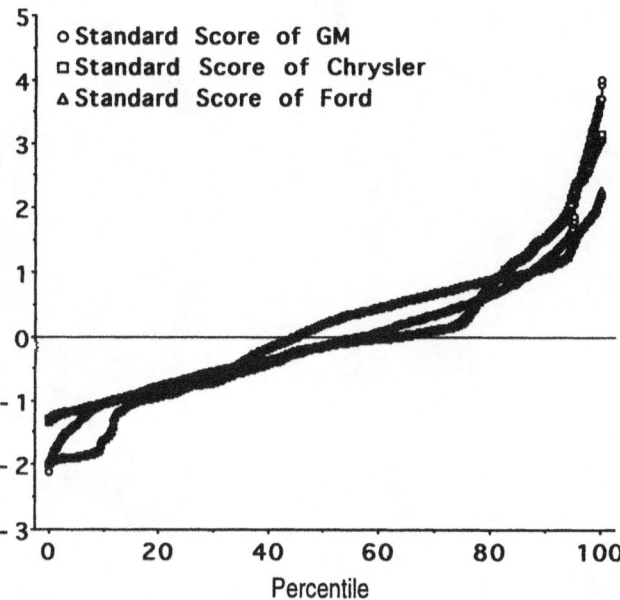

Figure 3.84

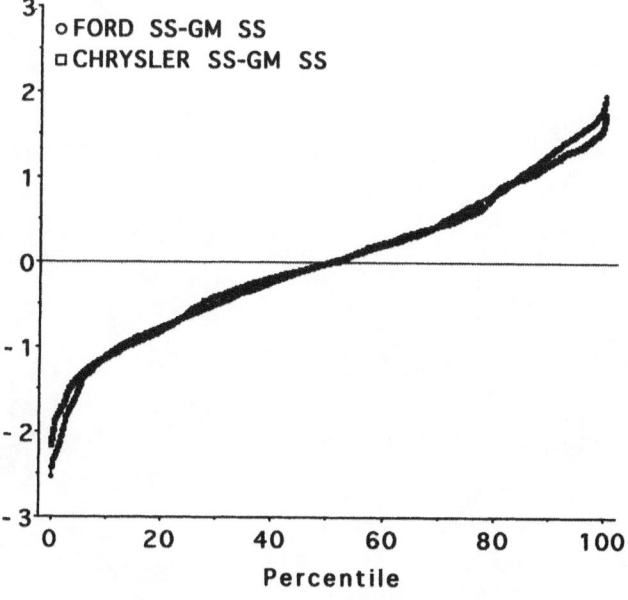

Microsoft   4

A plot of the closing prices of Microsoft is shown in Figure 4.1 and the cumulative probability density function is shown in Figure 4.2. The frequency histogram of prices with a bin size of $1.00 is shown in Figure 4.3, for a bin size of $0.50 in Figure 4.4, and a bin size of $0.25 in Figure 4.5. The multimodal (multiple 'peaks') character of these plots (Figures 4.3–4.5), with a major peak in the low teens and minor peaks in the mid $20s and low $30s, another around $45, another at around $60, and another broader peak around $80, suggests that the price tends to 'hover' around a price level, rather than oscillate widely. It also suggests that the prices contain a significant upward trend. The summary statistics are shown in Table 4.1.

The frequency histogram of Delta-1 price changes is shown in Figure 4.6 and the cumulative probability density function is shown in Figure 4.7, while the summary statistics are contained in Table 4.2. Although there is a significant upward trend in the prices, 80% of the price movement (i.e., the Delta-1 price change) is a dollar or less.

The Half1 and Half2 cumulative probability density functions are shown in Figure 4.8 and the summary statistics in Table 4.3 a-b. Although roughly parallel, these two cumulative probability density functions reflect the upward trend in prices and suggest that the price of Microsoft is not stationary. The Half1 and Half2 cumulative probability density functions for Delta-1 price changes are shown in Figure 4.9 and the summary statistics in Table 4.4a-b. The two cumu-

Figure 4.1

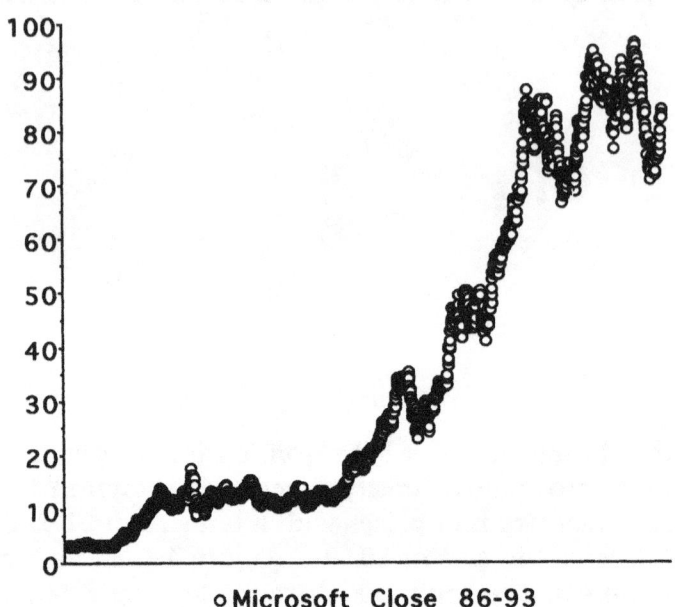

∘Microsoft Close 86-93

Figure 4.2

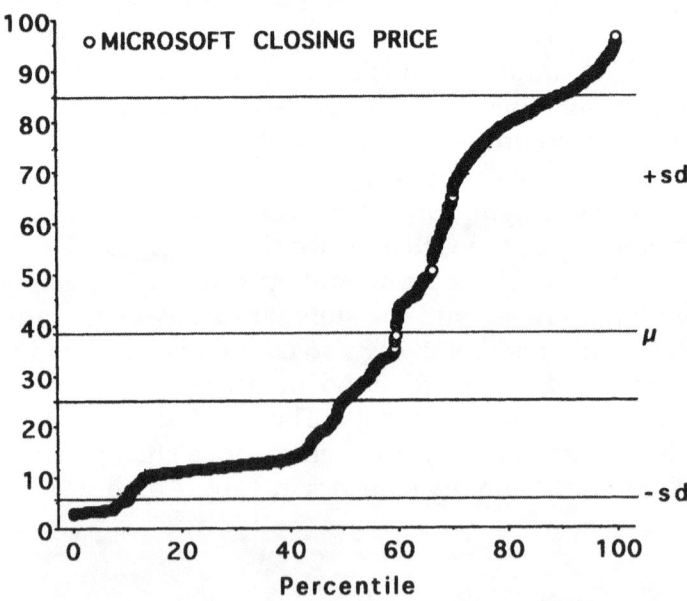

Figure 4.3

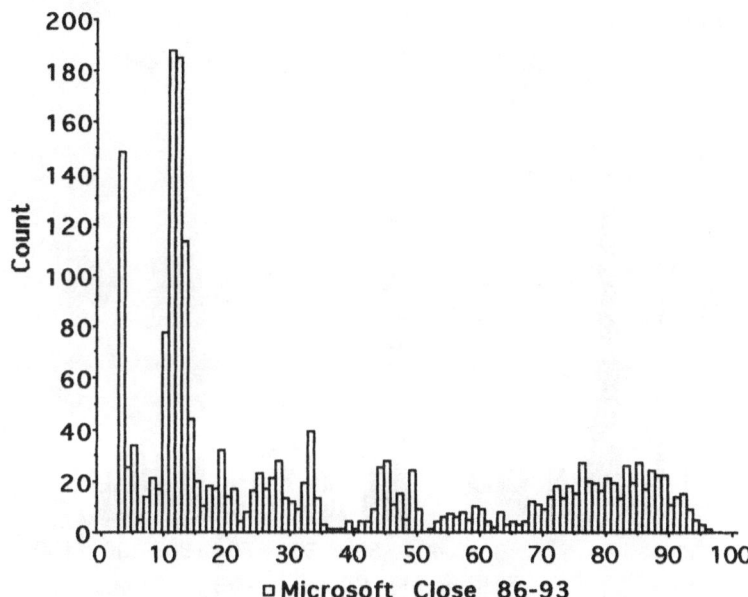

Figure 4.4

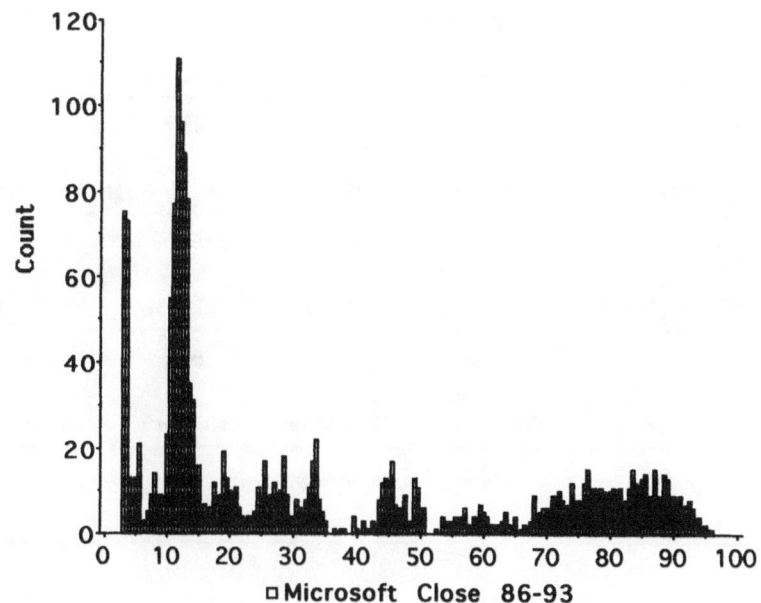

Figure 4.5

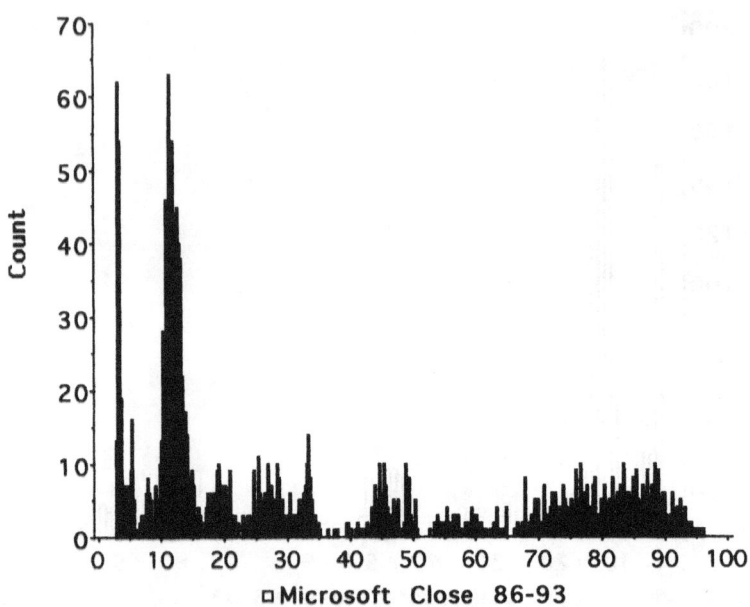

Table 4.1

X_1: *Microsoft Close 86–93*

Mean	Std. Dev.	Std. Error	Variance	Coef. Var.	Count
34.8	29.8	.7	888	85.7	1911

Minimum	Maximum	Range	Sum	Sum of Sqr.	# Missing
2.9	96.2	93.4	66485.2	4009199.7	0

# < 10th %	10th %	25th %	50th %	75th %	90th %
190	5.4	11.5	19.3	63.5	83.7

# > 90th %	Mode	Geo. Mean	Har. Mean	Kurtosis	Skewness
191	11.5	22.4	13.7	–1.1	.7

Figure 4.6

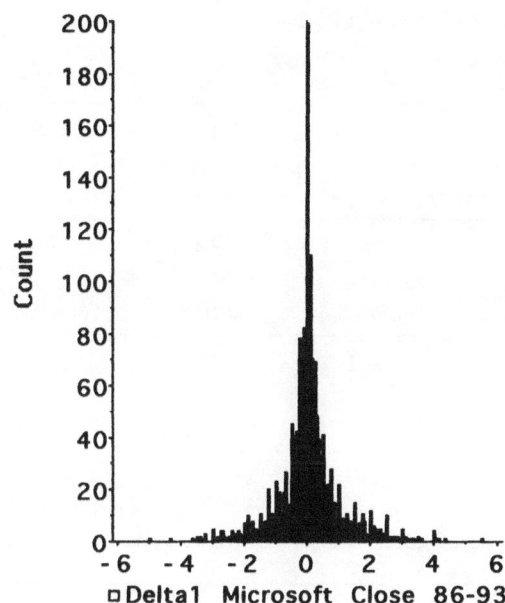

Figure 4.7

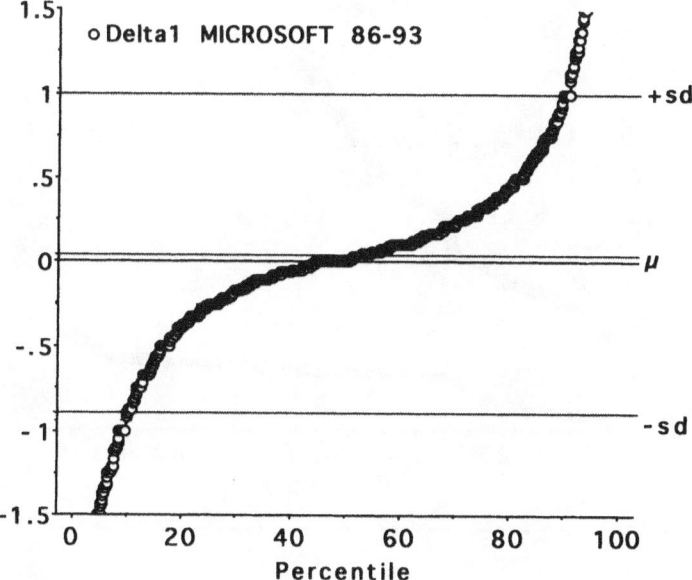

Table 4.2

X_1: *Delta-1 Microsoft Close 86–93*

Mean	Std. Dev.	Std. Error	Variance	Coef. Var.	Count
4.2E–2	1	2.2E–2	1	2351.9	1910

Minimum	Maximum	Range	Sum	Sum of Sqr.	# Missing
–6.5	7.9	14.4	79.4	1827.6	1

# < 10th %	10th %	25th %	50th %	75th %	90th %
191	–.9	–.2	0	.3	1

# > 90th %	Mode	Geo. Mean	Har. Mean	Kurtosis	Skewness
179	0	•	•	7.5	.2

Figure 4.8

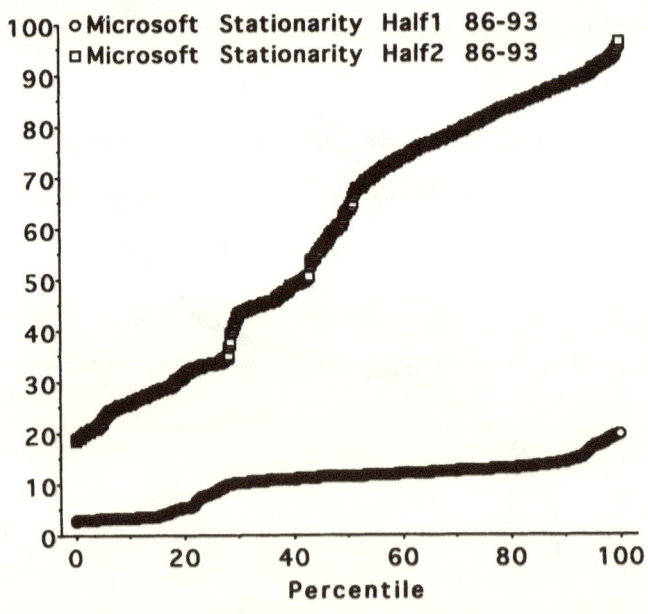

Table 4.3a-b

X_1: Microsoft Stationarity Half1 86–93

Mean	Std. Dev.	Std. Error	Variance	Coef. Var.	Count
10.5	4.1	.1	16.9	39.1	955

Minimum	Maximum	Range	Sum	Sum of Sqr.	# Missing
2.9	19.7	16.8	10024.3	121308.9	0

# < 10th %	10th %	25th %	50th %	75th %	90th %
92	3.4	8.3	11.5	12.8	14.4

# > 90th %	Mode	Geo. Mean	Har. Mean	Kurtosis	Skewness
94	11.5	9.4	8	–.4	–.5

X_2: Microsoft Stationarity Half2 86–93

Mean	Std. Dev.	Std. Error	Variance	Coef. Var.	Count
59.1	24	.8	578.3	40.7	955

Minimum	Maximum	Range	Sum	Sum of Sqr.	# Missing
18.5	96.2	77.7	56442.6	3887553.2	0

# < 10th %	10th %	25th %	50th %	75th %	90th %
95	25.9	33.4	63.5	81.3	88

# > 90th %	Mode	Geo. Mean	Har. Mean	Kurtosis	Skewness
95	89	53.3	47.2	–1.5	–.2

lative probability density functions cross, but do not overlap. This suggests that the Delta-1 price changes are also not stationary and this is confirmed by the box and whiskers plot in Figure 4.10 and the percentile comparison plot in Figure 4.11.

We will therefore decrease the size of our sample to 09/30/91 to 09/30/93 and the Half1 and Half2 cumulative probability density

Figure 4.9

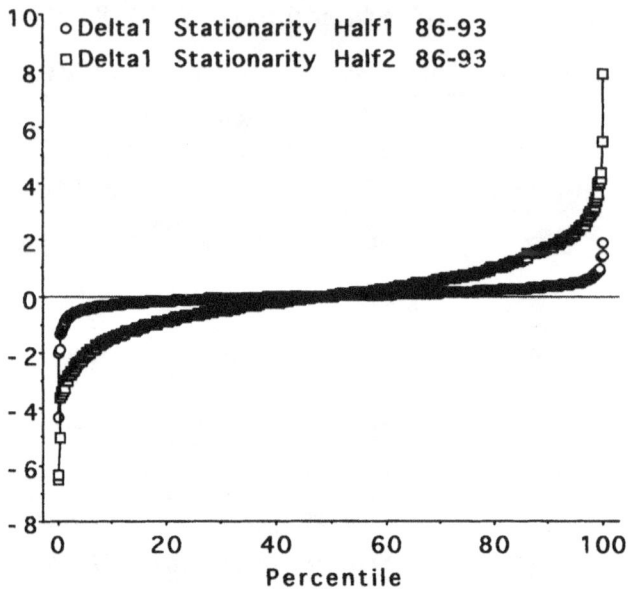

functions for the closing prices and Delta-1 price changes are shown in Figures 4.12 and 4.13, while the summary statistics are shown in Tables 4.5 a-b and 4.6 a-b, respectively. The parallel character of the plot of closing prices suggests an upward trend, but one that is stationary. The overlap of the cumulative density functions for Delta-1 price changes clearly suggests that the price changes are stationary. The overlap of the Odd and Even cumulative probability density functions strongly suggest that the prices (Figure 4.14, summary statistics, Table 4.7 a-b) and the Delta-1 price changes (Figure 4.15, summary statistics, Table 4.8 a-b) are random.

The differential spectrum with a bin width of $0.01 is shown in Figure 4.16 (summary statistics Table 4.9) and is clearly asymmetrical. This suggests that Microsoft contains significant serial dependencies. This is confirmed by the fact that the digram relative price changes significantly diverge from independence (see Table 4.10 for Chi-Square evaluations and Table 4.11 a-b for the summary statistics for the positive and negative price changes).

We can divide the High-Low distribution of price changes (summary statistics Table 4.12) into roughly equal quarters, where we

Table 4.4a-b

X_1: Delta-1 Stationarity Half1 86–93

Mean	Std. Dev.	Std. Error	Variance	Coef. Var.	Count
1.5E–2	.3	1.1E–2	.1	2226	954

Minimum	Maximum	Range	Sum	Sum of Sqr.	# Missing
–4.3	1.9	6.2	14.8	113.1	1

# < 10th %	10th %	25th %	50th %	75th %	90th %
93	–.3	–.1	0	.2	.4

# > 90th %	Mode	Geo. Mean	Har. Mean	Kurtosis	Skewness
95	0	•	•	30.5	–2.3

X_2: Delta-1 Stationarity Half2 86–93

Mean	Std. Dev.	Std. Error	Variance	Coef. Var.	Count
.1	1.3	4.3E–2	1.8	1993.8	955

Minimum	Maximum	Range	Sum	Sum of Sqr.	# Missing
–6.5	7.9	14.4	64.1	1714.2	0

# < 10th %	10th %	25th %	50th %	75th %	90th %
87	–1.5	–.7	0	.8	1.8

# > 90th %	Mode	Geo. Mean	Har. Mean	Kurtosis	Skewness
87	0	•	•	2.8	.1

code the lowest quarter ($0.70 to $1.50) as a '1', the second quarter ($1.501 to $2.20) as a '2', the third quarter ($2.201 to $2.80) as a '3', and the highest quarter ($2.801 to $7.20) as a '4' (see summary statistics for each of the quarters in Table 4.13 a-d). We can then generate a digram and trigram transition matrix (see Tables 4.14 a and b, respectively) and compare them with a similar matrix generated under the

Figure 4.10

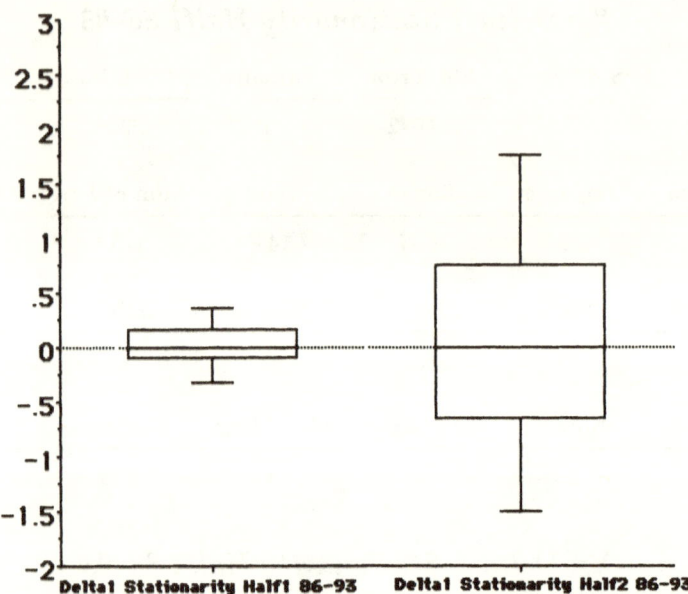

Figure 4.11

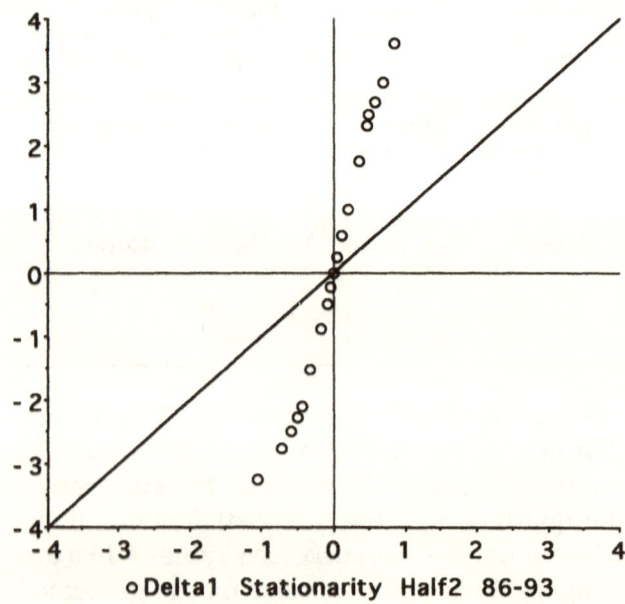

Figure 4.12

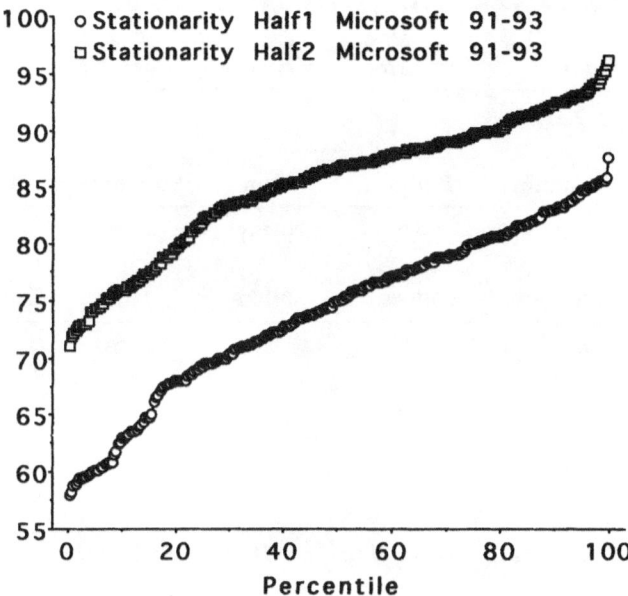

Figure 4.13

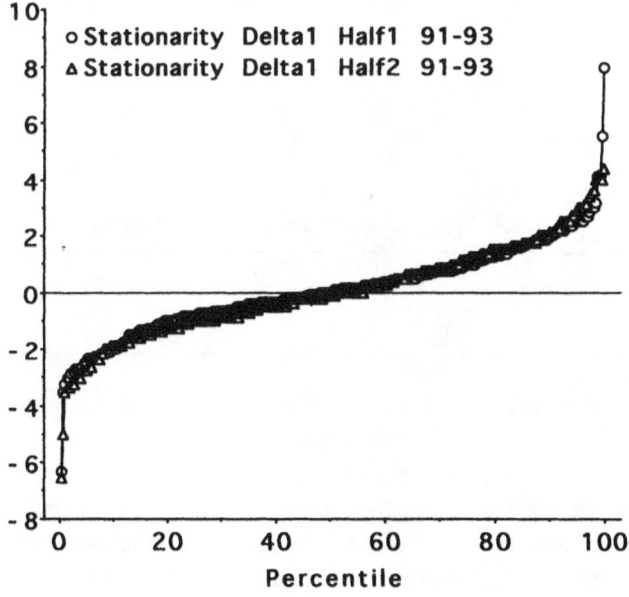

Table 4.5a-b

X_1: Stationarity Half1 Microsoft 91–93

Mean	Std. Dev.	Std. Error	Variance	Coef. Var.	Count
74	7.3	.5	53.7	9.9	253

Minimum	Maximum	Range	Sum	Sum of Sqr.	# Missing
58	87.7	29.7	18731	1400303.3	0

# < 10th %	10th %	25th %	50th %	75th %	90th %
25	62.9	69.2	75	80	83.1

# > 90th %	Mode	Geo. Mean	Har. Mean	Kurtosis	Skewness
25	•	73.7	73.3	–.7	–.4

X_2: Stationarity Half2 Microsoft 91–93

Mean	Std. Dev.	Std. Error	Variance	Coef. Var.	Count
85.4	5.9	.4	34.7	6.9	254

Minimum	Maximum	Range	Sum	Sum of Sqr.	# Missing
71	96.2	25.2	21683.6	1859872.5	0

# < 10th %	10th %	25th %	50th %	75th %	90th %
24	76	81.8	86.6	89.6	92.5

# > 90th %	Mode	Geo. Mean	Har. Mean	Kurtosis	Skewness
21	89	85.2	84.9	–.5	–.6

assumption of independence. As shown in these tables, the High-Low distribution contains significant serial dependencies.

We can look at the relative values in the observed and expected cells in these matrices to determine if certain 'patterns' contribute more to the ultimate Chi-Square than others. Some 'patterns' can occur more frequently than expected, while others can occur less

Table 4.6a-b

X_1: *Stationarity Delta-1 Half1 91–93*

Mean	Std. Dev.	Std. Error	Variance	Coef. Var.	Count
.1	1.6	.1	2.5	2205.8	253

Minimum	Maximum	Range	Sum	Sum of Sqr.	# Missing
–6.3	7.9	14.2	18.2	633.8	0

# < 10th %	10th %	25th %	50th %	75th %	90th %
24	–2	–.8	0	1	2

# > 90th %	Mode	Geo. Mean	Har. Mean	Kurtosis	Skewness
24	.5	•	•	2.8	.4

X_2: *Stationarity Delta-1 Half2 91–93*

Mean	Std. Dev.	Std. Error	Variance	Coef. Var.	Count
1.5E–2	1.7	.1	2.8	11333.6	254

Minimum	Maximum	Range	Sum	Sum of Sqr.	# Missing
–6.5	4.4	10.9	3.8	708.4	0

# < 10th %	10th %	25th %	50th %	75th %	90th %
23	–1.9	–1	–.1	1.2	2.1

# > 90th %	Mode	Geo. Mean	Har. Mean	Kurtosis	Skewness
23	•	•	•	.6	–4.7E–2

frequently than expected. Consider the matrix shown in Table 4.14a. Four patterns (1,1; 1,4; 4,1; and 4.4) contribute highly to the ultimate Chi-Square value. There are seven possible trigrams that contain the digram 1,1, they are 1,1,1; 1,1,2; 1,1,3; 1,1,4; 2,1,1; 3,1,1; and 4,1,1. Only one of these trigrams, 1,1,1, makes a large contribution to the trigram Chi-Square value. Several of the trigrams that contain the 1,1 digram

Figure 4.14

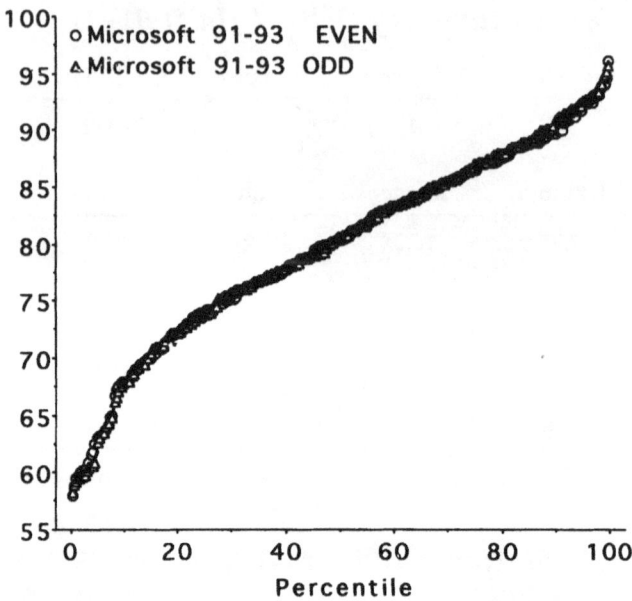

Figure 4.15

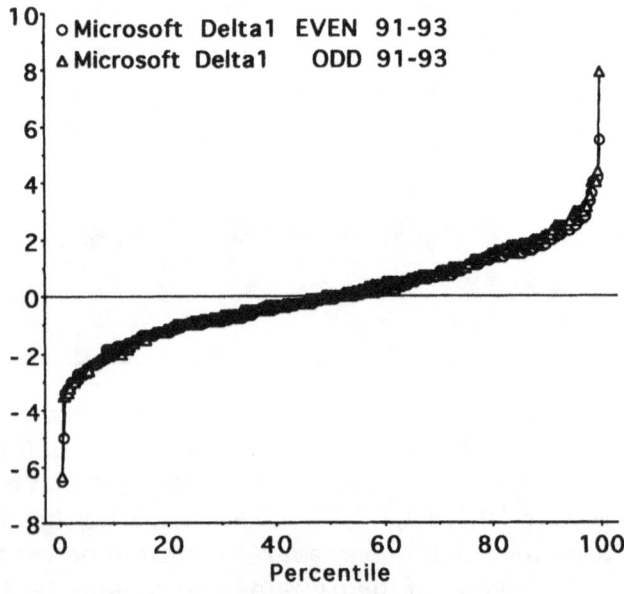

Table 4.7a-b

X_1: Microsoft 91–93 Even

Mean	Std. Dev.	Std. Error	Variance	Coef. Var.	Count
79.7	8.7	.5	75	10.9	252

Minimum	Maximum	Range	Sum	Sum of Sqr.	# Missing
58	96.2	38.2	20080.1	1618865.1	0

# < 10th %	10th %	25th %	50th %	75th %	90th %
24	68	74.1	80.4	86.5	89.8

# > 90th %	Mode	Geo. Mean	Har. Mean	Kurtosis	Skewness
25	•	79.2	78.7	–.5	–.5

X_2: Microsoft 91–93 Odd

Mean	Std. Dev.	Std. Error	Variance	Coef. Var.	Count
79.8	8.8	.6	78.3	11.1	252

Minimum	Maximum	Range	Sum	Sum of Sqr.	# Missing
58.3	95.8	37.4	20107.7	1624081.7	0

# < 10th %	10th %	25th %	50th %	75th %	90th %
25	67.9	73.9	80.7	87	91

# > 90th %	Mode	Geo. Mean	Har. Mean	Kurtosis	Skewness
25	88	79.3	78.7	–.5	–.5

occur less frequently than would be expected (i.e., 1,1,2; 1,1,3; 2,1,1; and 3,1,1). None of the eight trigrams that contain the digram 1,4 make a large contribution to the trigram Chi-Square, but all of these trigrams (1,1,4; 1,4,1; 1,4,2; 1,4,3; 1,4,4; 2,1,4; 3,1,4; and 4,1,4) occur more often than would be expected. This is also the case for the trigrams that contain the digram 4,1. The digram 4,4 occurs less often

Table 4.8a-b

X_1: *Microsoft Delta-1 Even 91–93*

Mean	Std. Dev.	Std. Error	Variance	Coef. Var.	Count
–2.6E–2	1.6	.1	2.5	–6171.1	252

Minimum	Maximum	Range	Sum	Sum of Sqr.	# Missing
–6.5	5.5	12	–6.5	626.9	0

# < 10th %	10th %	25th %	50th %	75th %	90th %
21	–1.9	–1	–.1	1	1.9

# > 90th %	Mode	Geo. Mean	Har. Mean	Kurtosis	Skewness
25	•	•	•	1.3	–2.0E–2

X_2: *Microsoft Delta-1 Odd 91–93*

Mean	Std. Dev.	Std. Error	Variance	Coef. Var.	Count
.1	1.7	.1	2.8	1598.7	252

Minimum	Maximum	Range	Sum	Sum of Sqr.	# Missing
–6.3	7.9	14.2	26.5	712.2	0

# < 10th %	10th %	25th %	50th %	75th %	90th %
23	–2	–.9	0	1.1	2.2

# > 90th %	Mode	Geo. Mean	Har. Mean	Kurtosis	Skewness
25	.5	•	•	1.8	.3

than would be expected. Three of the trigrams (1,4,4; 4,4,1; and 4,4,2) occur more often than would be expected, while the trigram 4,4,4 occurs less often than would be expected.

We can divide the Close-Open distribution (summary statistics, Table 4.15) into four roughly equal quarters and encode the lowest quarter (–$4.20 to –$0.90) as a '1', the next quarter (–$0.899 to 0) as a

Figure 4.16

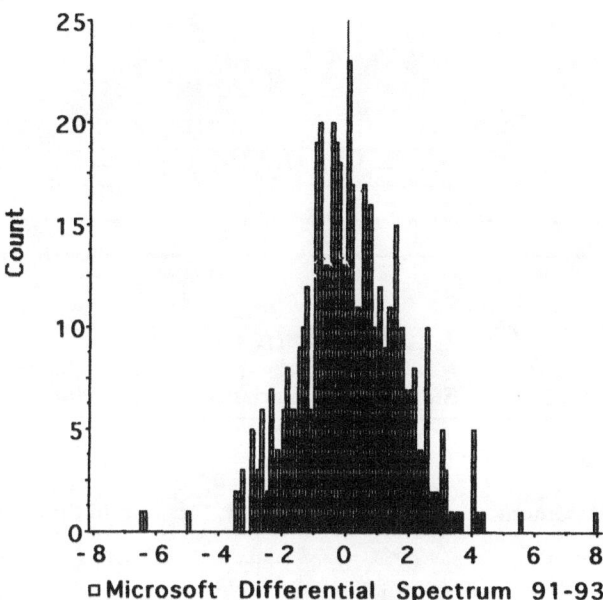

□Microsoft Differential Spectrum 91–93

Table 4.9

X_1: *Microsoft Differential Spectrum 91–93*

Mean	Std. Dev.	Std. Error	Variance	Coef. Var.	Count
4.3E–2	1.6	.1	2.7	3765.5	507

Minimum	Maximum	Range	Sum	Sum of Sqr.	# Missing
–6.5	7.9	14.4	21.9	1342.2	0

# < 10th %	10th %	25th %	50th %	75th %	90th %
50	–1.9	–1	0	1.1	2

# > 90th %	Mode	Geo. Mean	Har. Mean	Kurtosis	Skewness
50	0	•	•	1.6	.1

Table 4.10

	A	B	C	D	E	F	G	H	I
1	n	Digram	Expec Prob	(E)xpec Freq	(O)bser Freq	(E-O)	(E-O)^2	((E-O)^2)/E	SUM
2	504	11	0.1667	84.0168	146	-61.9832	3841.91708	45.7279625	
3		12	0.3333	167.9832	122	45.9832	2114.45468	12.5872985	58.315261
4		21	0.3333	167.9832	122	45.9832	2114.45468	12.5872985	70.9025595
5		22	0.1667	84.0168	114	-29.9832	898.992282	10.700149	83.489858
6									
7									
8								Chi-Square	94.1900071

Table 4.11a-b

X_1: *Column 1*

Mean	Std. Dev.	Std. Error	Variance	Coef. Var.	Count
−1.1	1	.1	1	−89.3	270

Minimum	Maximum	Range	Sum	Sum of Sqr.	# Missing
−6.5	0	6.5	−304.5	616.1	0

# < 10th %	10th %	25th %	50th %	75th %	90th %
26	−2.5	−1.6	−.9	−.4	−.1

# > 90th %	Mode	Geo. Mean	Har. Mean	Kurtosis	Skewness
24	0	•	•	5	−1.7

X_2: *Column 2*

Mean	Std. Dev.	Std. Error	Variance	Coef. Var.	Count
1.4	1.1	.1	1.2	78.6	237

Minimum	Maximum	Range	Sum	Sum of Sqr.	# Missing
.1	7.9	7.8	326.4	726.1	33

# < 10th %	10th %	25th %	50th %	75th %	90th %
21	.2	.6	1.2	1.9	2.7

# > 90th %	Mode	Geo. Mean	Har. Mean	Kurtosis	Skewness
24	.8	1	.6	5.5	1.7

Table 4.12

X_1: *Microsoft High–Low 91–93*

Mean	Std. Dev.	Std. Error	Variance	Coef. Var.	Count
2.3	1	4.5E–2	1	42.8	508

Minimum	Maximum	Range	Sum	Sum of Sqr.	# Missing
.7	7.2	6.6	1191.1	3303.4	0

# < 10th %	10th %	25th %	50th %	75th %	90th %
42	1.2	1.5	2.2	2.8	3.8

# > 90th %	Mode	Geo. Mean	Har. Mean	Kurtosis	Skewness
46	2.2	2.1	2	1.8	1.1

Table 4.13a

X_1: *1–Microsoft High–Low 91–93*

Mean	Std. Dev.	Std. Error	Variance	Coef. Var.	Count
1.3	.2	1.9E–2	4.7E–2	17.1	127

Minimum	Maximum	Range	Sum	Sum of Sqr.	# Missing
.7	1.5	.8	160.1	207.7	8

# < 10th %	10th %	25th %	50th %	75th %	90th %
7	1	1	1.2	1.5	1.5

# > 90th %	Mode	Geo. Mean	Har. Mean	Kurtosis	Skewness
0	1.5	1.2	1.2	–.6	–.5

'2', the next quarter ($0.001 to $1.00) as a '3', and the highest quarter ($1.001 to $4.90) as a '4' (summary statistics for each of the quarters are shown in Table 4.16a-d). The Chi-Square calculations for the di-

Table 4.13b-c

X_2: 2–Microsoft High–Low 91–93

Mean	Std. Dev.	Std. Error	Variance	Coef. Var.	Count
1.9	.2	1.5E–2	2.6E–2	8.5	118

Minimum	Maximum	Range	Sum	Sum of Sqr.	# Missing
1.6	2.2	.6	223.7	427	17

# < 10th %	10th %	25th %	50th %	75th %	90th %
11	1.7	1.8	2	2	2.1

# > 90th %	Mode	Geo. Mean	Har. Mean	Kurtosis	Skewness
11	2	1.9	1.9	–1.2	–1.5E–2

X_3: 3–Microsoft High–Low 91–93

Mean	Std. Dev.	Std. Error	Variance	Coef. Var.	Count
2.5	.2	1.6E–2	3.4E–2	7.5	135

Minimum	Maximum	Range	Sum	Sum of Sqr.	# Missing
2.2	2.8	.5	331.7	819.5	0

# < 10th %	10th %	25th %	50th %	75th %	90th %
0	2.2	2.2	2.5	2.6	2.8

# > 90th %	Mode	Geo. Mean	Har. Mean	Kurtosis	Skewness
0	2.2	2.5	2.4	–1.3	.3

gram category price changes are shown in Table 4.17. The Chi-Square approaches significance (the minimum 'z' > 1.96), but it is not significant. So, we will stop the analysis at this point.

The prices for 1986 to 1988 also appear to be stationary (see Figure 4.17 and summary statistics, Table 4.18 a-b) and the differen-

Table 4.13d

X₄: 4 – Microsoft High–Low 91–93

Mean	Std. Dev.	Std. Error	Variance	Coef. Var.	Count
3.7	.8	.1	.6	21.6	128

Minimum	Maximum	Range	Sum	Sum of Sqr.	# Missing
2.8	7.2	4.4	475.6	1849.2	7

# < 10th %	10th %	25th %	50th %	75th %	90th %
11	3	3.2	3.5	4	4.8

# > 90th %	Mode	Geo. Mean	Har. Mean	Kurtosis	Skewness
12	•	3.6	3.6	2.9	1.5

tial spectrum appears to be asymmetrical (see Figure 4.18 and summary statistics in Table 4.19).

We can divide the High-Low distribution (summary statistics in Table 4.20) into 4 approximately equal quarters. We can code the lowest quarter ($0.016 to $0.10) as a '1', the next quarter ($0.101 to $0.30) as a '2', the next quarter ($0.301 to $0.50) as a '3', and the highest quarter ($0.501 to $4.10) as a '4' (summary statistics for each of the four quarters are shown in Table 4.21 a-d). The digram category price change Chi-Square calculations are shown in Table 4.22 and the Chi-Square is statistically significant. In this case, the digrams 1,1, 4,3 and 4,4 occurred more frequently than expected and the digrams 1,3 and 3,1 occurred less frequently than expected and the digrams 1,4 and 4,1 did not occur at all.

We can divide the Close-Open distribution (summary statistics Table 4.23) into four approximately equal quarters and encode the lowest quarter (–$3.60 to –$0.10) as a '1', the next quarter (–$0.099 to $0) as a '2', the next quarter ($0.001 to $0.10) as a '3', and the highest quarter ($0.011 to $1.50) as a '4' (summary statistics for each quarter are shown in Table 4.24a-d). The digram category price change Chi-Square calculation is shown in Table 4.25 and the Chi-Square is significant. Three digrams, 2,2, 3,3, and 4,4 all occur more often than predicted.

Table 4.14a

	A	B	C	D	E	F	G	H
1	Symbols	Frequency	Probability					
2	1	126	0.2485					
3	2	118	0.2327					
4	3	135	0.2663					
5	4	128	0.2525					
6	Sum	507						
7								
8								
9								
10	Digrams	(E)xpected	(O)bserved	(E-O)	(E-O)^2	((E-O)^2)/E	Sum	
11	11	31.314	52.000	-20.686	427.927	13.666		
12	12	29.325	31.000	-1.675	2.804	0.096	13.761	
13	13	33.550	24.000	9.550	91.208	2.719	16.480	
14	14	31.811	19.000	12.811	164.113	5.159	21.639	
15	21	29.325	27.000	2.325	5.408	0.184	16.664	
16	22	27.464	32.000	-4.536	20.580	0.749	17.414	
17	23	31.420	29.000	2.420	5.857	0.186	17.600	
18	24	29.791	30.000	-0.209	0.044	0.001	17.602	
19	31	33.550	29.000	4.550	20.705	0.617	18.217	
20	32	31.420	27.000	4.420	19.537	0.622	18.839	
21	33	35.947	44.000	-8.053	64.855	1.804	20.643	
22	34	34.083	35.000	-0.917	0.841	0.025	20.668	
23	41	31.811	19.000	12.811	164.113	5.159	25.827	
24	42	29.791	27.000	2.791	7.789	0.261	26.089	
25	43	34.083	38.000	-3.917	15.344	0.450	26.539	
26	44	32.316	44.000	-11.684	136.526	4.225	30.763	
27								
28		n=		507		Chi-Square	30.7634752	
29								
30								
31						2=8 3666	8.36660027	
32								
33								
34								
35								
36								
37								
38								
39								
40								
41								
42								
43								
44								
45								
46								
47								
48								
49								
50								
51								

A regression of the Close-Open distribution on the High-Low distribution, as shown in Figure 4.19, indicates that there is a poor relationship between these two distributions and this is confirmed by the low correlation coefficient ($r = 0.069$). The shape of the plot in the scatter diagram suggests that there is a nonlinear relationship between High-Low and Close-Open.

Table 4.14b

Symbols	Frequency	Probability								
1	126	0.2490								
2	118	0.2332								
3	135	0.2668								
4	127	0.2510								
Sum	506									

Digrams	(E)xpected	(O)bserved	(E-O)	(E-O)^2	((E-O)^2)/E	Sum
111	7.813	27.000	-19.187	368.146	47 120	
112	7.317	9.000	-1 683	2.833	0.387	47.508
113	8.371	10.000	-1.629	2.654	0.317	47.825
114	7.875	6.000	1.875	3.515	0.446	48.271
121	7.317	10.000	-2.683	7.199	0.984	49.255
122	6.852	8.000	-1.148	1.317	0 192	49.447
123	7.839	6.000	1.839	3.384	0.432	49.879
124	7.375	7.000	0 375	0.141	0.019	49.898
131	8.371	8.000	0 371	0.138	0.016	49.914
132	7.839	6.000	1 839	3.384	0.432	50.346
133	8.969	8.000	0 969	0.939	0.105	50.451
134	8 437	2.000	6 437	41.440	4.911	55.362
141	7 875	5.000	2 875	8.265	1.050	56.412
142	7.375	5.000	2.375	5.640	0.765	57.176
143	8.437	4.000	4.437	19.690	2.334	59.510
144	7.937	5.000	2 937	8.628	1.087	60.597
211	7.317	9.000	-1.683	2.833	0.387	60.984
212	6 852	8.000	-1 148	1.317	0.192	61.177
213	7.839	6.000	1.839	3.384	0.432	61.608
214	7.375	4.000	3.375	11.390	1.544	63.153
221	6.852	10.000	-3.148	9.908	1.446	64.599
222	6 417	9.000	-2.583	6.671	1.040	65.638
223	7.342	7.000	0.342	0.117	0.016	65.654
224	6.907	6.000	0.907	0.822	0.119	65.773
231	7.839	7.000	0.839	0.705	0.090	65.863
232	7 342	4.000	3.342	11.167	1.521	67.384
233	8.399	8.000	0.399	0.160	0.019	67.403
234	7.902	10.000	-2.098	4.403	0.557	67.960
241	7.375	4.000	3.375	11.390	1.544	69.505
242	6.907	9.000	-2.093	4.382	0.634	70.139
243	7.902	9.000	-1.098	1.206	0.153	70.292
244	7.433	8.000	-0.567	0.321	0.043	70.335
311	8 371	10.000	-1.629	2.654	0.317	70.652
312	7.839	9.000	-1 161	1.347	0.172	70.824
313	8.969	5.000	3.969	15.752	1 756	72.580
314	8.437	5.000	3 437	11.816	1.400	73.980
321	7.839	2.000	5.839	34.099	4.350	78.330
322	7.342	9.000	-1.658	2.750	0.375	78.705
323	8.399	8.000	0.399	0.160	0.019	78.724
324	7.902	8.000	-0.098	0.010	0.001	78.725
331	8 969	8.000	0.969	0.939	0.105	78.830
332	8.399	8.000	0 399	0.160	0.019	78.849
333	9.609	19.000	-9.391	88.182	9 177	88.025
334	9.040	9.000	0.040	0.002	0.000	88.025
341	8 437	6.000	2.437	5.941	0.704	88.729

Table 4.14b continues

The relative price change (summary statistics Table 4.26a-b) digram, trigram, and tetragram Chi-Square calculations also detect significant serial dependencies in the closing prices for Microsoft for 1986–1988, as shown in Tables 4.27 to 4.29, respectively.

The closing prices for Microsoft for 05/04/93 to 05/04/94 are shown in Figure 4.20, while the frequency distribution of prices is shown in Figure 4.21 and the cumulative probability density function of prices is shown in Figure 4.22 (summary statistics Table 4.30). Compare this cumulative probability density function to those shown in Figures 4.23 and 4.24, which show the cumulative probability functions for 1986–88 and 1991–93. It is clear that these cumulative probability den-

Table 4.14b (Continued)

342	7 902	7.000	0 902	0 813	0 103	88.832		
343	9 040	11.000	-1 960	3 841	0 425	89 257		
344	8 504	11.000	-2 496	6 228	0 732	89.990		
411	7 875	6.000	1 875	3 515	0 446	90.436		
412	7 375	5.000	2 375	5 640	0 765	91 201		
413	8 437	3.000	5 437	29 565	3 504	94.705		
414	7 937	4 000	3 937	15 503	1 953	96 658		
421	7 375	4.000	3 375	11 390	1 544	98.202		
422	6 907	6.000	0 907	0.822	0 119	98.321		
423	7 902	8 000	-0 098	0 010	0 001	98.323		
424	7 433	9.000	-1 567	2 454	0 330	98.653		
431	8 437	6.000	2 437	5 941	0 704	99.357		
432	7 902	9.000	-1 098	1 206	0 153	99.509		
433	9 040	9.000	0 040	0.002	0 000	99 510		
434	8 504	14.000	-5 496	30.202	3 551	103.061	sqrt(2*ChiSquare)	sqrt(2*(df-1))
441	7 937	4 000	3 937	15 503	1 953	105.014	15 92734071	11 22497216
442	7 433	6.000	1 433	2 055	0 276	105.291		
443	8 504	14.000	-5 496	30.202	3 551	108.842		
444	8 000	20.000	-12 000	143 991	17 998	126.840	z	4.702368545
	n=	506			Chi-Square	126 840		
expected=	506							

Table 4.15

X_1: *Microsoft Close–Open 91–93*

Mean	Std. Dev.	Std. Error	Variance	Coef. Var.	Count
5.3E–3	1.5	.1	2.1	27435.3	508

Minimum	Maximum	Range	Sum	Sum of Sqr.	# Missing
–4.2	4.9	9.1	2.7	1080.7	0

# < 10th %	10th %	25th %	50th %	75th %	90th %
50	–1.8	–.9	0	1	1.8

# > 90th %	Mode	Geo. Mean	Har. Mean	Kurtosis	Skewness
48	•	•	•	.4	.2

Table 4.16a-b

X_1: 1–Microsoft Close–Open 91–93

Mean	Std. Dev.	Std. Error	Variance	Coef. Var.	Count
−1.8	.7	.1	.6	−41.5	126

Minimum	Maximum	Range	Sum	Sum of Sqr.	# Missing
−4.2	−.9	3.3	−225.3	471.6	23

# < 10th %	10th %	25th %	50th %	75th %	90th %
12	−2.9	−2.2	−1.6	−1.2	−1

# > 90th %	Mode	Geo. Mean	Har. Mean	Kurtosis	Skewness
3	−1	•	•	.2	−.9

X_2: 2–Microsoft Close–Open 91–93

Mean	Std. Dev.	Std. Error	Variance	Coef. Var.	Count
−.4	.3	2.4E−2	.1	−67.8	149

Minimum	Maximum	Range	Sum	Sum of Sqr.	# Missing
−.9	0	.9	−63.2	39	0

# < 10th %	10th %	25th %	50th %	75th %	90th %
15	−.8	−.7	−.5	−.2	0

# > 90th %	Mode	Geo. Mean	Har. Mean	Kurtosis	Skewness
0	•	•	•	−1.3	2.4E−2

sity functions show very different distributions. This is confirmed by the box and whiskers plots shown in Figure 4.25 (prices for 87–93) and Figure 4.26 (Delta-1 price changes for the same time period).

The cumulative probability density functions for Half1 and Half2 are shown in Figure 4.27. These two probability density functions cross, but do not overlap. This suggests that Microsoft prices for

Table 4.16c-d

X_3: 3–Microsoft Close–Open 91–93

Mean	Std. Dev.	Std. Error	Variance	Coef. Var.	Count
.6	.3	2.6E–2	.1	50.3	120

Minimum	Maximum	Range	Sum	Sum of Sqr.	# Missing
.1	1	.9	67.5	47.5	29

# < 10th %	10th %	25th %	50th %	75th %	90th %
12	.2	.2	.5	.8	1

# > 90th %	Mode	Geo. Mean	Har. Mean	Kurtosis	Skewness
0	.2	.5	.4	–1.2	3.2E–2

X_4: 4–Microsoft Close–Open 91–93

Mean	Std. Dev.	Std. Error	Variance	Coef. Var.	Count
2	.8	.1	.7	42.6	113

Minimum	Maximum	Range	Sum	Sum of Sqr.	# Missing
1.1	4.9	3.8	223.7	522.6	36

# < 10th %	10th %	25th %	50th %	75th %	90th %
5	1.2	1.4	1.8	2.3	3.2

# > 90th %	Mode	Geo. Mean	Har. Mean	Kurtosis	Skewness
10	1.2	1.8	1.7	1.8	1.5

1993 to 1994 may not be stationary and this is confirmed by the box and whiskers plot shown in Figure 4.28. This plot shows that the two distributions are different. This is confirmed by the box and whiskers plot shown in Figure 4.29, which shows the weekly prices for 1993–1994. The summary statistics are shown in Table 4.31a-b. The cumulative probability density functions for Delta-1 Half1 and Half2 are

Table 4.17

	A	B	C	D	E	F	G	H
1	Symbols	Frequency	Probability					
2	1	126	0.2485					
3	2	148	0.2919					
4	3	120	0.2367					
5	4	113	0.2229					
6	Sum	507						
7								
8								
9								
10	Digrams	(E)xpected	(O)bserved	(E-O)	(E-O)^2	((E-O)^2)/E	Sum	
11	11	31.314	28.000	3.314	10.980	0.351		
12	12	36.781	34.000	2.781	7.734	0.210	0.561	
13	13	29.822	34.000	-4.178	17.452	0.585	1.146	
14	14	28.083	30.000	-1.917	3.676	0.131	1.277	
15	21	36.781	40.000	-3.219	10.362	0.282	1.428	
16	22	43.203	53.000	-9.797	95.978	2.222	3.649	
17	23	35.030	29.000	6.030	36.356	1.038	4.687	
18	24	32.986	26.000	6.986	48.807	1.480	6.167	
19	31	29.822	28.000	1.822	3.321	0.111	4.799	
20	32	35.030	34.000	1.030	1.060	0.030	4.829	
21	33	28.402	29.000	-0.598	0.357	0.013	4.841	
22	34	26.746	29.000	-2.254	5.082	0.190	5.031	
23	41	28.083	30.000	-1.917	3.676	0.131	5.162	
24	42	32.986	28.000	4.986	24.862	0.754	5.916	
25	43	26.746	27.000	-0.254	0.065	0.002	5.918	
26	44	25.185	28.000	-2.815	7.922	0.315	6.233	
27								
28	Sum=	507	507			Chi-Square	6.23303725	
29								
30								
31						3.53073286	5.47722558	
32								
33					z=	-1.94649272		
34								
35								
36								
37								
38								
39								
40								
41								
42								
43								
44								
45								
46								
47								
48								
49								
50								
51								

shown in Figure 4.30 and the summary statistics are shown in Table 4.32 a-b. This suggests that the Delta-1 price changes are probably stationary. This is confirmed by the box and whiskers plot of weekly Delta-1 price changes shown in Figure 4.31. The differential spectrum is shown in Figure 4.32 (summary statistics Table 4.33). The differential spectrum looks asymmetrical and this is confirmed by the fact that the digram RPC transition matrix is significantly different from

Figure 4.17

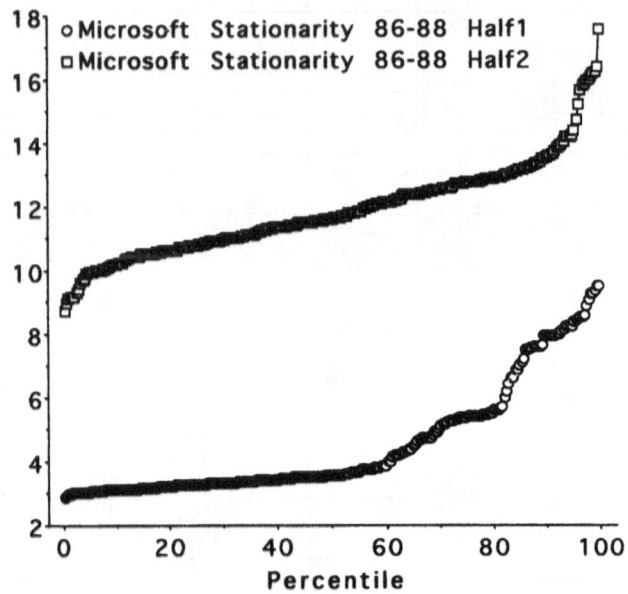

Table 4.18a

X_1: *Microsoft Stationarity 86–88 Half1*

Mean	Std. Dev.	Std. Error	Variance	Coef. Var.	Count
4.5	1.8	.1	3.2	39.5	253

Minimum	Maximum	Range	Sum	Sum of Sqr.	# Missing
2.9	9.5	6.7	1146.9	6006.5	0

# < 10th %	10th %	25th %	50th %	75th %	90th %
20	3.1	3.3	3.5	5.4	7.9

# > 90th %	Mode	Geo. Mean	Har. Mean	Kurtosis	Skewness
25	3.5	4.3	4	.4	1.3

Table 4.18b

X_2: *Microsoft Stationarity 86–88 Half2*

Mean	Std. Dev.	Std. Error	Variance	Coef. Var.	Count
11.9	1.5	.1	2.2	12.5	253

Minimum	Maximum	Range	Sum	Sum of Sqr.	# Missing
8.7	17.5	8.8	3003.8	36217.4	0

# < 10th %	10th %	25th %	50th %	75th %	90th %
25	10.2	10.8	11.6	12.8	13.6

# > 90th %	Mode	Geo. Mean	Har. Mean	Kurtosis	Skewness
25	•	11.8	11.7	1.2	.9

Figure 4.18

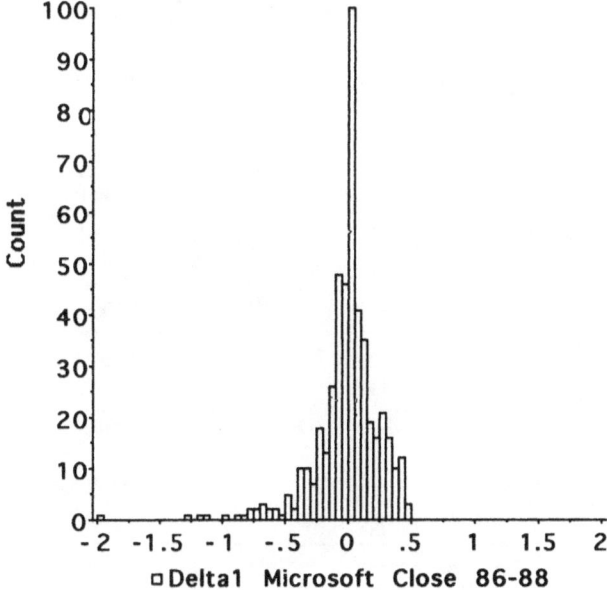

Table 4.19

X_1: Delta-1 Microsoft Close 86–88

Mean	Std. Dev.	Std. Error	Variance	Coef. Var.	Count
2.2E–2	.4	1.7E–2	.1	1713.4	506

Minimum	Maximum	Range	Sum	Sum of Sqr.	# Missing
–4.3	1.9	6.2	11	70.5	1

# < 10th %	10th %	25th %	50th %	75th %	90th %
49	–.3	–.1	7.8E–3	.1	.4

# > 90th %	Mode	Geo. Mean	Har. Mean	Kurtosis	Skewness
51	0	•	•	39.6	–3.1

Table 4.20

X_1: Microsoft High–Low 86–88

Mean	Std. Dev.	Std. Error	Variance	Coef. Var.	Count
.4	.4	1.6E–2	.1	101.8	507

Minimum	Maximum	Range	Sum	Sum of Sqr.	# Missing
1.6E–2	4.1	4.1	183.3	134.9	0

# < 10th %	10th %	25th %	50th %	75th %	90th %
51	.1	.1	.3	.5	.8

# > 90th %	Mode	Geo. Mean	Har. Mean	Kurtosis	Skewness
50	.4	.2	.2	25.6	3.7

independence (Table 4.34) and the summary statistics for the price decreases and increases are shown in Table 4.35a-b. The trigram transition matrix is shown in Table 4.36.

Table 4.21a-b

X_1: 1–Microsoft High–Low 86–88

Mean	Std. Dev.	Std. Error	Variance	Coef. Var.	Count
.1	2.1E–2	2.1E–3	4.5E–4	33.2	101

Minimum	Maximum	Range	Sum	Sum of Sqr.	# Missing
1.6E–2	.1	.1	6.5	.5	70

# < 10th %	10th %	25th %	50th %	75th %	90th %
9	3.1E–2	4.7E–2	.1	.1	.1

# > 90th %	Mode	Geo. Mean	Har. Mean	Kurtosis	Skewness
7	.1	.1	.1	–.8	–.4

X_2: 2–Microsoft High–Low 86–88

Mean	Std. Dev.	Std. Error	Variance	Coef. Var.	Count
.2	.1	4.4E–3	3.4E–3	31.5	171

Minimum	Maximum	Range	Sum	Sum of Sqr.	# Missing
.1	.3	.2	31.5	6.4	0

# < 10th %	10th %	25th %	50th %	75th %	90th %
17	.1	.1	.2	.2	.3

# > 90th %	Mode	Geo. Mean	Har. Mean	Kurtosis	Skewness
14	.1	.2	.2	–1	.5

If we parse the Delta-1 price changes into three unequal parts (<10%, 11–90%, >91%), as shown in Table 4.37 a-c, the digram transition matrix does not diverge from independence (see Table 4.38). Parsing the Delta-1 price changes into four approximately equal parts (see Table 4.39 a-d), also does not yield a digram transition matrix that diverges from independence (Table 4.40).

Table 4.21c-d

X_3: 3–Microsoft High–Low 86–88

Mean	Std. Dev.	Std. Error	Variance	Coef. Var.	Count
.4	.1	4.7E–3	2.8E–3	13.3	125

Minimum	Maximum	Range	Sum	Sum of Sqr.	# Missing
.3	.5	.2	49.4	19.9	46

# < 10th %	10th %	25th %	50th %	75th %	90th %
7	.3	.4	.4	.4	.5

# > 90th %	Mode	Geo. Mean	Har. Mean	Kurtosis	Skewness
11	.4	.4	.4	–.8	.3

X_4: 4–Microsoft High–Low 86–88

Mean	Std. Dev.	Std. Error	Variance	Coef. Var.	Count
.9	.5	4.5E–2	.2	54.4	110

Minimum	Maximum	Range	Sum	Sum of Sqr.	# Missing
.5	4.1	3.6	95.9	108.1	61

# < 10th %	10th %	25th %	50th %	75th %	90th %
10	.6	.6	.7	1	1.2

# > 90th %	Mode	Geo. Mean	Har. Mean	Kurtosis	Skewness
11	.6	.8	.8	20.1	3.9

Table 4.22

	A	B	C	D	E	F	G	H
1	Symbols	Frequency	Probability					
2	1	101	0.1996					
3	2	171	0.3379					
4	3	124	0.2451					
5	4	110	0.2174					
6	Sum	506						
7								
8								
9								
10	Digrams	(E)xpected	(O)bserved	(E-O)	(E-O)^2	((E-O)^2)/E	Sum	
11	11	20.160	60.000	-39.840	1587.219	78.731		
12	12	34.132	40.000	-5.868	34.429	1.009	79.739	
13	13	24.751	1.000	23.751	564.109	22.791	102.531	
14	14	21.957	0.000	21.957	482.089	21.957	124.487	
15	21	34.132	39.000	-4.868	23.693	0.694	103.225	
16	22	57.789	80.000	-22.211	493.349	8.537	111.762	
17	23	41.905	31.000	10.905	118.922	2.838	114.600	
18	24	37.174	21.000	16.174	261.595	7.037	121.637	
19	31	24.751	2.000	22.751	517.607	20.913	135.513	
20	32	41.905	42.000	-0.095	0.009	0.000	135.513	
21	33	30.387	47.000	-16.613	275.980	9.082	144.595	
22	34	26.957	33.000	-6.043	36.524	1.355	145.950	
23	41	21.957	0.000	21.957	482.089	21.957	167.906	
24	42	37.174	9.000	28.174	793.769	21.353	189.259	
25	43	26.957	45.000	-18.043	325.567	12.077	201.337	
26	44	23.913	56.000	-32.087	1029.573	43.055	244.392	
27								
28		n=	506			Chi-Square	244.391594	
29								
30		506						
31						22.1084416	5.47722558	
32								
33					z=	16.631216		
34								
35								
36								
37								
38								
39								
40								
41								
42								
43								
44								
45								
46								
47								
48								
49								
50								
51								

Table 4.23

X_1: Microsoft Close–Open 86–88

Mean	Std. Dev.	Std. Error	Variance	Coef. Var.	Count
1.6E–2	.3	1.5E–2	.1	2049.1	507

Minimum	Maximum	Range	Sum	Sum of Sqr.	# Missing
–3.6	1.5	5.1	8.4	57.9	0

# < 10th %	10th %	25th %	50th %	75th %	90th %
48	–.3	–.1	0	.1	.3

# > 90th %	Mode	Geo. Mean	Har. Mean	Kurtosis	Skewness
48	0	•	•	28.2	–2.5

Table 4.24a

X_1: 1–Microsoft Close–Open 86–88

Mean	Std. Dev.	Std. Error	Variance	Coef. Var.	Count
–.4	.4	3.7E–2	.2	–113.6	118

Minimum	Maximum	Range	Sum	Sum of Sqr.	# Missing
–3.6	–.1	3.5	–41.4	33.2	39

# < 10th %	10th %	25th %	50th %	75th %	90th %
12	–.7	–.4	–.2	–.1	–.1

# > 90th %	Mode	Geo. Mean	Har. Mean	Kurtosis	Skewness
12	–.1	•	•	36.8	–5.2

Table 4.24b-c

X_2: 2–Microsoft Close–Open 86–88

Mean	Std. Dev.	Std. Error	Variance	Coef. Var.	Count
–3.1E–2	3.1E–2	2.6E–3	9.9E–4	–101.5	143

Minimum	Maximum	Range	Sum	Sum of Sqr.	# Missing
–.1	0	.1	–4.4	.3	14

# < 10th %	10th %	25th %	50th %	75th %	90th %
10	–.1	–.1	–3.1E–2	0	0

# > 90th %	Mode	Geo. Mean	Har. Mean	Kurtosis	Skewness
0	0	•	•	–1.1	–.5

X_3: 3–Microsoft Close–Open 86–88

Mean	Std. Dev.	Std. Error	Variance	Coef. Var.	Count
.1	2.3E–2	2.5E–3	5.5E–4	45.3	89

Minimum	Maximum	Range	Sum	Sum of Sqr.	# Missing
1.6E–2	.1	.1	4.6	.3	68

# < 10th %	10th %	25th %	50th %	75th %	90th %
8	3.1E–2	3.1E–2	4.7E–2	.1	.1

# > 90th %	Mode	Geo. Mean	Har. Mean	Kurtosis	Skewness
9	.1	4.6E–2	4.0E–2	–1.1	.2

Table 4.24d

X_4: 4–Microsoft Close–Open 86–88

Mean	Std. Dev.	Std. Error	Variance	Coef. Var.	Count
.3	.2	1.9E–2	.1	73.9	157

Minimum	Maximum	Range	Sum	Sum of Sqr.	# Missing
.1	1.5	1.3	49.6	24.2	0

# < 10th %	10th %	25th %	50th %	75th %	90th %
4	.1	.2	.2	.4	.6

# > 90th %	Mode	Geo. Mean	Har. Mean	Kurtosis	Skewness
16	.1	.3	.2	4.7	2

Figure 4.19

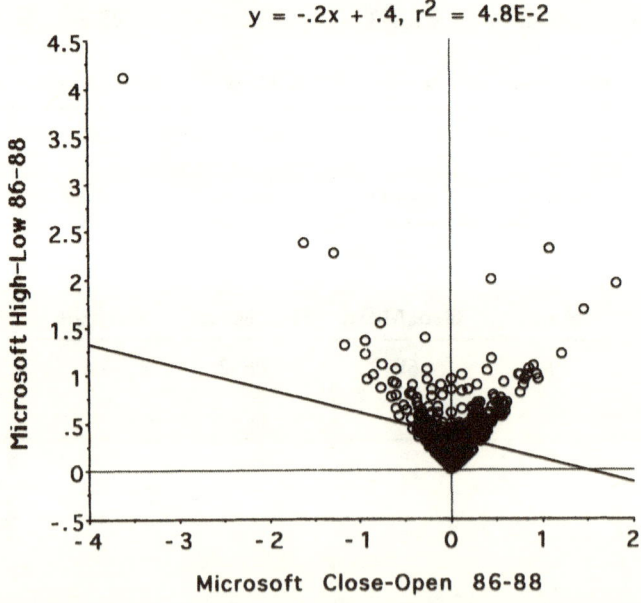

$$y = -.2x + .4, \ r^2 = 4.8\text{E-}2$$

Microsoft High–Low 86–88

Microsoft Close–Open 86-88

Table 4.25

	A	B	C	D	E	F	G	H
1	Symbols	Frequency	Probability					
2	1	118	0.2332					
3	2	143	0.2826					
4	3	89	0.1759					
5	4	156	0.3083					
6	Sum	506						
7								
8								
9								
10	Digrams	(E)xpected	(O)bserved	(E-O)	(E-O)^2	((E-O)^2)/E	Sum	
11	11	27.518	37.000	-9.482	89.912	3.267		
12	12	33.348	28.000	5.348	28.599	0.858	4.125	
13	13	20.755	13.000	7.755	60.139	2.898	7.023	
14	14	36.379	40.000	-3.621	13.108	0.360	7.383	
15	21	33.348	22.000	11.348	128.773	3.862	10.884	
16	22	40.413	62.000	-21.587	465.997	11.531	22.415	
17	23	25.152	31.000	-5.848	34.197	1.360	23.775	
18	24	44.087	28.000	16.087	258.790	5.870	29.645	
19	31	20.755	15.000	5.755	33.119	1.596	25.370	
20	32	25.152	27.000	-1.848	3.414	0.136	25.506	
21	33	15.654	29.000	-13.346	178.112	11.378	36.884	
22	34	27.439	18.000	9.439	89.090	3.247	40.131	
23	41	36.379	44.000	-7.621	58.073	1.596	41.727	
24	42	44.087	26.000	18.087	327.138	7.420	49.147	
25	43	27.439	16.000	11.439	130.845	4.769	53.916	
26	44	48.095	70.000	-21.905	479.835	9.977	63.893	
27								
28		n=	506			Chi-Square	63.892908	
29								
30		506						
31						11.3042388	5.47722558	
32								
33					z=	5.82701327		
34								
35								
36								
37								
38								
39								
40								
41								
42								
43								
44								
45								
46								
47								
48								
49								
50								
51								5.82701327

Table 4.26a-b

X_1: 1–Delta-1 Microsoft Close 86–88

Mean	Std. Dev.	Std. Error	Variance	Coef. Var.	Count
–.2	.4	2.3E–2	.1	–193	253

Minimum	Maximum	Range	Sum	Sum of Sqr.	# Missing
–4.3	0	4.3	–47.2	41.4	0

# < 10th %	10th %	25th %	50th %	75th %	90th %
25	–.4	–.2	–.1	–2.7E–2	0

# > 90th %	Mode	Geo. Mean	Har. Mean	Kurtosis	Skewness
0	0	•	•	70.7	–7

X_2: 2–Delta-1 Microsoft Close 86–88

Mean	Std. Dev.	Std. Error	Variance	Coef. Var.	Count
.2	.2	1.6E–2	.1	108.6	253

Minimum	Maximum	Range	Sum	Sum of Sqr.	# Missing
1.6E–2	1.9	1.9	58.2	29.1	0

# < 10th %	10th %	25th %	50th %	75th %	90th %
21	3.1E–2	.1	.1	.3	.5

# > 90th %	Mode	Geo. Mean	Har. Mean	Kurtosis	Skewness
23	.1	.1	.1	11.8	2.8

Table 4.27

	A	B	C	D	E	F	G	H	I
1	n	Digram	Expec Prob	(E)xpec Freq	(O)bser Freq	(E-O)	(E-O)^2	((E-O)^2)/E	SUM
2	505	11	0.1667	84.1835	142	-57.8165	3342.74767	39.7078724	
3		12	0.3333	168.3165	111	57.3165	3285.18117	19.5178795	59.2257519
4		21	0.3333	168.3165	111	57.3165	3285.18117	19.5178795	78.7436314
5		22	0.1667	84.1835	141	-56.8165	3228.11467	38.3461685	98.261511
6									
7				505					
8								Chi-Square	136.607679

Table 4.28

	A	B	C	D	E	F	G	H	I
1	n	Trigram	Expec Prob	(E)xpec Freq	(O)bser Freq	(E-O)	(E-O)^2	((E-O)^2)/E	SUM
2	504	111	0.042	21.002	77.000	-55.998	3135.812	149.312	
3		112	0.125	63.000	65.000	-2.000	4.000	0.063	149.376
4		121	0.208	104.998	46.000	58.998	3480.802	33.151	182.527
5		122	0.125	63.000	64.000	-1.000	1.000	0.016	182.543
6		211	0.125	63.000	65.000	-2.000	4.000	0.063	182.606
7		212	0.208	104.998	46.000	58.998	3480.802	33.151	215.757
8		221	0.125	63.000	65.000	-2.000	4.000	0.063	215.821
9		222	0.042	21.002	76.000	-54.998	3024.815	144.027	
10									
11				504.000				Chi-Square	359.848
12									
13									
14						Sqrt(2*Chi)	26.827		
15					-				
16						Sqrt(2*(df-1))	3.742		
17									
18						Z	23.085		

Table 4.29

	A	B	C	D	E	F	G	H	I
1	n	Tetragram	Expected P	(E)xpected P	(O)bser P	(E-O)	(E-O)^2	((E-O)^2)/E	Sum
2	503	1111	0.008	4.190	41.000	-36.810	1354.977	323.384	
3		1112	0.033	16.765	36.000	-19.235	369.986	22.069	345.453
4		1121	0.075	37.725	30.000	7.725	59.676	1.582	347.035
5		1122	0.050	25.150	34.000	-8.850	78.322	3.114	350.149
6		1211	0.075	37.725	24.000	13.725	188.376	4.993	355.143
7		1212	0.133	67.065	22.000	45.065	2030.853	30.282	385.425
8		1221	0.096	48.371	32.000	16.371	268.026	5.541	390.966
9		1222	0.033	16.765	32.000	-15.235	232.106	13.845	404.810
10		2111	0.033	16.765	36.000	-19.235	369.986	22.069	426.879
11		2112	0.092	46.105	29.000	17.105	292.580	6.346	433.225
12		2121	0.133	67.065	16.000	51.065	2607.633	38.882	472.107
13		2122	0.075	37.725	30.000	7.725	59.676	1.582	473.689
14		2211	0.050	25.150	41.000	-15.850	251.222	9.989	483.678
15		2212	0.075	37.725	24.000	13.725	188.376	4.993	488.671
16		2221	0.033	16.765	32.000	-15.235	232.106	13.845	502.516
17		2222	0.008	4.190	44.000	-39.810	1584.837	378.244	880.760
18									
19									
20				505.246				Chi-Square	880.760
21									
22									
23						Sqrt(2* Chi)	41.970		
24						Sqrt(2*(df-1))	5.477		
25									
26						Z=	36.493		

Figure 4.20

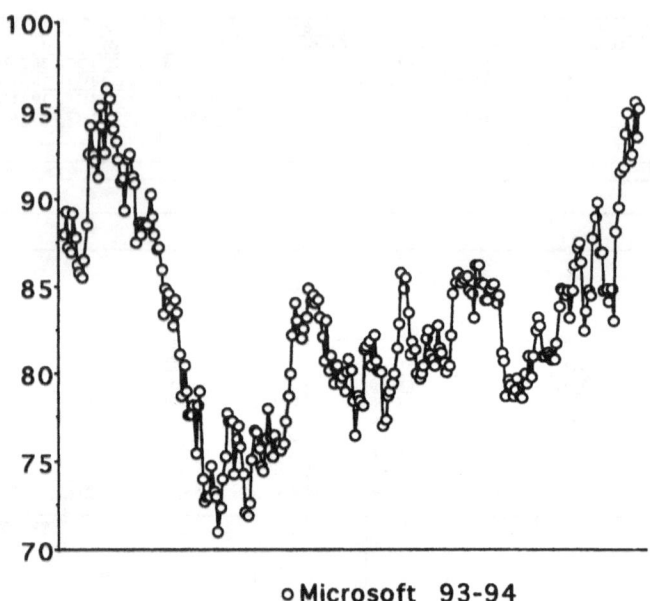

o Microsoft 93-94

Figure 4.21

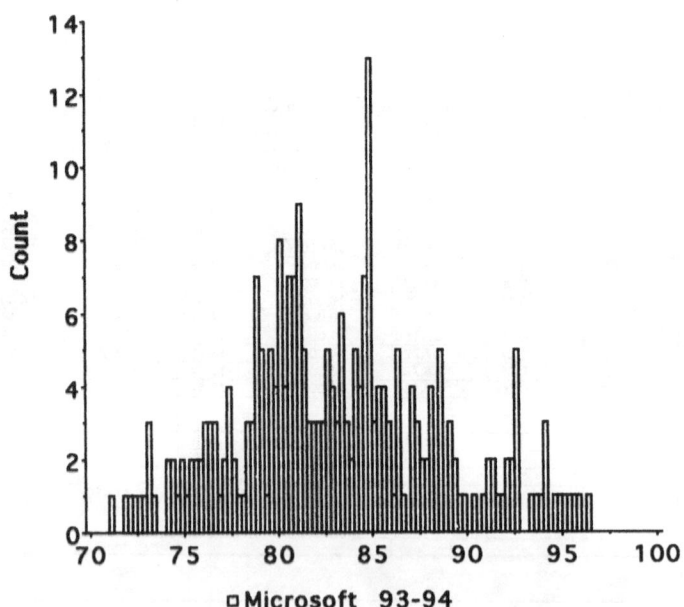

□ Microsoft 93-94

Figure 4.22

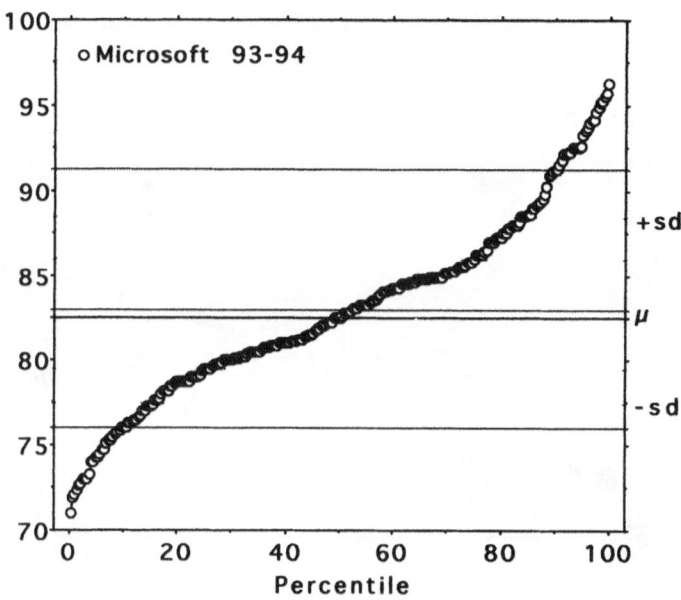

Table 4.30

X_1: Microsoft 93–94

Mean	Std. Dev.	Std. Error	Variance	Coef. Var.	Count
82.9	5.5	.3	29.8	6.6	254

Minimum	Maximum	Range	Sum	Sum of Sqr.	# Missing
71	96.2	25.2	21066.9	1754837.9	0

# < 10th %	10th %	25th %	50th %	75th %	90th %
24	76	79.5	82.5	86	91.2

# > 90th %	Mode	Geo. Mean	Har. Mean	Kurtosis	Skewness
24	•	82.8	82.6	−.3	.3

Figure 4.23

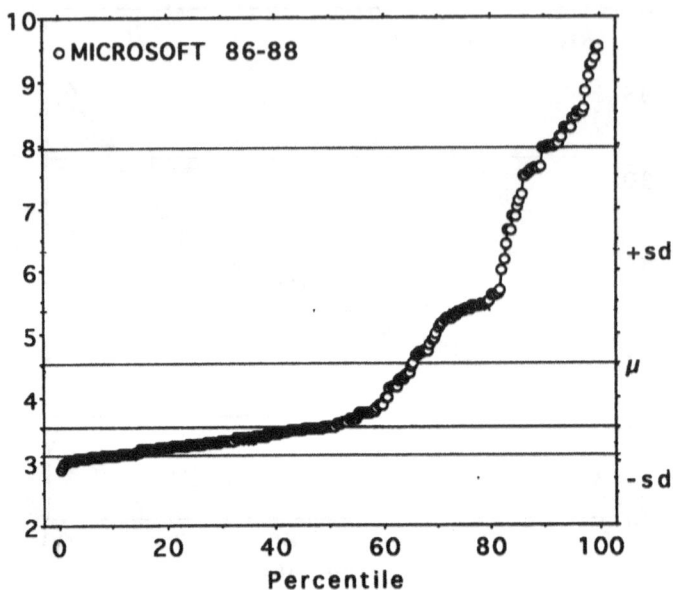

Figure 4.24

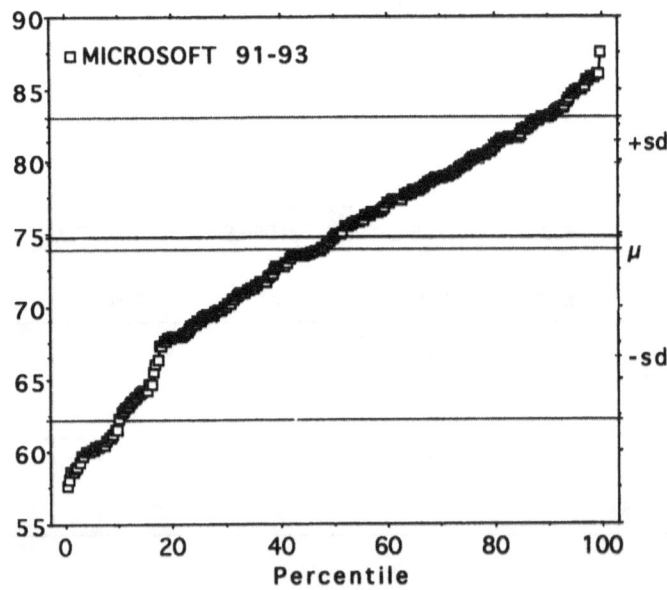

Figure 4.25

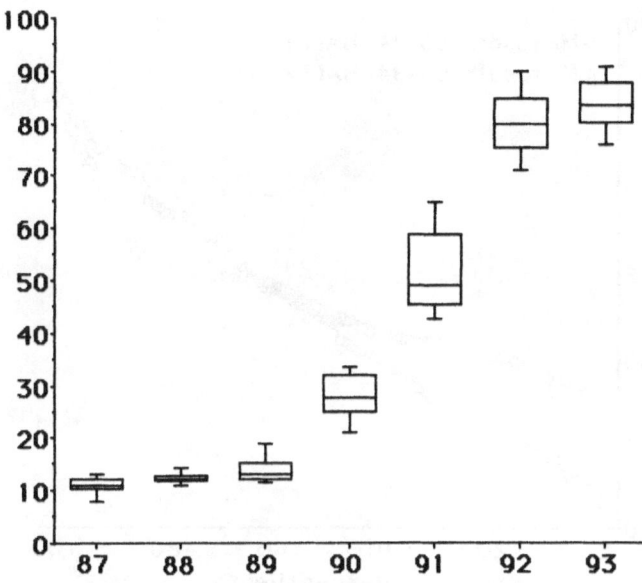

Figure 4.26

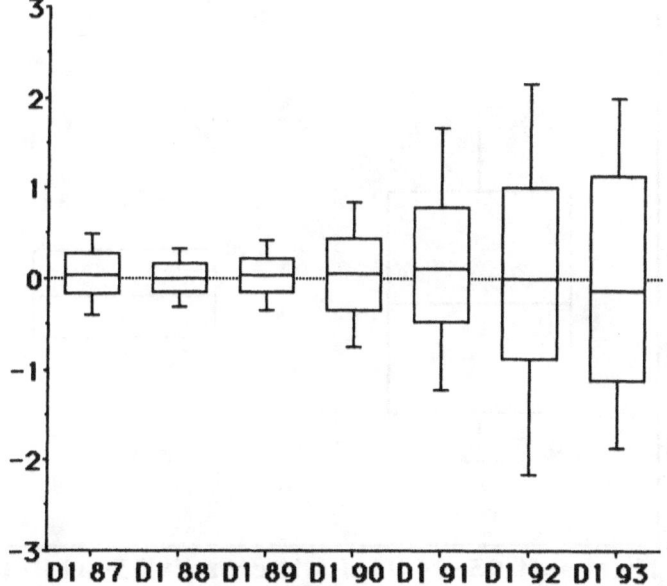

Figure 4.27

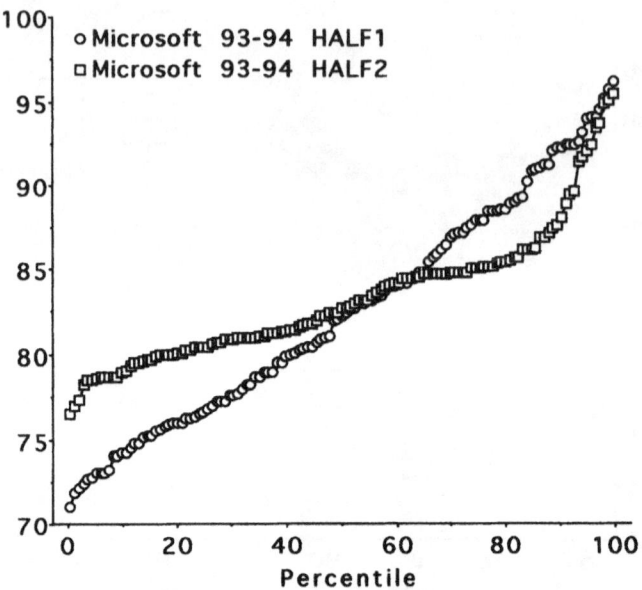

Figure 4.28

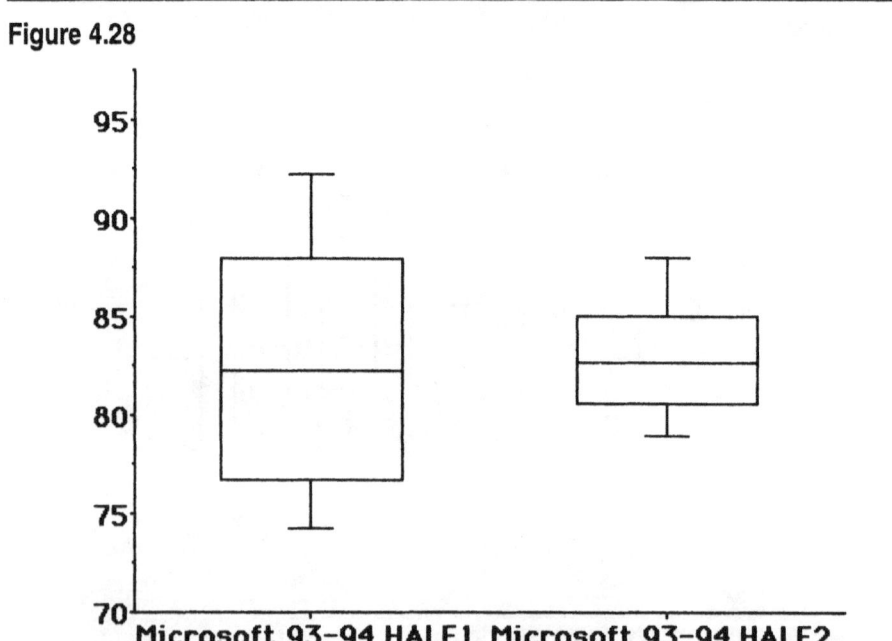

Figure 4.29

Table 4.31a

X_1: *Microsoft 93–94 Half1*

Mean	Std. Dev.	Std. Error	Variance	Coef. Var.	Count
82.5	6.6	.6	43.9	8	127

Minimum	Maximum	Range	Sum	Sum of Sqr.	# Missing
71	96.2	25.2	10479.9	870314.3	0

# < 10th %	10th %	25th %	50th %	75th %	90th %
12	74.2	76.7	82.2	88	92.2

# > 90th %	Mode	Geo. Mean	Har. Mean	Kurtosis	Skewness
12	•	82.3	82	−1	.2

Table 4.31b

X₂: *Microsoft 93–94 Half2*

Mean	Std. Dev.	Std. Error	Variance	Coef. Var.	Count
83.4	3.9	.3	15.5	4.7	127

Minimum	Maximum	Range	Sum	Sum of Sqr.	# Missing
76.5	95.5	19	10587.1	884523.6	0

# < 10th %	10th %	25th %	50th %	75th %	90th %
13	79	80.5	82.8	85.1	88.1

# > 90th %	Mode	Geo. Mean	Har. Mean	Kurtosis	Skewness
13	84.8	83.3	83.2	1.1	1.1

Figure 4.30

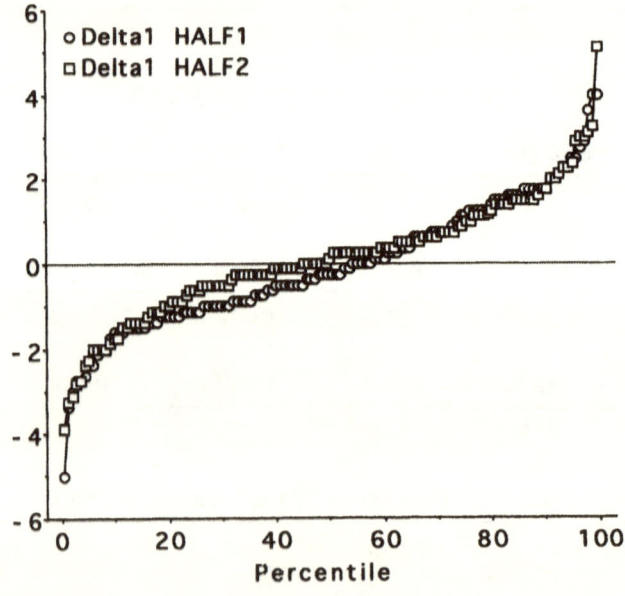

Table 4.32a-b

X_1: Delta-1 Half1

Mean	Std. Dev.	Std. Error	Variance	Coef. Var.	Count
−.1	1.5	.1	2.4	−2737.6	126

Minimum	Maximum	Range	Sum	Sum of Sqr.	# Missing
−5	4	9	−7.1	300	1

# < 10th %	10th %	25th %	50th %	75th %	90th %
12	−1.6	−1.1	−.2	1.1	1.8

# > 90th %	Mode	Geo. Mean	Har. Mean	Kurtosis	Skewness
12	•	•	•	.3	.1

X_2: Delta-1 Half2

Mean	Std. Dev.	Std. Error	Variance	Coef. Var.	Count
.1	1.4	.1	2.1	1278.8	127

Minimum	Maximum	Range	Sum	Sum of Sqr.	# Missing
−3.9	5.1	9	14.2	261	0

# < 10th %	10th %	25th %	50th %	75th %	90th %
12	−1.8	−.6	.1	1	1.8

# > 90th %	Mode	Geo. Mean	Har. Mean	Kurtosis	Skewness
12	.2	•	•	.9	.1

Figure 4.31

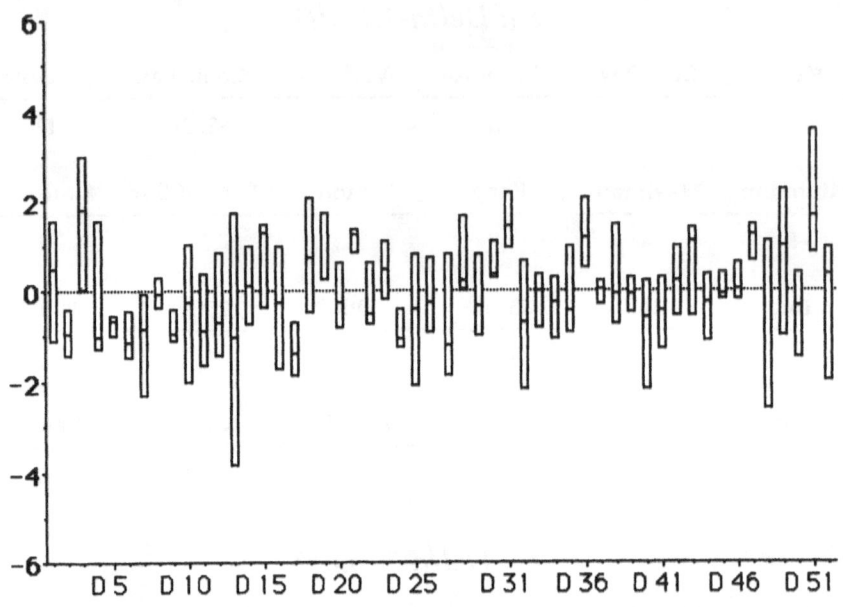

Figure 4.32

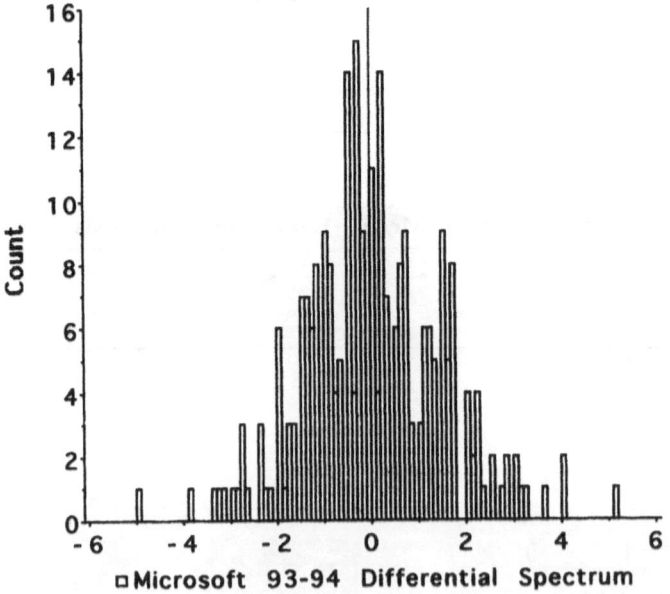

Table 4.33

X_1: *Microsoft 93–94 Differential Spectrum*

Mean	Std. Dev.	Std. Error	Variance	Coef. Var.	Count
2.8E–2	1.5	.1	2.2	5297	253

Minimum	Maximum	Range	Sum	Sum of Sqr.	# Missing
–5	5.1	10.1	7.1	561	1

# < 10th %	10th %	25th %	50th %	75th %	90th %
23	–1.8	–1	0	1	1.8

# > 90th %	Mode	Geo. Mean	Har. Mean	Kurtosis	Skewness
24	–.2	•	•	.5	.1

Table 4.34

	A	B	C	D	E	F	G	H	I
1	n	Digram	Expec Prob	(E)xpec Freq	(O)bser Freq	(E-O)	(E-O)^2	((E-O)^2)/E	SUM
2	252	11	0.1667	42.0084	75	-32.9916	1088.44567	25.9101911	
3		12	0.3333	83.9916	61	22.9916	528.613671	6.29364925	32.2038403
4		21	0.3333	83.9916	61	22.9916	528.613671	6.29364925	38.4974896
5		22	0.1667	42.0084	55	-12.9916	168.781671	4.01780764	44.7911388
6			SUM=	252	252				
7									
8								Chi-Square	48.8089465

Table 4.35a-b

X_1: – Delta-1 Price Changes

Mean	Std. Dev.	Std. Error	Variance	Coef. Var.	Count
−1.1	.9	.1	.8	−86.7	136

Minimum	Maximum	Range	Sum	Sum of Sqr.	# Missing
−5	0	5	−142.9	262.3	0

# < 10th %	10th %	25th %	50th %	75th %	90th %
14	−2.4	−1.5	−.9	−.2	−.1

# > 90th %	Mode	Geo. Mean	Har. Mean	Kurtosis	Skewness
11	−.2	•	•	2.2	−1.3

X_2: + Delta-1 Price Changes

Mean	Std. Dev.	Std. Error	Variance	Coef. Var.	Count
1.3	1	.1	.9	74.6	117

Minimum	Maximum	Range	Sum	Sum of Sqr.	# Missing
.1	5.1	5	150.1	298.7	19

# < 10th %	10th %	25th %	50th %	75th %	90th %
4	.2	.5	1.1	1.8	2.5

# > 90th %	Mode	Geo. Mean	Har. Mean	Kurtosis	Skewness
11	.2	.9	.6	1.8	1.2

Table 4.36

	A	B	C	D	E	F	G	H	I
1	n	Trigram	Expec Prob	(E)xpec Freq	(O)bser Freq	(E-O)	(E-O)^2	((E-O)^2)/E	SUM
2	251	111	0.042	10.459	36.000	-25.541	652.334	62.370	
3		112	0.125	31.375	39.000	-7.625	58.141	1.853	64.223
4		121	0.208	52.291	36.000	16.291	265.391	5.075	69.298
5		122	0.125	31.375	24.000	7.375	54.391	1.734	71.032
6		211	0.125	31.375	39.000	-7.625	58.141	1.853	72.885
7		212	0.208	52.291	22.000	30.291	917.534	17.547	90.431
8		221	0.125	31.375	24.000	7.375	54.391	1.734	92.165
9		222	0.042	10.459	31.000	-20.541	421.926	40.340	
10									
11			n=	251.000	251.000			Chi-Square	132.505

Table 4.37a

$X_1: 1$

Mean	Std. Dev.	Std. Error	Variance	Coef. Var.	Count
−2.6	.7	.2	.6	−28.1	23

Minimum	Maximum	Range	Sum	Sum of Sqr.	# Missing
−5	−1.9	3.1	−60.7	172.2	183

# < 10th %	10th %	25th %	50th %	75th %	90th %
2	−3.5	−3	−2.4	−2	−2

# > 90th %	Mode	Geo. Mean	Har. Mean	Kurtosis	Skewness
1	−2	•	•	2.6	−1.5

Table 4.37b-c

X_2: 2

Mean	Std. Dev.	Std. Error	Variance	Coef. Var.	Count
7.6E–3	1	.1	.9	12737.7	206

Minimum	Maximum	Range	Sum	Sum of Sqr.	# Missing
–1.8	1.8	3.5	1.6	191.5	0

# < 10th %	10th %	25th %	50th %	75th %	90th %
20	–1.2	–.8	0	.8	1.5

# > 90th %	Mode	Geo. Mean	Har. Mean	Kurtosis	Skewness
13	–.2	•	•	–1	.1

X_3: 3

Mean	Std. Dev.	Std. Error	Variance	Coef. Var.	Count
2.8	.8	.2	.6	28.6	24

Minimum	Maximum	Range	Sum	Sum of Sqr.	# Missing
2	5.1	3.1	66.2	197.2	182

# < 10th %	10th %	25th %	50th %	75th %	90th %
0	2	2.2	2.5	3.1	4

# > 90th %	Mode	Geo. Mean	Har. Mean	Kurtosis	Skewness
1	•	2.7	2.6	1.5	1.3

Table 4.38

	A	B	C	D	E	F	G
1	Symbols	Frequency	Probability				
2	1	23	0.0913				
3	2	205	0.8135				
4	3	24	0.0952				
5							
6	Sum	252					
7							
8							
9							
10	Digrams	(E)xpected	(O)bserved	(E-O)	(E-O)^2	((E-O)^2)/E	Sum
11	11	2.10	2	0.10	0.010	0.005	
12	12	18.71	17	1.71	2.925	0.156	0.161
13	13	2.19	4	-1.81	3.274	1.495	1.656
14	21	18.71	19	-0.29	0.084	0.004	1.660
15	22	166.77	167	-0.23	0.055	0.000	1.661
16	23	19.52	19	0.52	0.274	0.014	1.675
17	31	2.19	2	0.19	0.036	0.017	1.691
18	32	19.52	21	-1.48	2.179	0.112	1.803
19	33	2.29	1	1.29	1.653	0.723	
20							
21	SUM=	252	252			Chi-Square	2.526
22					2.24771536		
23					4		
24				Z=	-1.75228464		

Table 4.39a

$X_1: 1$

Mean	Std. Dev.	Std. Error	Variance	Coef. Var.	Count
-1.8	.8	.1	.6	-45.2	66

Minimum	Maximum	Range	Sum	Sum of Sqr.	# Missing
-5	-1	4	-116.4	246.8	27

# < 10th %	10th %	25th %	50th %	75th %	90th %
7	-2.8	-2	-1.5	-1.1	-1

# > 90th %	Mode	Geo. Mean	Har. Mean	Kurtosis	Skewness
0	-1	•	•	3	-1.6

Table 4.39b-c

X_2: 2

Mean	Std. Dev.	Std. Error	Variance	Coef. Var.	Count
−.4	.3	3.4E−2	.1	−74.1	70

Minimum	Maximum	Range	Sum	Sum of Sqr.	# Missing
−.9	0	.9	−26.5	15.5	23

# < 10th %	10th %	25th %	50th %	75th %	90th %
0	−.9	−.5	−.3	−.1	0

# > 90th %	Mode	Geo. Mean	Har. Mean	Kurtosis	Skewness
0	−.2	•	•	−1	−.3

X_3: 3

Mean	Std. Dev.	Std. Error	Variance	Coef. Var.	Count
.9	.5	.1	.3	58.9	93

Minimum	Maximum	Range	Sum	Sum of Sqr.	# Missing
.1	1.8	1.6	83.8	101.4	0

# < 10th %	10th %	25th %	50th %	75th %	90th %
4	.2	.4	.8	1.4	1.6

# > 90th %	Mode	Geo. Mean	Har. Mean	Kurtosis	Skewness
8	.2	.7	.5	−1.4	.2

Table 4.39d

X_4: 4

Mean	Std. Dev.	Std. Error	Variance	Coef. Var.	Count
2.8	.8	.2	.6	28.6	24

Minimum	Maximum	Range	Sum	Sum of Sqr.	# Missing
2	5.1	3.1	66.2	197.2	69

# < 10th %	10th %	25th %	50th %	75th %	90th %
0	2	2.2	2.5	3.1	4

# > 90th %	Mode	Geo. Mean	Har. Mean	Kurtosis	Skewness
1	•	2.7	2.6	1.5	1.3

Table 4.40

	A	B	C	D	E	F	G	H
1	Symbols	Frequency	Probability					
2	1	66	0.2619					
3	2	70	0.2778					
4	3	92	0.3651					
5	4	24	0.0952					
6	Sum	252						
7								
8								
9								
10	Digrams	(E)xpected	(O)bserved	(E-O)	(E-O)^2	((E-O)^2)/E	Sum	
11	11	17.286	16.000	1.286	1.653	0.096		
12	12	18.333	20.000	-1.667	2.778	0.152	0.247	
13	13	24.095	22.000	2.095	4.390	0.182	0.429	
14	14	6.286	8.000	-1.714	2.939	0.468	0.897	
15	21	18.333	17.000	1.333	1.778	0.097	0.526	
16	22	19.444	22.000	-2.556	6.531	0.336	0.862	
17	23	25.556	26.000	-0.444	0.198	0.008	0.870	
18	24	6.667	5.000	1.667	2.778	0.417	1.287	
19	31	24.095	28.000	-3.905	15.247	0.633	1.503	
20	32	25.556	23.000	2.556	6.531	0.256	1.758	
21	33	33.587	31.000	2.587	6.694	0.199	1.958	
22	34	8.762	10.000	-1.238	1.533	0.175	2.133	
23	41	6.286	5.000	1.286	1.653	0.263	2.395	
24	42	6.667	5.000	1.667	2.778	0.417	2.812	
25	43	8.762	13.000	-4.238	17.961	2.050	4.862	
26	44	2.286	1.000	1.286	1.653	0.723	5.585	
27								
28	n=	252	252			Chi-Square	5.58532646	
29								
30								
31						3.34225267	5.47722558	
32								
33					z=	-2.1349729		
34								
35								
36								
37								
38								
39								
40								
41								
42								
43								
44								
45								
46								
47								
48								
49								
50								
51								-2.1349729

IBM   5

The closing prices for IBM for 1962 to 1993 are shown in Figure 5.1 and the cumulative probability density function is shown in Figure 5.2 (summary statistics, Table 5.1). The frequency histogram for prices with a bin width of $1.00 is shown in Figure 5.3, while the frequency histogram with a bin width of $0.50 is shown in Figure 5.4. The frequency distribution of Delta-1 price changes with a bin width of $0.25 is shown in Figure 5.5 and $0.10 in Figure 5.6, while the cumulative probability density function for Delta-1 price changes is shown in Figure 5.7 (summary statistics Table 5.2). The cumulative probability density function for Delta-30 and Delta-60 price changes is shown in Figure 5.8. The frequency histogram for Delta-30 and Delta-60 price changes is shown in Figures 5.9 and 5.10, respectively.

The cumulative probability density functions for Half1 and Half2 are shown in Figure 5.11. These two probability density functions are essentially parallel and thus suggest that IBM prices are stationary for 1962–1993. This is confirmed by the compare percentile plot shown in Figure 5.12 and in the summary statistics (Table 5.3a-b). Note for example, that the mean ± 1 standard deviation for Half1 and the mean ± 1 standard deviation for Half2 overlap, which means that they are not statistically significantly different. The cumulative probability density functions for Half1 and Half2 for Delta-1 price changes is shown in Figure 5.13 (summary statistics Table 5.4a-b).

Figure 5.1

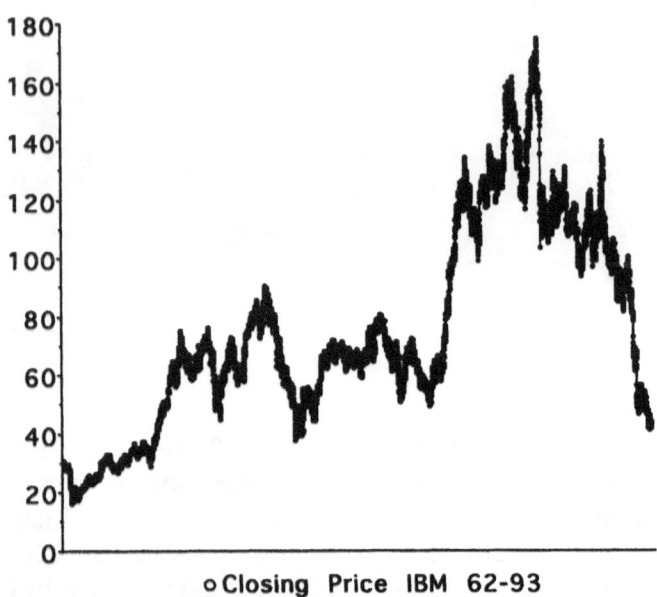

Figure 5.2

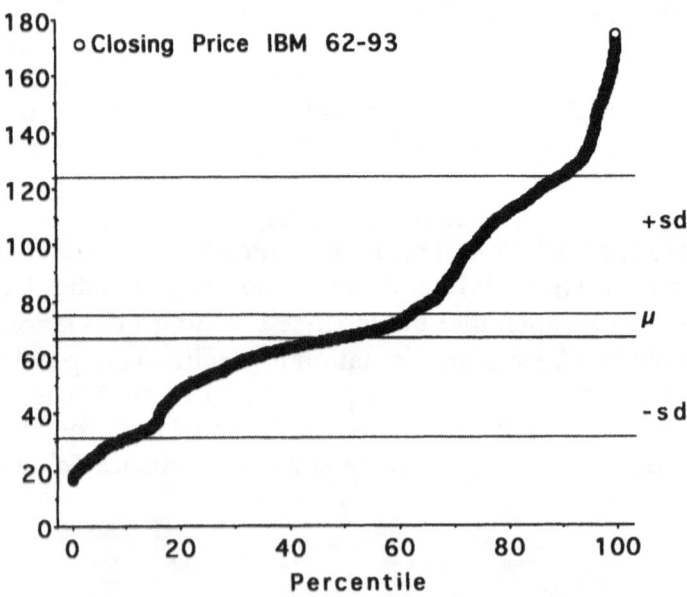

Table 5.1

X_1: *Closing Price IBM 62–93*

Mean	Std. Dev.	Std. Error	Variance	Coef. Var.	Count
74.9	34.1	.4	1166.1	45.6	7987

Minimum	Maximum	Range	Sum	Sum of Sqr.	# Missing
16.3	174.8	158.4	598586.5	54173257.4	0

# < 10th %	10th %	25th %	50th %	75th %	90th %
796	31.2	53	66.9	102.2	123.9

# > 90th %	Mode	Geo. Mean	Har. Mean	Kurtosis	Skewness
793	•	66.9	58.7	–.4	.5

Figure 5.3

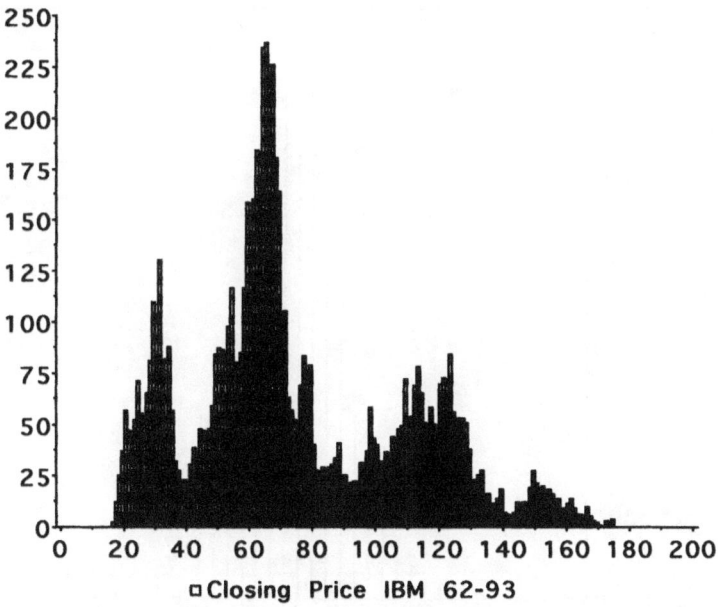

□ Closing Price IBM 62-93

Figure 5.4

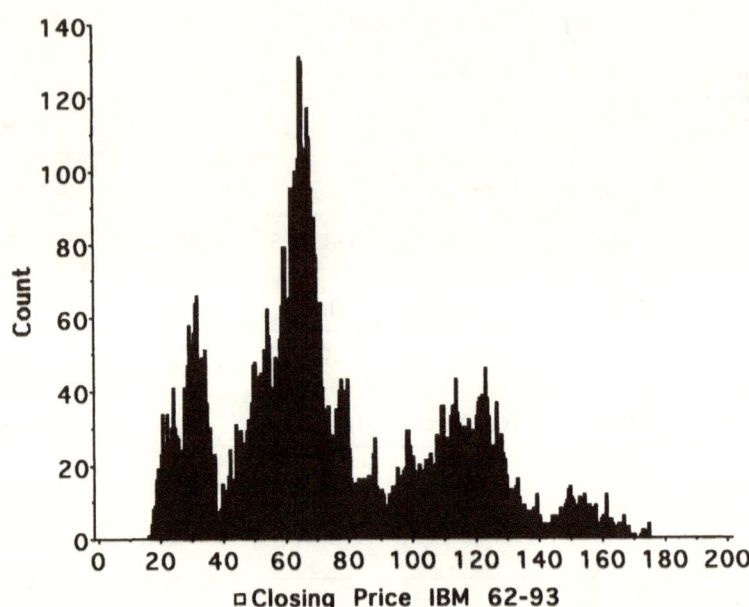

Figure 5.5

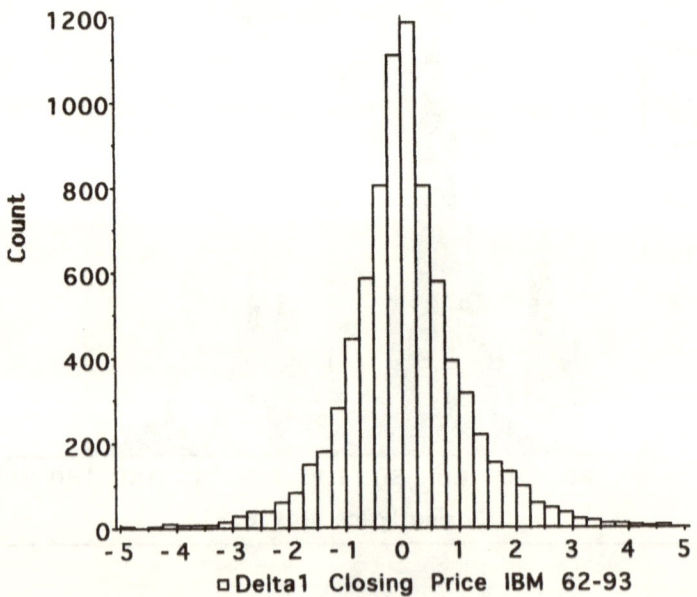

Figure 5.6

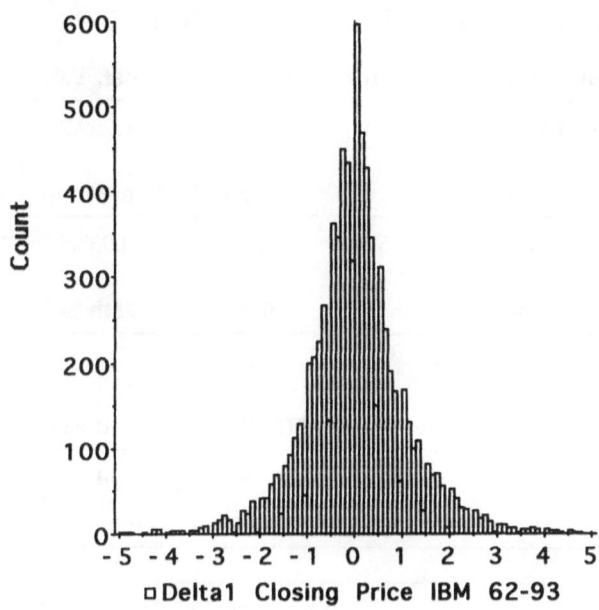

□ Delta1 Closing Price IBM 62-93

Figure 5.7

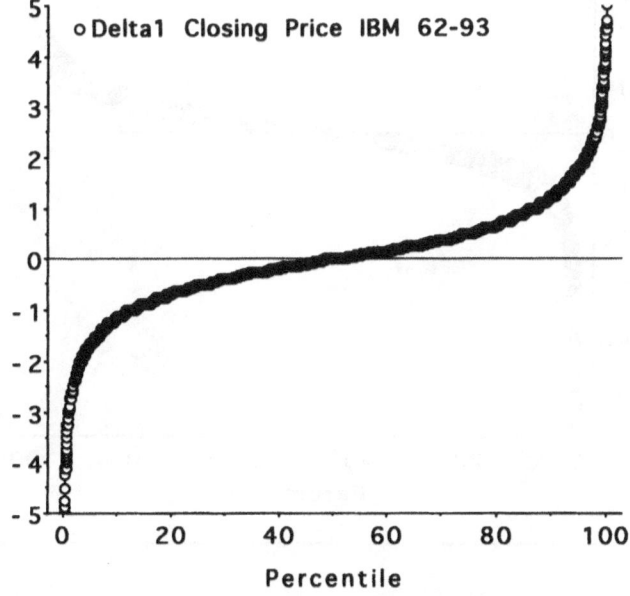

Table 5.2

X_1: Delta-1 Closing Price IBM 62–93

Mean	Std. Dev.	Std. Error	Variance	Coef. Var.	Count
1.4E–3	1.2	1.3E–2	1.4	81259.7	7986

Minimum	Maximum	Range	Sum	Sum of Sqr.	# Missing
–31.8	11.8	43.5	11.5	10933.6	1

# < 10th %	10th %	25th %	50th %	75th %	90th %
783	–1.1	–.5	0	.5	1.2

# > 90th %	Mode	Geo. Mean	Har. Mean	Kurtosis	Skewness
755	0	•	•	74.1	–2.6

Figure 5.8

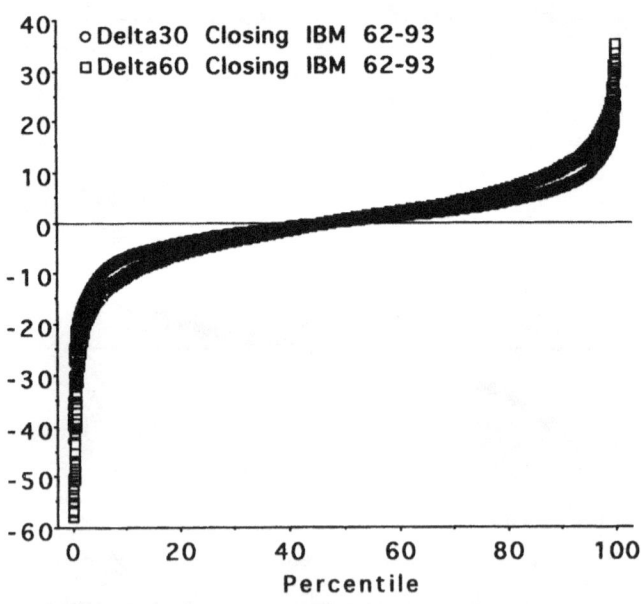

Figure 5.9

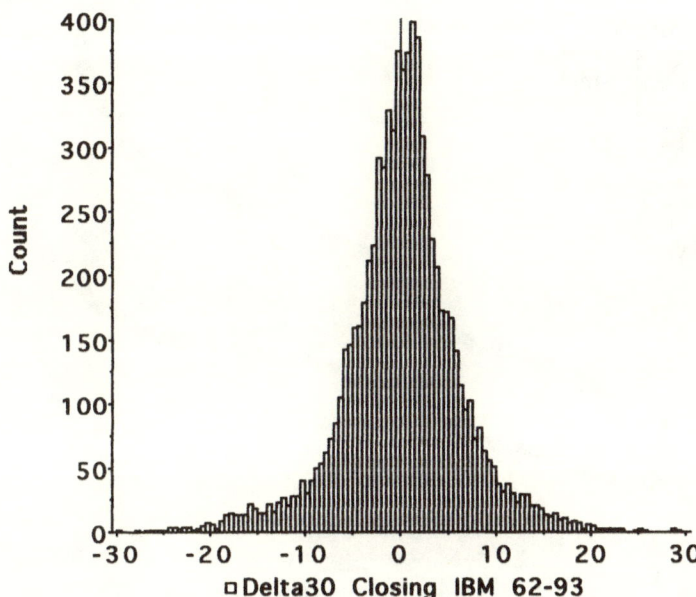

□Delta30 Closing IBM 62-93

Figure 5.10

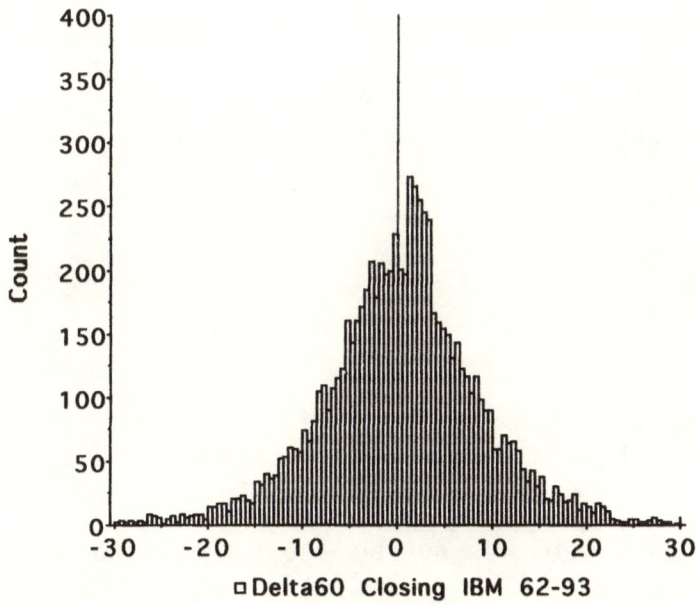

□Delta60 Closing IBM 62-93

Figure 5.11

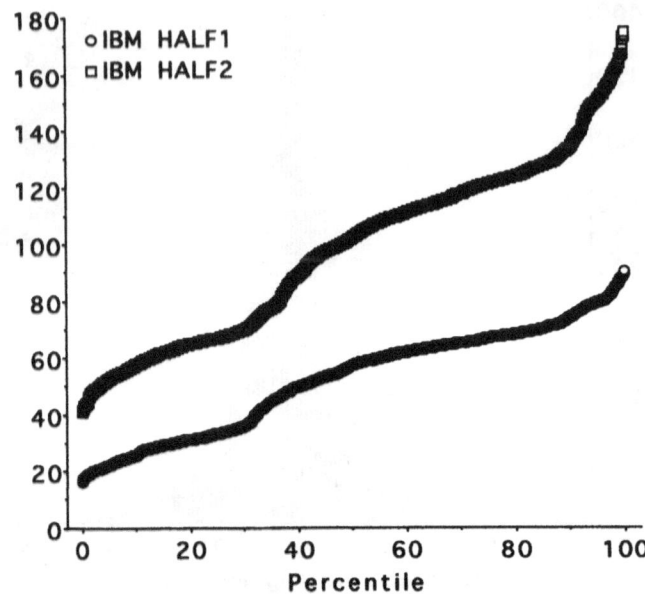

Figure 5.12

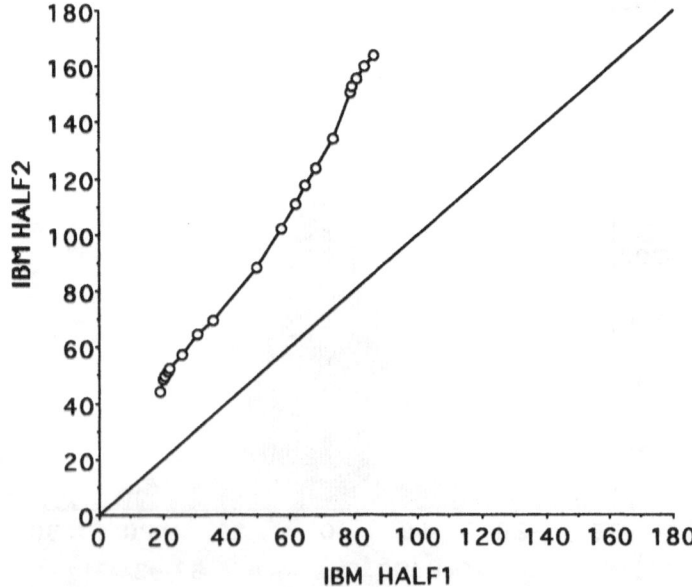

Table 5.3a-b

X_1: IBM Half1

Mean	Std. Dev.	Std. Error	Variance	Coef. Var.	Count
52.2356	18.4871	.2926	341.7715	35.3917	3993

Minimum	Maximum	Range	Sum	Sum of Sqr.	# Missing
16.3125	90.1875	73.875	208576.6012	1.2259E7	0

# < 10th %	10th %	25th %	50th %	75th %	90th %
399	25.8875	33.2656	57.3438	66.8789	73.9375

# > 90th %	Mode	Geo. Mean	Har. Mean	Kurtosis	Skewness
397	64.5	48.4285	44.2489	−1.1764	−.2214

X_2: IBM Half2

Mean	Std. Dev.	Std. Error	Variance	Coef. Var.	Count
97.6629	30.9609	.49	958.5771	31.7018	3993

Minimum	Maximum	Range	Sum	Sum of Sqr.	# Missing
41	174.75	133.75	389967.8775	4.1912E7	0

# < 10th %	10th %	25th %	50th %	75th %	90th %
399	57.6	66.875	102.25	121.25	134

# > 90th %	Mode	Geo. Mean	Har. Mean	Kurtosis	Skewness
398	•	92.4785	87.1938	−.9871	.0961

The closing prices for IBM for 05/04/93 to 05/04/94 are shown in Figure 5.14 and the cumulative probability density functions are shown in Figure 5.15. The summary statistics prices are shown in Table 5.5, while the Delta-1 price change summary statistics are shown in Table 5.6. The cumulative probability density functions for 05/04/92 to 05/04/93 are shown in Figure 5.16. The scales (i.e., the

Figure 5.13

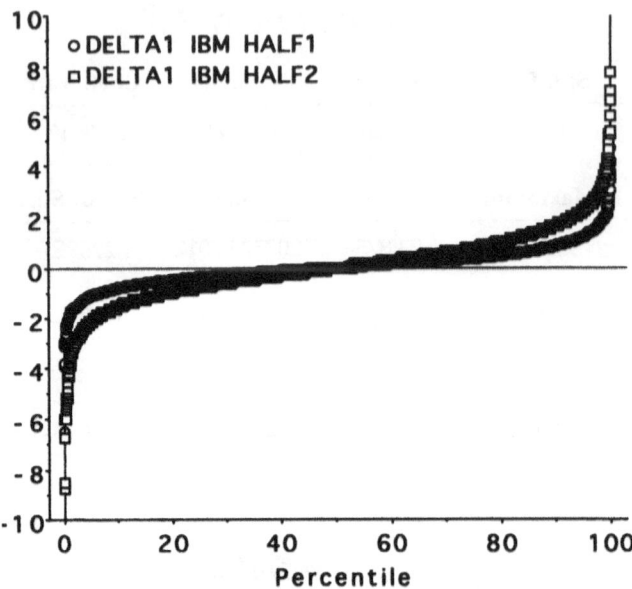

Table 5.4a

X_1: *Delta-1 IBM Half1*

Mean	Std. Dev.	Std. Error	Variance	Coef. Var.	Count
.0088	.7235	.0115	.5235	8179.082	3992

Minimum	Maximum	Range	Sum	Sum of Sqr.	# Missing
−6.5	4.7031	11.2031	35.3125	2089.4484	1

# < 10th %	10th %	25th %	50th %	75th %	90th %
397	−.7969	−.3594	0	.3438	.8437

# > 90th %	Mode	Geo. Mean	Har. Mean	Kurtosis	Skewness
398	0	•	•	4.5962	.05

Table 5.4b

X₂: Delta-1 IBM Half2

Mean	Std. Dev.	Std. Error	Variance	Coef. Var.	Count
–.0061	1.4886	.0236	2.2159	–24317.1247	3992

Minimum	Maximum	Range	Sum	Sum of Sqr.	# Missing
–31.75	11.75	43.5	–24.4375	8843.9443	1

# < 10th %	10th %	25th %	50th %	75th %	90th %
386	–1.5	–.75	0	.75	1.625

# > 90th %	Mode	Geo. Mean	Har. Mean	Kurtosis	Skewness
382	•	•	•	55.4243	–2.5508

Figure 5.14

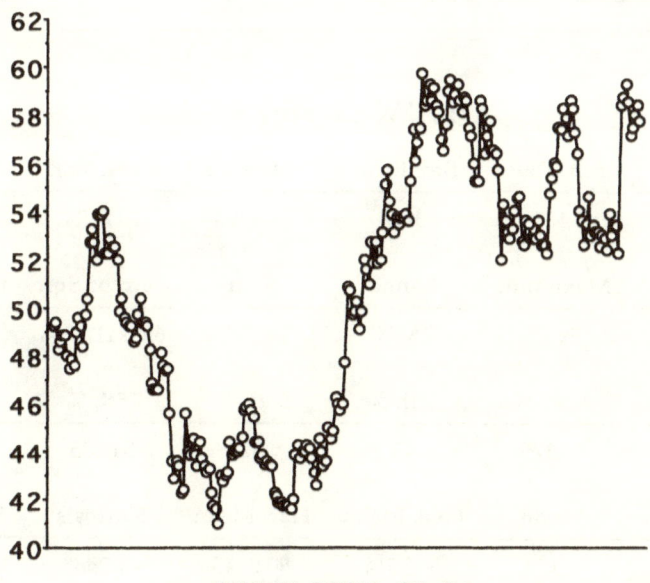

○ IBM CLOSING 93-94

Figure 5.15

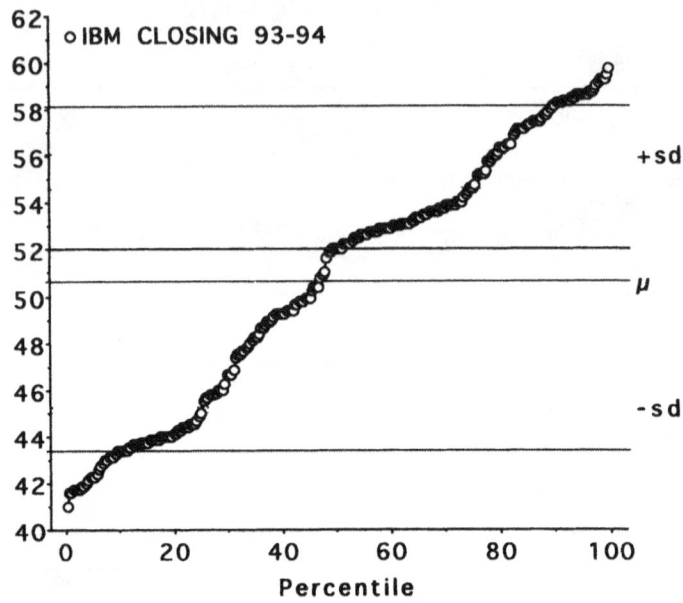

Table 5.5

X_1: IBM Closing 93–94

Mean	Std. Dev.	Std. Error	Variance	Coef. Var.	Count
50.6452	5.4258	.3404	29.4394	10.7134	254

Minimum	Maximum	Range	Sum	Sum of Sqr.	# Missing
41	59.75	18.75	12863.875	658941.3906	0

# < 10th %	10th %	25th %	50th %	75th %	90th %
24	43.375	45	52	54.625	58.1375

# > 90th %	Mode	Geo. Mean	Har. Mean	Kurtosis	Skewness
25	•	50.3514	50.0545	−1.2585	−.1073

Table 5.6

X_1: Delta-1 IBM Closing 93–94

Mean	Std. Dev.	Std. Error	Variance	Coef. Var.	Count
.0335	1.0335	.0648	1.068	3088.2304	254

Minimum	Maximum	Range	Sum	Sum of Sqr.	# Missing
–3.75	6.125	9.875	8.5	270.5	0

# < 10th %	10th %	25th %	50th %	75th %	90th %
25	–1.0125	–.5	0	.5	1.375

# > 90th %	Mode	Geo. Mean	Har. Mean	Kurtosis	Skewness
22	0	•	•	5.327	1.1111

vertical axis) are different, but the size of the plots is identical, so if the cumulative probability density functions were similar, you should be able to overlay the two plots and they should be similar or identical. If you trace Figure 5.16 and place the tracing over Figure 5.15, you will note that they are not similar. Therefore, not only are the absolute prices different, but the distribution of prices within each individual time series is also different.

The cumulative probability density functions for Half1 and Half2 IBM closing prices and Delta-1 price changes for 1993–1994 are shown in Figures 5.17 (summary statistics Table 5.7a-b) and 5.18 (summary statistics Table 5.8a-b). The differential spectrum is shown in Figure 5.19. The differential spectrum is asymmetrical and this suggests that the IBM 1993–1994 time series contains serial dependencies and this is confirmed by the fact that the digram transition matrix diverges from independence (Table 5.9). The trigram transition matrix is shown in Table 5.10. The Chi-Square is also statistically significant. This suggests that the time series contains serial dependencies involving four sequential prices. For example, the trigram 1,1,1 occurs much more often than would be expected if the time series did not contain serial dependencies. Note that this trigram encodes the serial relationship Price1>Price2>Price3>Price4. The other tri-

Figure 5.16

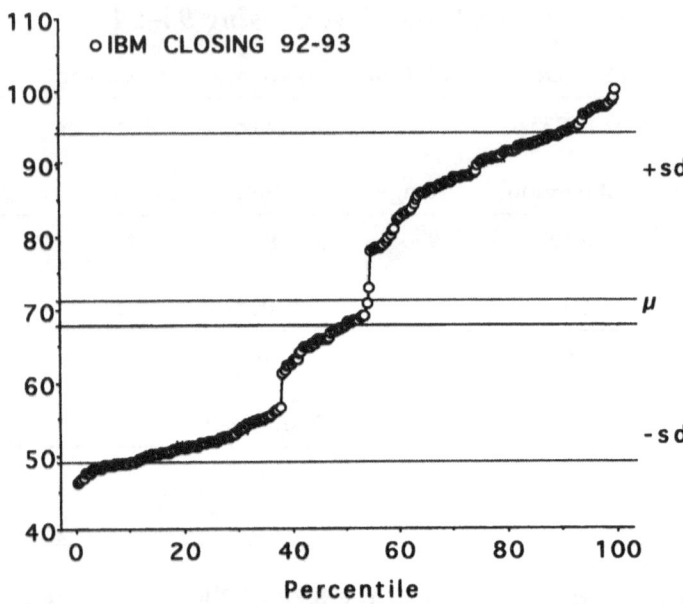

Figure 5.17

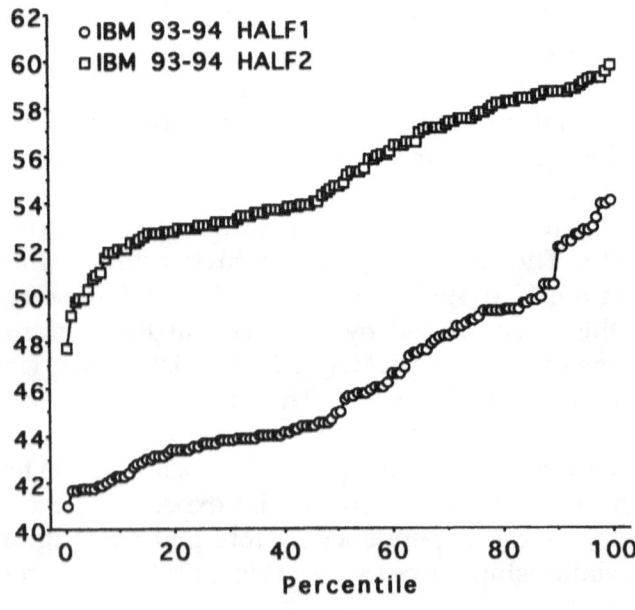

Table 5.7a-b

X_1: IBM 93–94 Half1

Mean	Std. Dev.	Std. Error	Variance	Coef. Var.	Count
46.1478	3.3872	.3018	11.4729	7.3398	126

Minimum	Maximum	Range	Sum	Sum of Sqr.	# Missing
41	54	13	5814.625	269766.3594	381

# < 10th %	10th %	25th %	50th %	75th %	90th %
11	42.25	43.625	44.9375	48.875	51.8375

# > 90th %	Mode	Geo. Mean	Har. Mean	Kurtosis	Skewness
13	44	46.0277	45.9108	−.6752	.6199

X_2: IBM 93–94 Half2

Mean	Std. Dev.	Std. Error	Variance	Coef. Var.	Count
55.0723	2.7449	.2426	7.5344	4.9841	128

Minimum	Maximum	Range	Sum	Sum of Sqr.	# Missing
47.75	59.75	12	7049.25	389175.0312	379

# < 10th %	10th %	25th %	50th %	75th %	90th %
12	52	53	54.625	57.5625	58.625

# > 90th %	Mode	Geo. Mean	Har. Mean	Kurtosis	Skewness
10	58.625	55.004	54.9353	−.909	−.1267

gram that makes a large contribution to the Chi-Square is 2,1,2, which encodes the serial relationship Price1<Price2>Price3<Price4.

The RPC methodology parses the time series into two classes, i.e., Price1>Price2, which is encoded as a '1' (a price increase) and Price1<Price2, which is encoded as a '2' (a price decrease), as shown in Table 5.11. To increase sensitivity of this method, we can parse

Figure 5.18

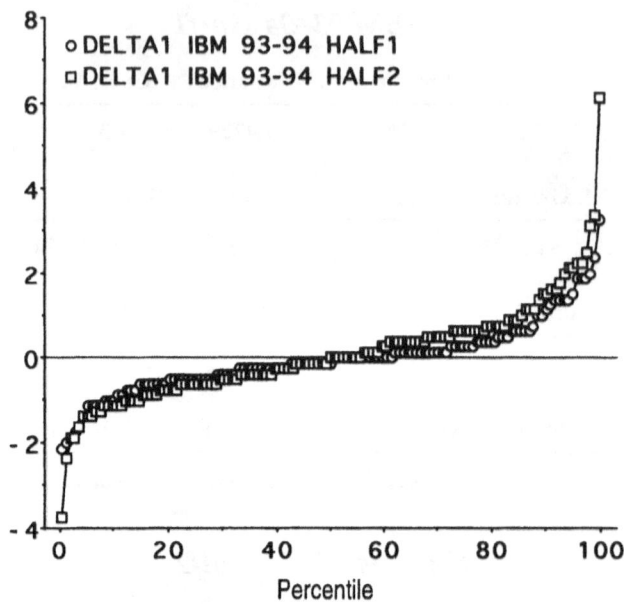

Table 5.8a

X_1: *Delta-1 IBM 93–94 Half1*

Mean	Std. Dev.	Std. Error	Variance	Coef. Var.	Count
−.026	.8417	.0753	.7085	−3237.3832	125

Minimum	Maximum	Range	Sum	Sum of Sqr.	# Missing
−2.125	3.25	5.375	−3.25	87.9375	382

# < 10th %	10th %	25th %	50th %	75th %	90th %
10	−1	−.5	−.125	.25	1.125

# > 90th %	Mode	Geo. Mean	Har. Mean	Kurtosis	Skewness
12	0	•	•	2.0284	.7143

Table 5.8b

X_2: Delta-1 IBM 93–94 Half2

Mean	Std. Dev.	Std. Error	Variance	Coef. Var.	Count
.0787	1.1909	.1057	1.4184	1512.5023	127

Minimum	Maximum	Range	Sum	Sum of Sqr.	# Missing
–3.75	6.125	9.875	10	179.5	380

# < 10th %	10th %	25th %	50th %	75th %	90th %
10	–1.125	–.625	0	.625	1.5

# > 90th %	Mode	Geo. Mean	Har. Mean	Kurtosis	Skewness
12	•	•	•	5.1077	1.1482

Figure 5.19

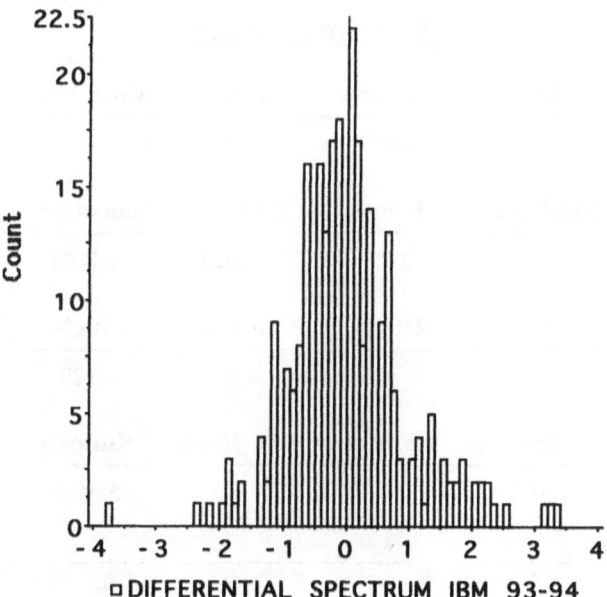

Table 5.9

	A	B	C	D	E	F	G	H	I
1	n	Digram	Expec Prob	(E)xpec Freq	(O)bser Freq	(E-O)	(E-O)^2	((E-O)^2)/E	SUM
2	252	11	0.1667	42.008	86.000	-43.992	1935.261	46.068	
3		12	0.3333	83.992	61.000	22.992	528.614	6.294	52.362
4		21	0.3333	83.992	62.000	21.992	483.630	5.758	58.120
5		22	0.1667	42.008	43.000	-0.992	0.983	0.023	63.878
6		N=		252.000	252.000				
7									
8								Chi-Square	63.902

Table 5.10

	A	B	C	D	E	F	G	H	I
1	n	Trigram	Expec Prob	(E)xpec Freq	(O)bser Freq	(E-O)	(E-O)^2	((E-O)^2)/E	SUM
2	251	111	0.042	10.459	49.000	-38.541	1485.396	142.018	
3		112	0.125	31.375	37.000	-5.625	31.641	1.008	143.027
4		121	0.208	52.291	36.000	16.291	265.391	5.075	148.102
5		122	0.125	31.375	25.000	6.375	40.641	1.295	149.398
6		211	0.125	31.375	37.000	-5.625	31.641	1.008	150.406
7		212	0.208	52.291	24.000	28.291	800.371	15.306	165.712
8		221	0.125	31.375	25.000	6.375	40.641	1.295	167.008
9		222	0.042	10.459	18.000	-7.541	56.864	5.437	
10									
11		n=		251.000	251.000			Chi-Square	172.444

Table 5.11a

X_1: 1–IBM 93–94

Mean	Std. Dev.	Std. Error	Variance	Coef. Var.	Count
–.5853	.5714	.047	.3265	–97.6305	148

Minimum	Maximum	Range	Sum	Sum of Sqr.	# Missing
–3.75	0	3.75	–86.625	98.7031	106

# < 10th %	10th %	25th %	50th %	75th %	90th %
14	–1.25	–.875	–.5	–.125	0

# > 90th %	Mode	Geo. Mean	Har. Mean	Kurtosis	Skewness
0	0	•	•	5.9453	–1.8981

each of these classes into 2 classes: 'S1' and 'S2' represent the price decreases, where 'S1' encodes the prices in the lower half of the price decreases and 'S2' encodes the prices in the upper half of the price

Table 5.11b

X_2: 2–IBM 93–94

Mean	Std. Dev.	Std. Error	Variance	Coef. Var.	Count
.906	.9073	.0885	.8233	100.1523	105

Minimum	Maximum	Range	Sum	Sum of Sqr.	# Missing
.125	6.125	6	95.125	171.7969	149

# < 10th %	10th %	25th %	50th %	75th %	90th %
0	.125	.375	.625	1.375	2

# > 90th %	Mode	Geo. Mean	Har. Mean	Kurtosis	Skewness
10	.125	.5893	.3787	9.498	2.4818

decreases. 'L1' and 'L2' represent the price increases, where 'L1' en-codes the lower half of the price increases and 'L2' encodes the upper half of the price increase. The digram transition matrix is shown in Table 5.12. Note that the S1S1, S1S2, S2S1, and S2S2 cells in this matrix represent the same prices as in the 1,1 cell in the RPC digram transition matrix shown in Table 5.9. The digrams that begin with 'S2' (that is, the higher half of the price decreases) occur more often than would be expected, while the digrams that begin with 'S1' (that is, the lower half of the price decreases) occur less often than ex-pected.

We can use the mean and standard deviations as the method for parsing the time series. For example, we can encode the prices that fall into the class <mean – standard deviation as a '1', the mean + and – standard deviation as a '2', and the >mean + standard deviation as a '3'. The summary statistics for the three classes are shown in Table 5.13 and the transition matrix is shown in Table 5.14. When the time series is parsed in this way, it does not contain serial dependencies.

Table 5.12

	A	B	C	D	E	F
1						
2	S	0.584		S1	0.341	
3	L	0.415		S2	0.242	
4				L1	0.246	
5				L2	0.171	
6						
7	PATTERN	OBSERVED	EXPECTED			
8	S1S1	20	29.349	-9.349	87.408	2.978
9	S1S2	19	20.817	-1.817	3.303	0.159
10	S1L1	19	21.159	-2.159	4.660	0.220
11	S1L2	19	14.675	4.325	18.709	1.275
12	S2S1	25	20.817	4.183	17.494	0.840
13	S2S2	22	14.766	7.234	52.333	3.544
14	S2L1	12	15.008	-3.008	9.048	0.603
15	S2LS2	11	10.409	0.591	0.350	0.034
16	L1S1	14	21.159	-7.159	51.247	2.422
17	L1S2	13	15.008	-2.008	4.032	0.269
18	L1L1	10	15.254	-5.254	27.604	1.810
19	L1L2	11	10.579	0.421	0.177	0.017
20	L2S1	19	14.675	4.325	18.709	1.275
21	L2S2	16	10.409	5.591	31.262	3.003
22	L2L1	6	10.579	-4.579	20.971	1.982
23	L2L2	16	7.337	8.663	75.042	10.228
24						
25		252	252			30.658
26						
27						
28				7.830		
29				5.477		
30						
31			Z	2.353		

Table 5.13a-b

X_1: 1–<M–SD IBM 93–94

Mean	Std. Dev.	Std. Error	Variance	Coef. Var.	Count
−1.56	.5898	.118	.3478	−37.8049	25

Minimum	Maximum	Range	Sum	Sum of Sqr.	# Missing
−3.75	−1.125	2.625	−39	69.1875	229

# < 10th %	10th %	25th %	50th %	75th %	90th %
2	−2.125	−1.875	−1.375	−1.125	−1.125

# > 90th %	Mode	Geo. Mean	Har. Mean	Kurtosis	Skewness
0	−1.125	•	•	5.5961	−2.1946

X_2: 2–SD–M+SD IBM 93–94

Mean	Std. Dev.	Std. Error	Variance	Coef. Var.	Count
−.0791	.4911	.0351	.2411	−620.9662	196

Minimum	Maximum	Range	Sum	Sum of Sqr.	# Missing
−1	1	2	−15.5	48.25	58

# < 10th %	10th %	25th %	50th %	75th %	90th %
13	−.75	−.5	−.125	.25	.625

# > 90th %	Mode	Geo. Mean	Har. Mean	Kurtosis	Skewness
12	0	•	•	−.7472	.1173

Table 5.13c

X_3: 3–>M+SD IBM 93–94

Mean	Std. Dev.	Std. Error	Variance	Coef. Var.	Count
1.9688	.9677	.1711	.9365	49.1543	32

Minimum	Maximum	Range	Sum	Sum of Sqr.	# Missing
1.125	6.125	5	63	153.0625	222

# < 10th %	10th %	25th %	50th %	75th %	90th %
0	1.125	1.375	1.75	2.1875	3.1625

# > 90th %	Mode	Geo. Mean	Har. Mean	Kurtosis	Skewness
3	1.375	1.8177	1.7109	8.7706	2.6856

Table 5.14

	A	B	C	D	E	F	G
1	Symbols	Frequency	Probability				
2	1	25	0.0992				
3	2	195	0.7738				
4	3	32	0.1270				
5							
6	Sum	252					
7							
8							
9							
10	Digrams	(E)xpected	(O)bserved	(E-O)	(E-O)^2	((E-O)^2)/E	Sum
11	11	2.48	1	1.48	2.191	0.883	
12	12	19.35	22	-2.65	7.048	0.364	1.248
13	13	3.17	2	1.17	1.380	0.435	1.682
14	21	19.35	20	-0.65	0.429	0.022	1.704
15	22	150.89	149	1.89	3.583	0.024	1.728
16	23	24.76	26	-1.24	1.533	0.062	1.790
17	31	3.17	4	-0.83	0.681	0.215	2.005
18	32	24.76	24	0.76	0.580	0.023	2.028
19	33	4.06	4	0.06	0.004	0.001	
20							
21	Sum=	252	252			Chi-Square	2.029
22							
23							
24							2.01451038
25							4
26						Z=	-1.98548962

"Ma" and the "Baby" Bells

6

The closing prices for AT & T for 1970 to 1993 are shown in Figure 6.1. The Delta-100, Delta-200, and Delta-500 price change cumulative probability density functions are shown in Figures 6.2 to 6.4, respectively. Fifty percent of the price change activity occurs within a band that is $2.635 wide (i.e., 25th percentile is $-0.815 to the 75th percentile which is $1.82) for the Delta-100 price changes and this increases to $4.86 for the Delta-500 price changes, as shown in Table 6.1 a-c. The frequency histogram for closing prices for 1985–1994 is shown in Figure 6.5.

The cumulative probability density functions for closing prices for 1985–1994 are shown in Figures 6.6 to 6.15 for AT & T, Ameritech, Pacific Telesis, Southwestern Bell, Bell Atlantic, Bellsouth, NYNEX, USWEST, MCI, and Sprint, respectively. If we examine these figures, we will note that they fall into five different shaped cumulative probability density functions. AT & T does not closely resemble any of the baby Bells or their competitors, Sprint or MCI. Bellsouth, Bell Atlantic, and Southwestern Bell all resemble each other more than they resemble the other companies. NYNEX and USWEST also resemble each other more than they do the others, while SPRINT and MCI resemble each other more than the others. Pacific Telesis and

Figure 6.1

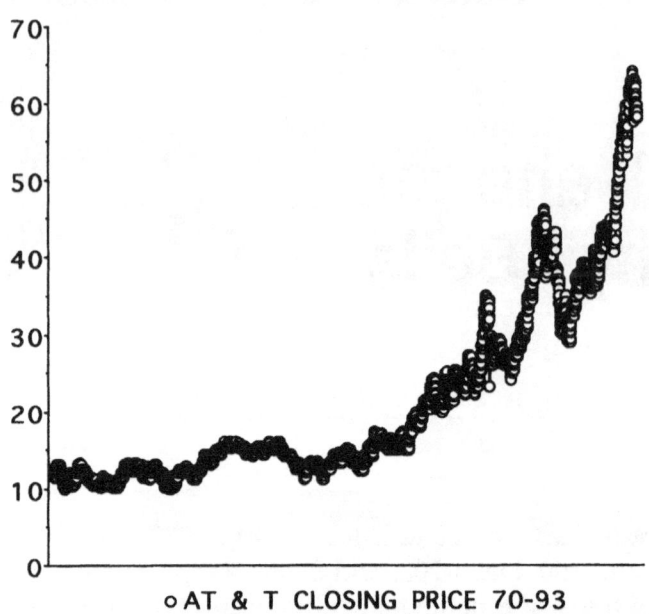

○ AT & T CLOSING PRICE 70-93

Figure 6.2

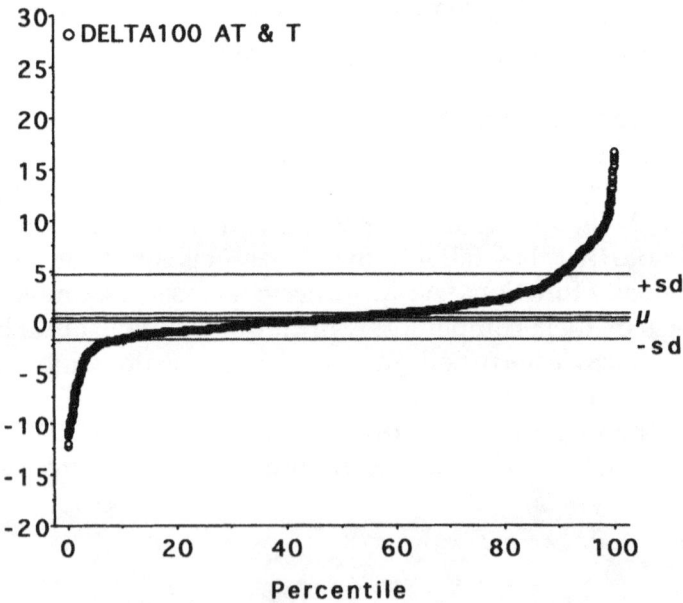

Table 6.1a-b

X_1: Delta-100 AT&T

Mean	Std. Dev.	Std. Error	Variance	Coef. Var.	Count
.8296	3.0669	.0399	9.4056	369.6748	5903

Minimum	Maximum	Range	Sum	Sum of Sqr.	# Missing
−12.24	16.56	28.8	4897.18	59574.5772	100

# < 10th %	10th %	25th %	50th %	75th %	90th %
588	−1.72	−.815	.32	1.82	4.56

# > 90th %	Mode	Geo. Mean	Har. Mean	Kurtosis	Skewness
589	0	•	•	4.0443	.8499

X_2: Delta-200 AT&T

Mean	Std. Dev.	Std. Error	Variance	Coef. Var.	Count
1.602	4.552	.0598	20.7205	284.1518	5803

Minimum	Maximum	Range	Sum	Sum of Sqr.	# Missing
−15.84	22.92	38.76	9296.14	135112.5516	200

# < 10th %	10th %	25th %	50th %	75th %	90th %
579	−1.82	−.62	.8	2.92	6.4

# > 90th %	Mode	Geo. Mean	Har. Mean	Kurtosis	Skewness
579	−.08	•	•	3.8475	.9497

Ameritech are also similar to each other. This similarity is confirmed by the frequency histograms for closing prices shown in Figures 6.16 and 6.17. The cumulative probability density functions for closing prices 1993–1994 for Half1 and Half2 for each of these time series are shown in Figures 6.18 to 6.27. With the possible exception of NYNEX (Figure 6.24) and MCI (Figure 6.26), these cumulative probability

Table 6.1c

X_3: Delta-500 AT&T

Mean	Std. Dev.	Std. Error	Variance	Coef. Var.	Count
3.3249	5.5245	.0745	30.5205	166.1566	5503

Minimum	Maximum	Range	Sum	Sum of Sqr.	# Missing
−9.4	26	35.4	18296.92	228759.3384	500

# < 10th %	10th %	25th %	50th %	75th %	90th %
550	−1.68	−.04	1.96	4.82	9.92

# > 90th %	Mode	Geo. Mean	Har. Mean	Kurtosis	Skewness
549	2	•	•	3.6906	1.7666

density functions either overlap or are essentially parallel, which suggests that the underlying time series are stationary.

The differential spectrums for each of these time series are shown in Figures 6.28 to 6.37. They all appear to be asymmetrical and this implies that each of these time series contains serial dependencies.

Figure 6.3

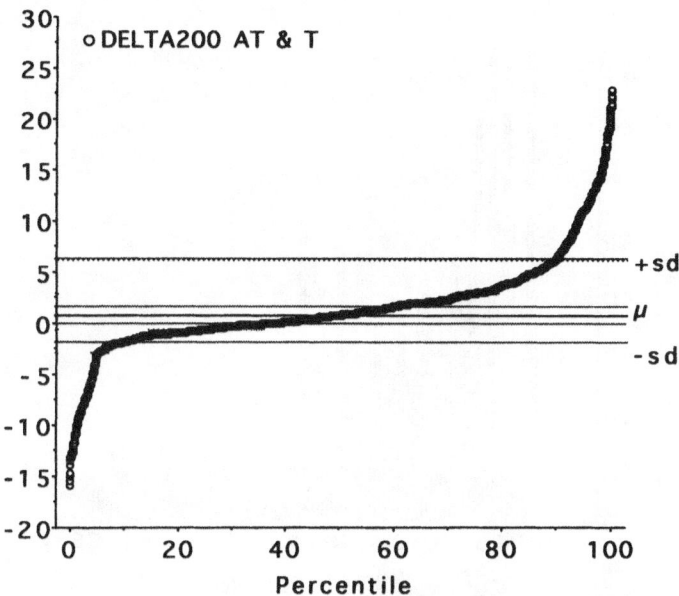

Figure 6.4

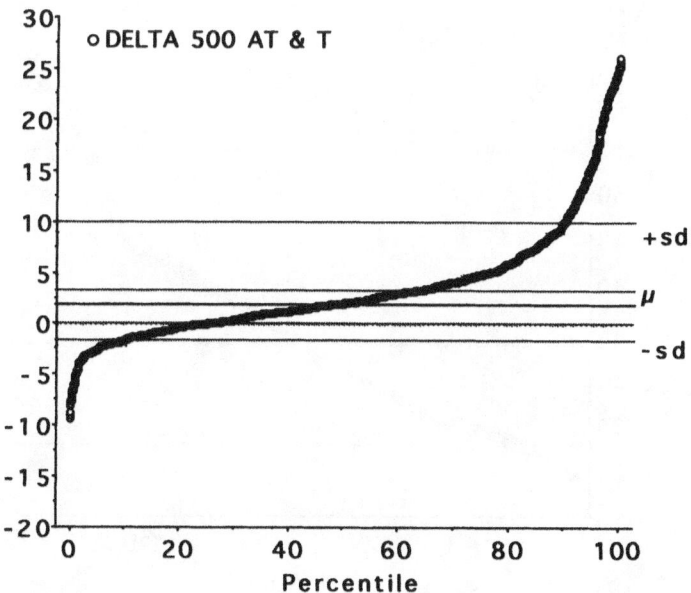

Figure 6.5

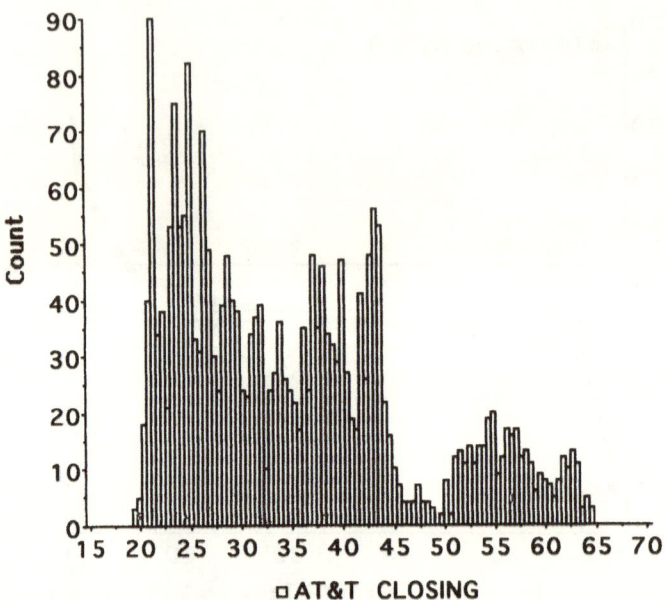

Figure 6.6

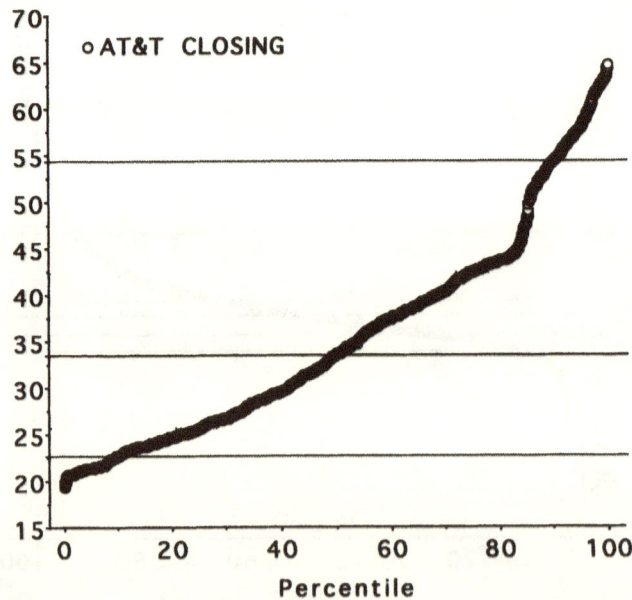

Figure 6.7

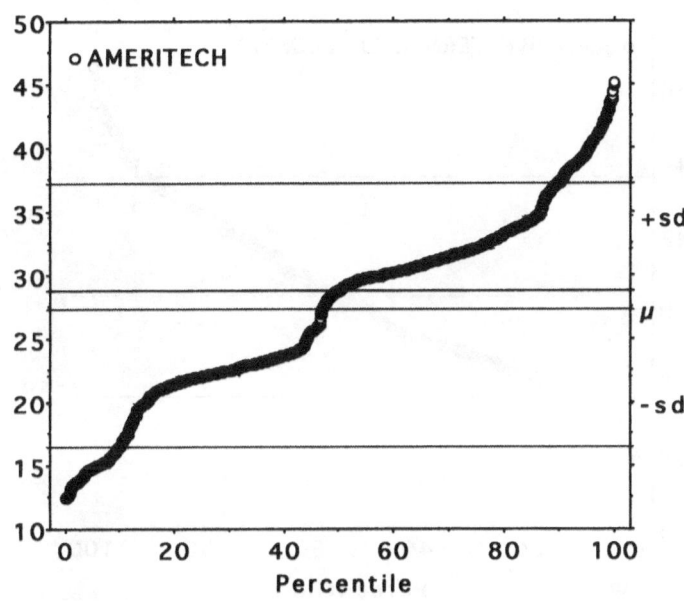

Figure 6.8

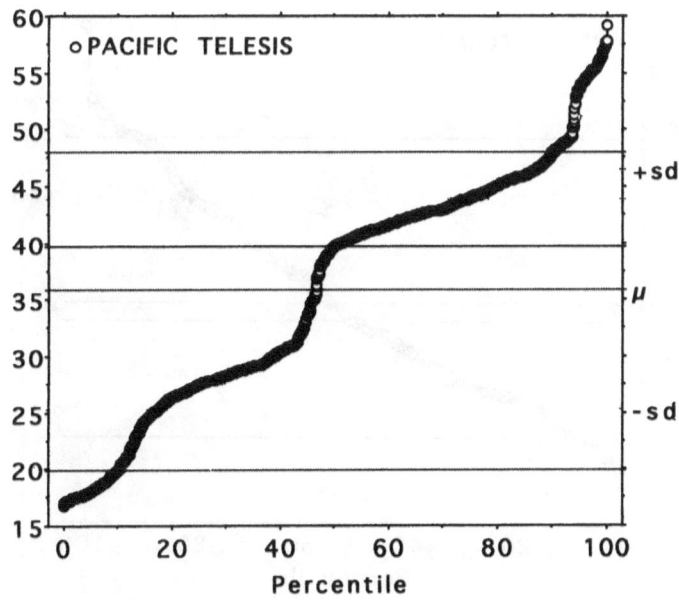

Figure 6.9

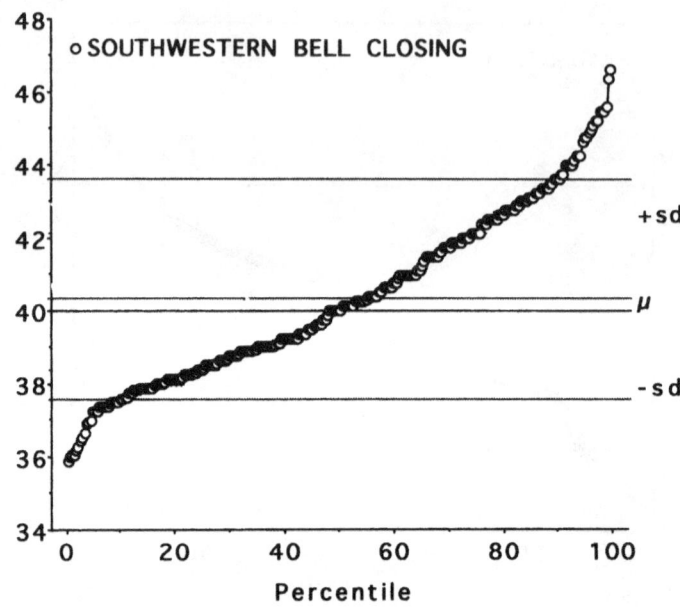

Percentile

Figure 6.10

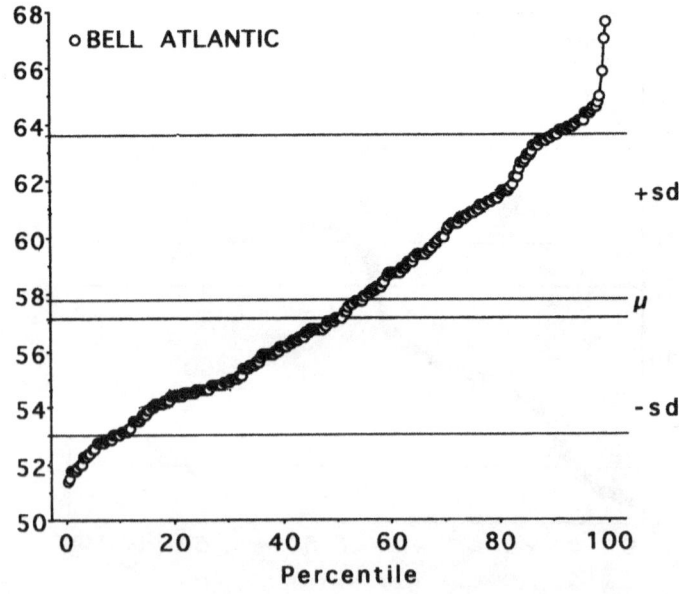

Percentile

Figure 6.11

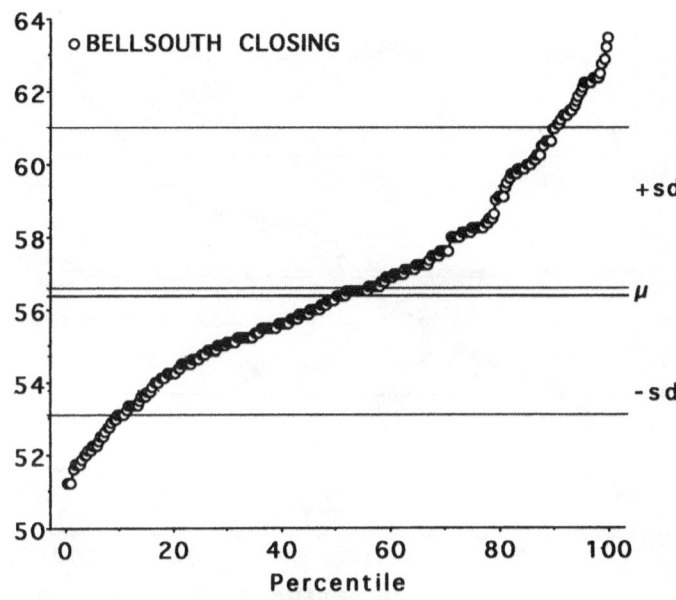

Figure 6.12

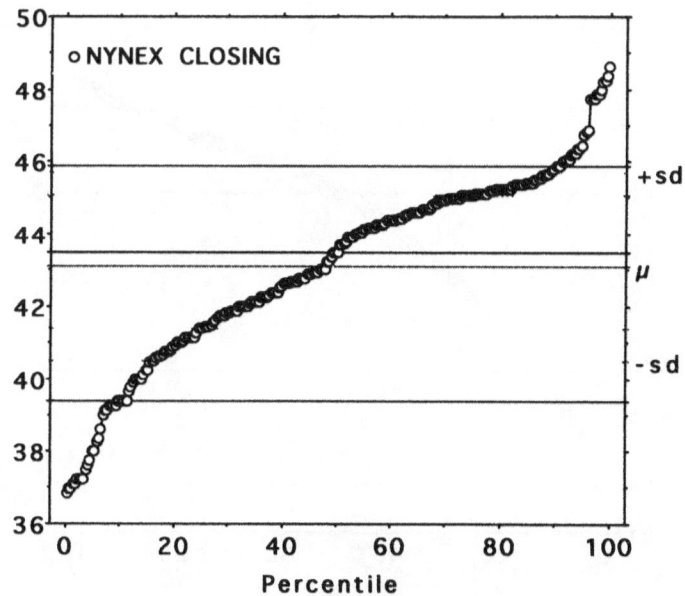

Figure 6.13

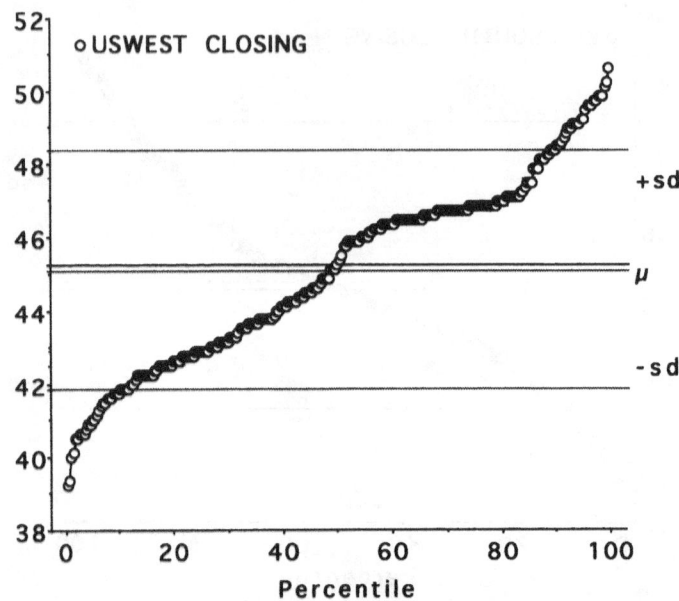

Figure 6.14

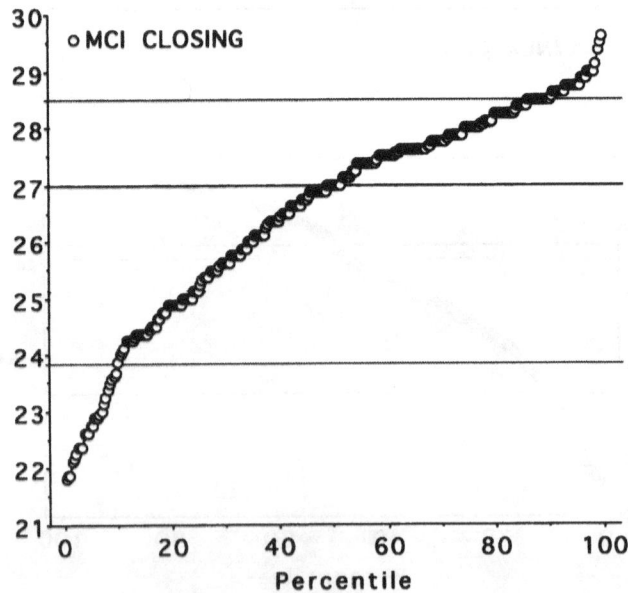

Figure 6.15

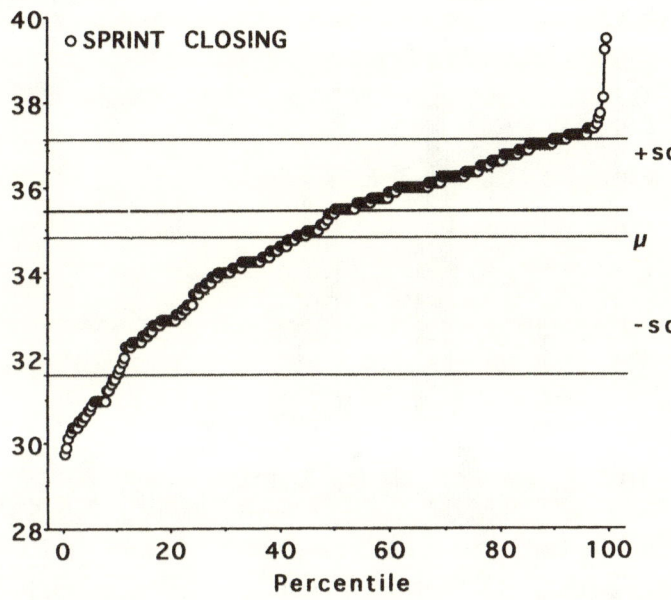

Figure 6.16

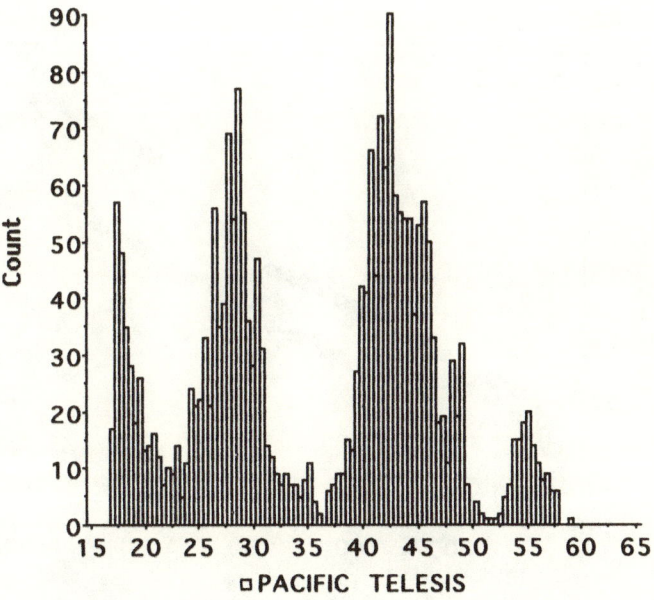

Figure 6.17

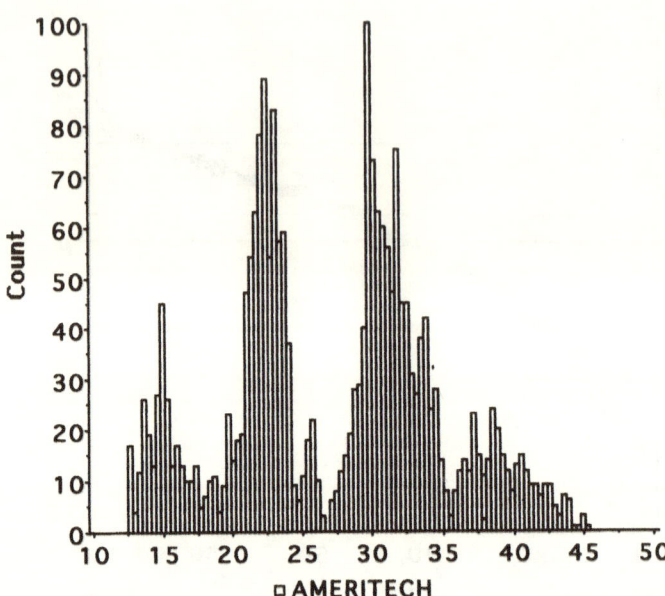

Figure 6.18

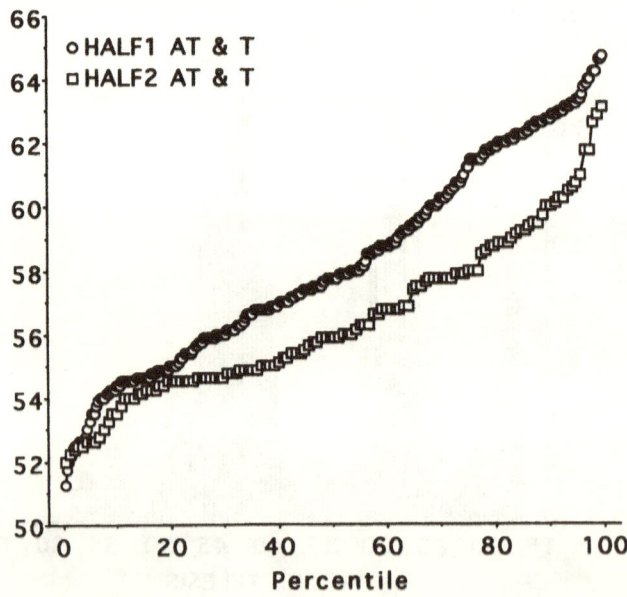

Figure 6.19

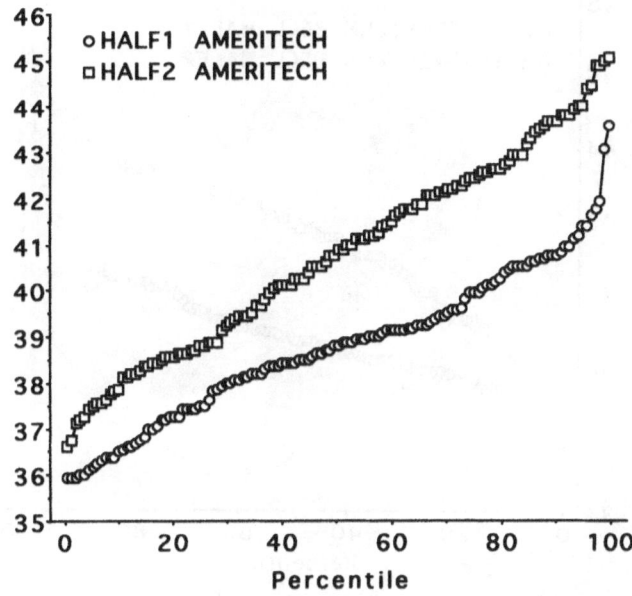

Figure 6.20

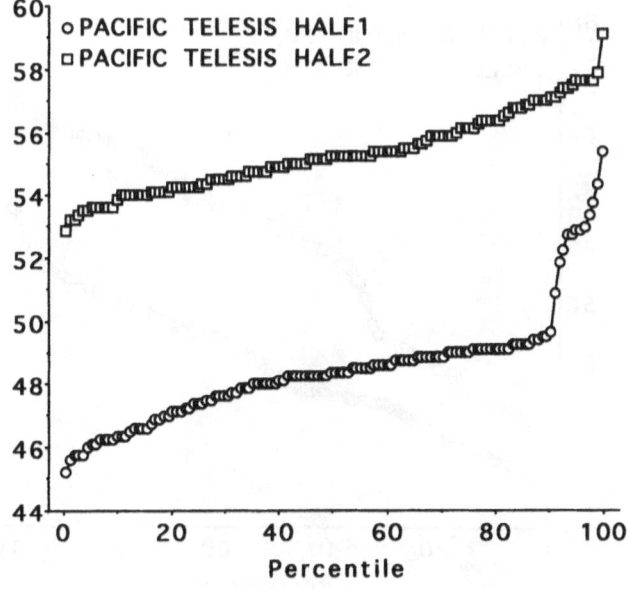

Figure 6.21

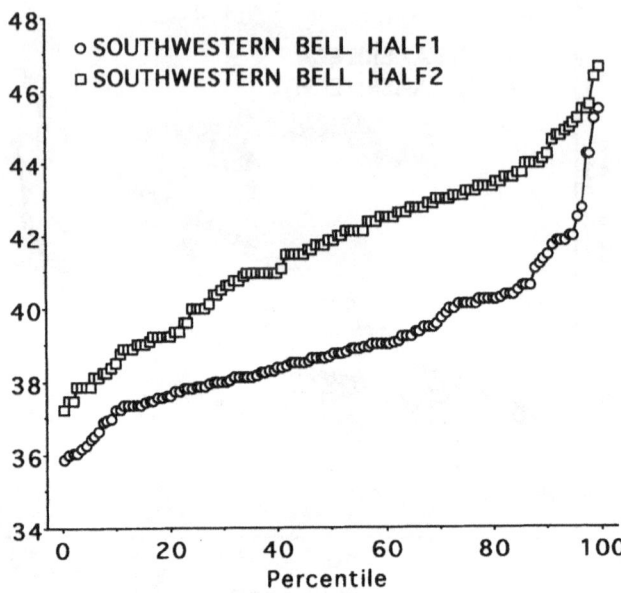

Figure 6.22

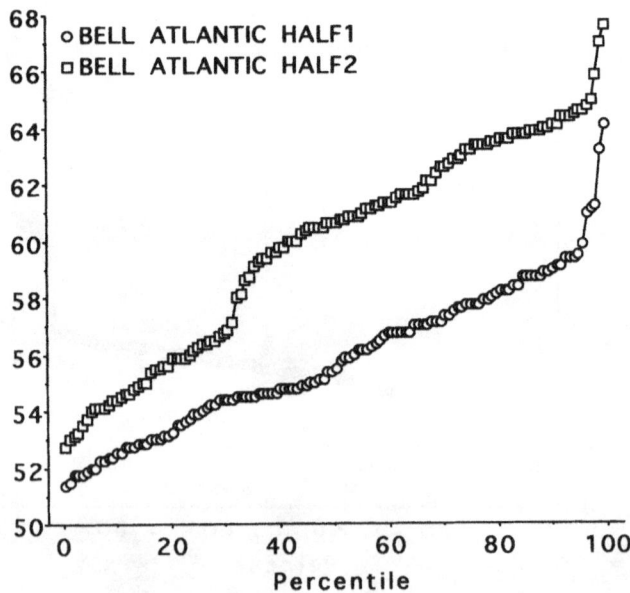

Figure 6.23

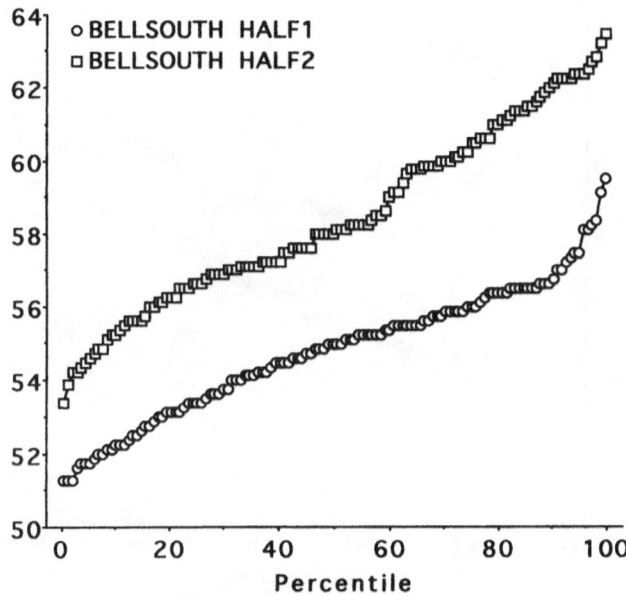

Figure 6.24

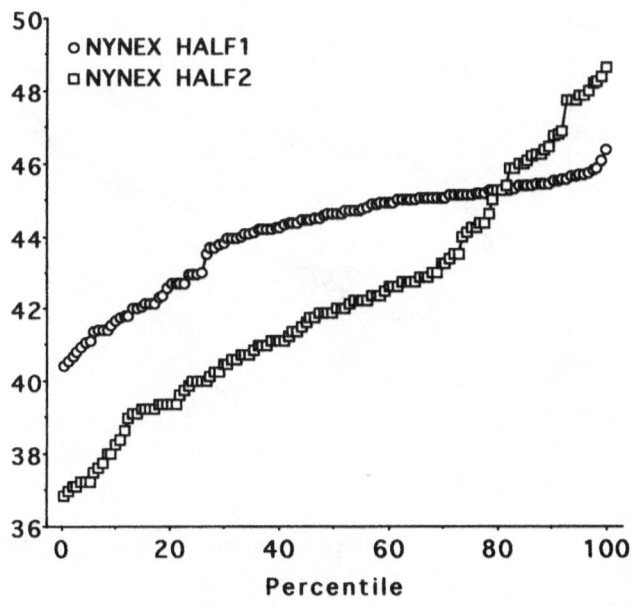

Figure 6.25

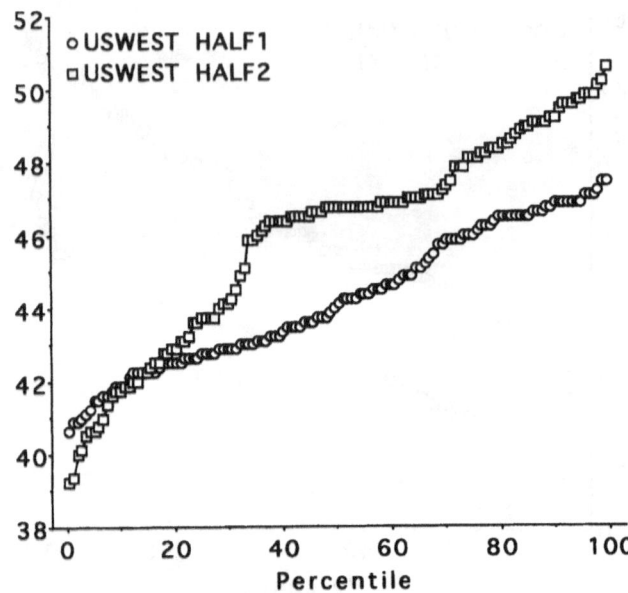

Figure 6.26

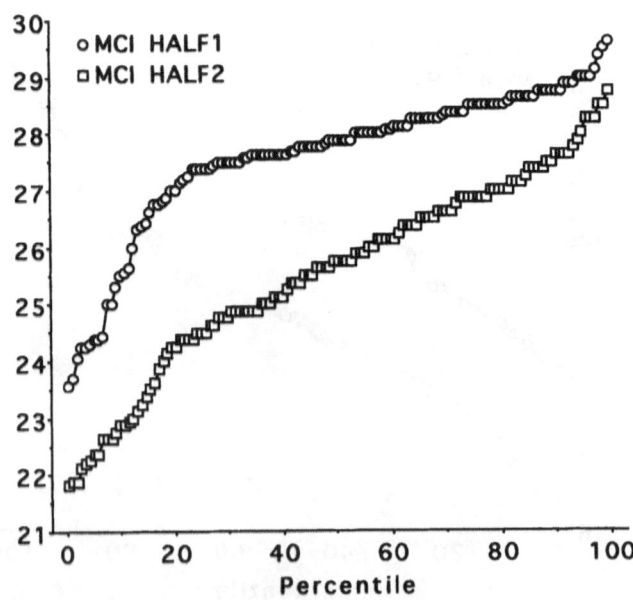

Figure 6.27

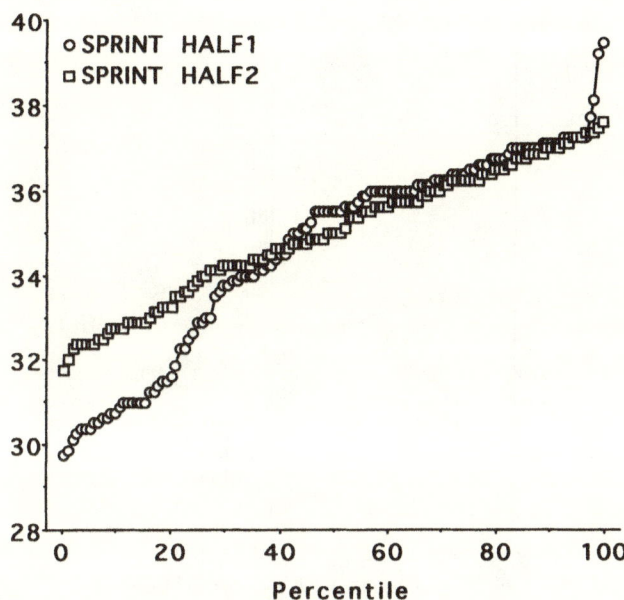

Figure 6.28

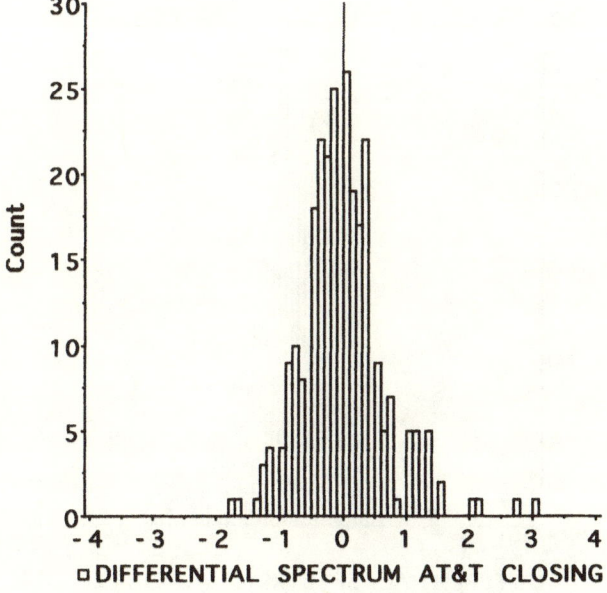

Figure 6.29

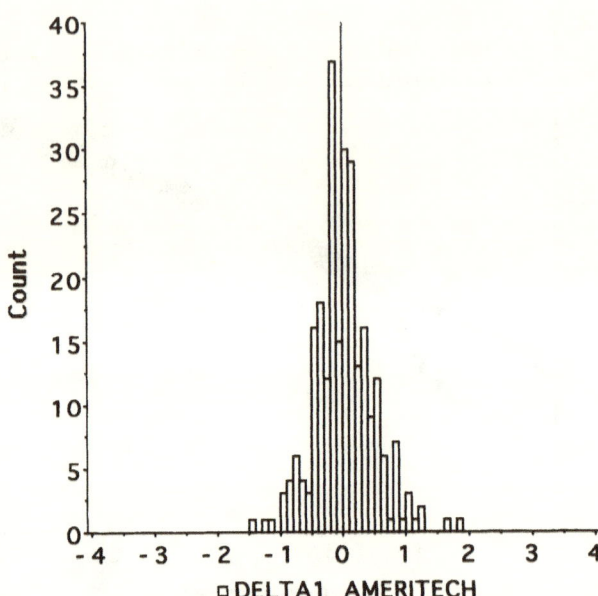

Figure 6.30

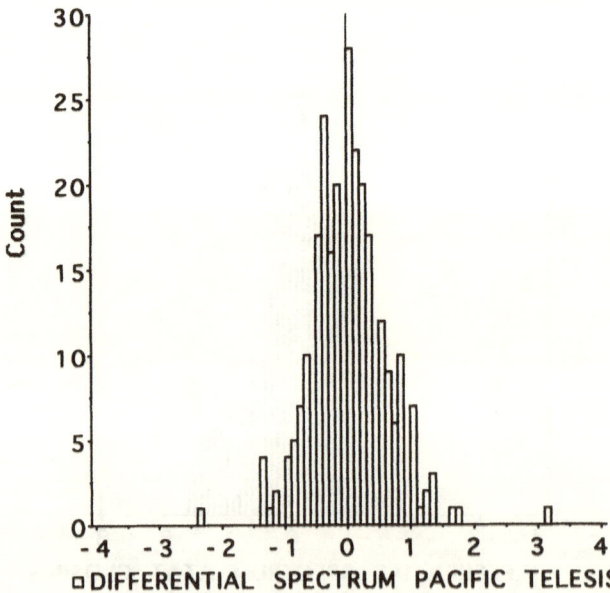

Figure 6.31

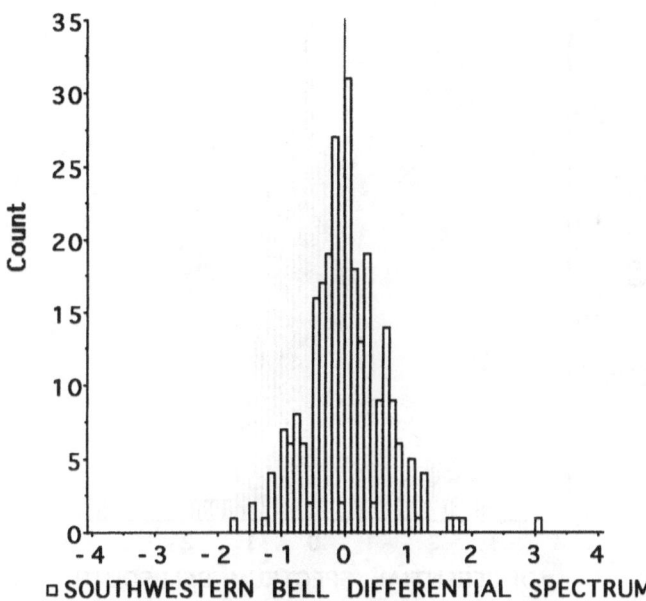

Figure 6.32

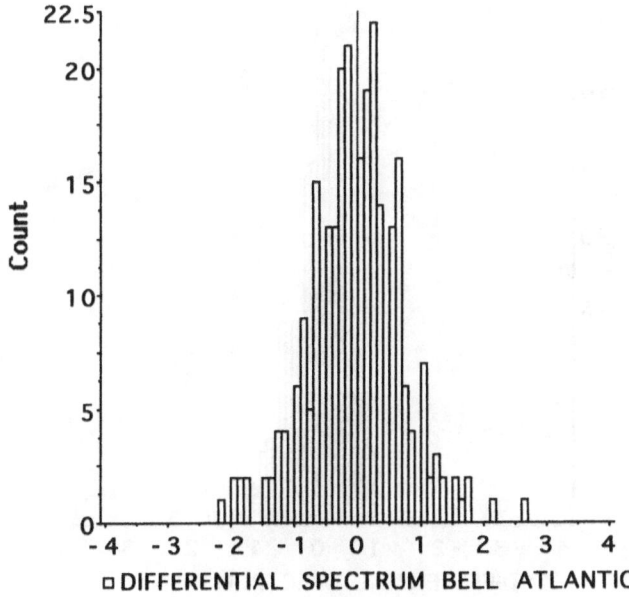

Figure 6.33

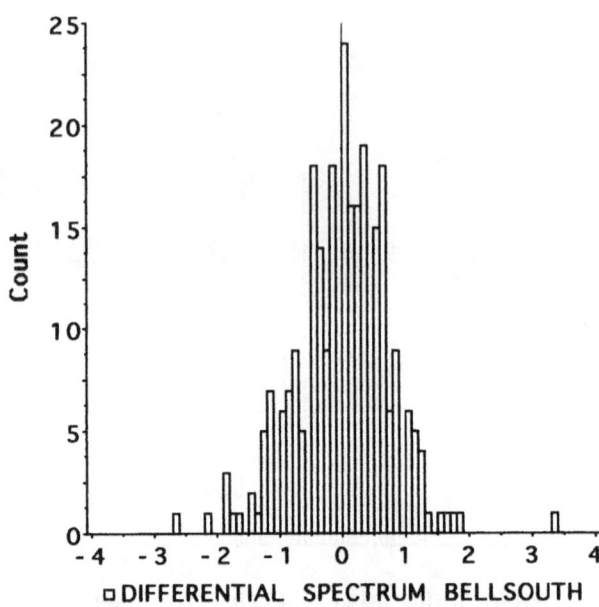

Figure 6.34

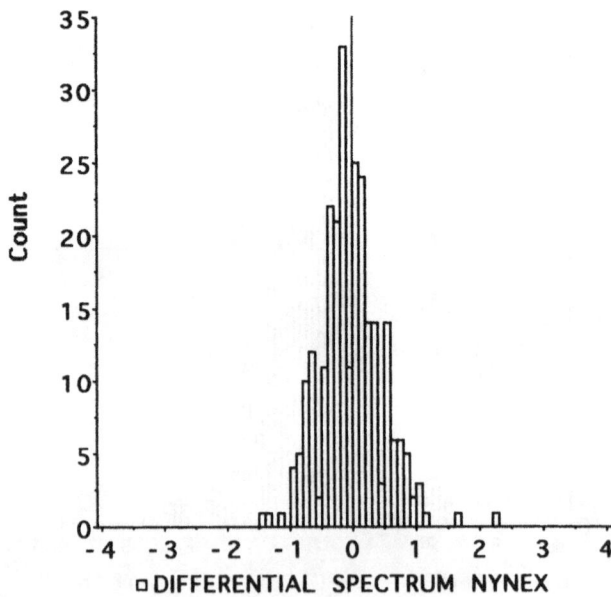

Figure 6.35

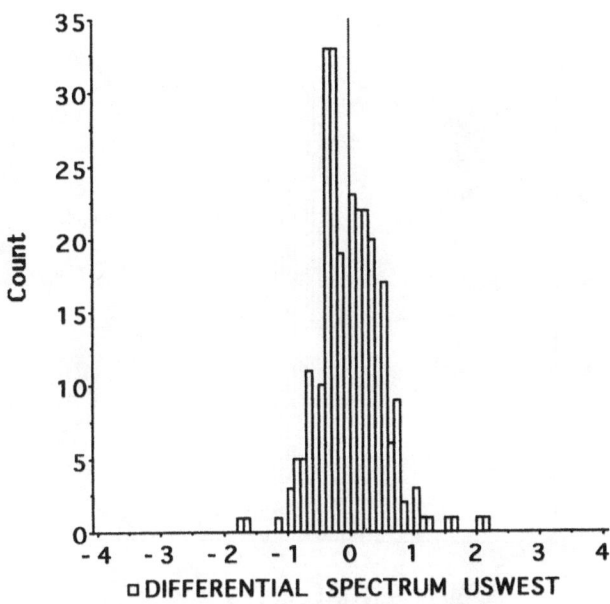

Figure 6.36

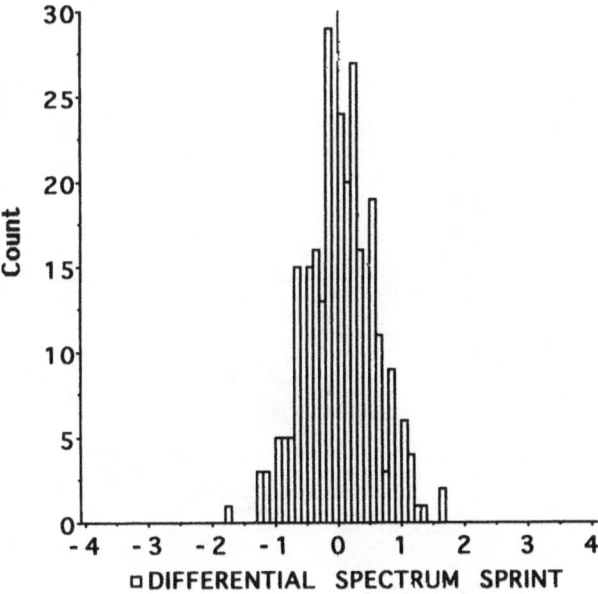

Figure 6.37

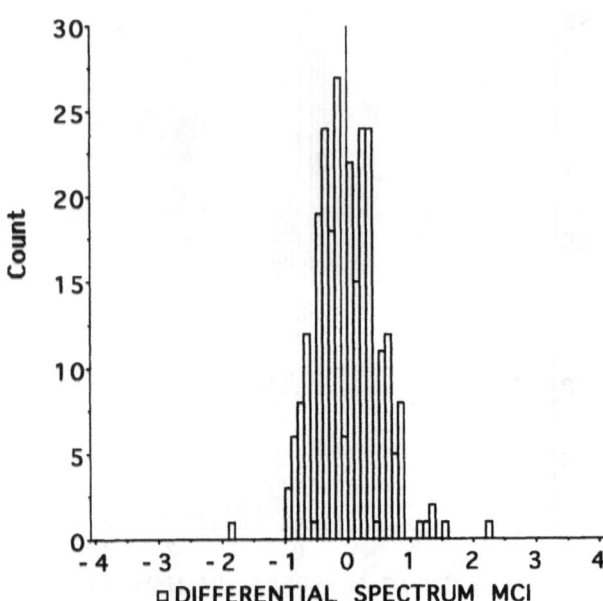

Futures

7

Grains

The weekly closing prices of soybeans for 09/04/81 to 05/13/94 are shown in Figure 7.1 and the cumulative probability density function is shown in Figure 7.2. The summary statistics for closing prices are shown in Table 7.1, while the Delta-1 price changes are shown in Table 7.2. The cumulative probability density functions for Half1 and Half2 for prices (Figure 7.3, summary statistics Table 7.3 a-b) and Delta-1 price changes (Figure 7.4 and Table 7.4 a-b) suggest that both prices and price changes are stationary. The differential spectrum, with a bin size of $0.01 is shown in Figure 7.5 is asymmetrical, which suggests that the soybean time series contains serial dependencies and this is confirmed by the fact that the RPC digrams diverge from independence, as shown in Table 7.5. The price increases and decreases are summarized in Table 7.6 a-b.

The closing prices and Delta-1 price changes for soybean meal also appear to be stationary (see Figures 7.6 and 7.7 and summary Tables 7.7 a-b and 7.8 a-b) and contain serial dependencies, as shown by the asymmetrical differential spectrum (Figure 7.8) and the significant Chi-Square value for the digram transition matrix (Table 7.9). The price increases and decreases are summarized in Table 7.10 a-b.

On the other hand, closing prices for soybean oil may not be stationary because the cumulative probability density functions cross

Figure 7.1

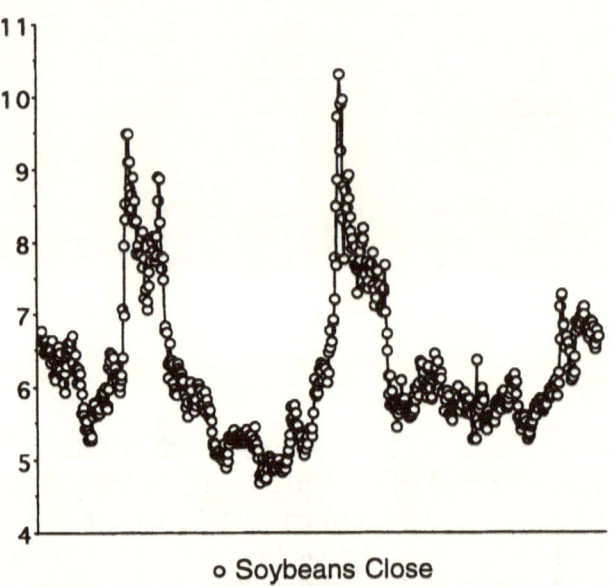

○ Soybeans Close

Figure 7.2

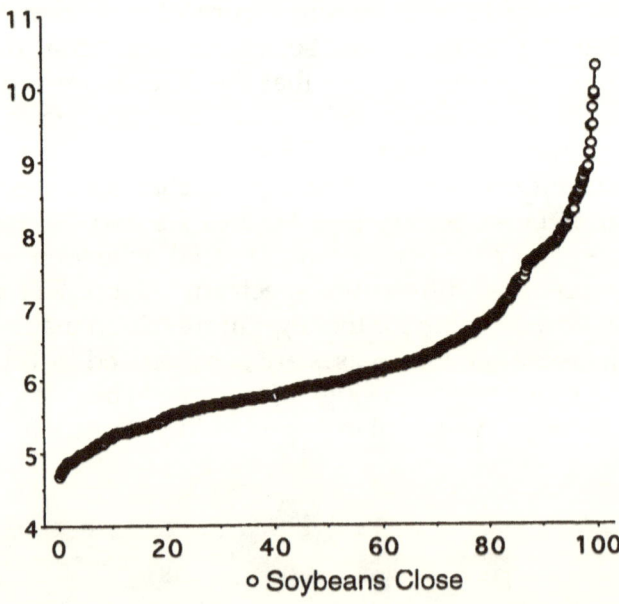

○ Soybeans Close

Table 7.1

X_1: Soybeans Close

Mean	Std. Dev.	Std. Error	Variance	Coef. Var.	Count
6.2	1	3.8E–2	1	15.7	664

Minimum	Maximum	Range	Sum	Sum of Sqr.	# Missing
4.7	10.3	5.6	4118	26169.7	0

# < 10th %	10th %	25th %	50th %	75th %	90th %
66	5.2	5.6	5.9	6.6	7.7

# > 90th %	Mode	Geo. Mean	Har. Mean	Kurtosis	Skewness
65	•	6.1	6.1	1.5	1.3

Table 7.2

X_1: Delta-1 Soybeans Close

Mean	Std. Dev.	Std. Error	Variance	Coef. Var.	Count
–1.2E–4	.2	8.3E–3	4.5E–2	–178599.6	663

Minimum	Maximum	Range	Sum	Sum of Sqr.	# Missing
–1.6	1.2	2.8	–.1	30	1

# < 10th %	10th %	25th %	50th %	75th %	90th %
65	–.2	–.1	5.0E–3	.1	.2

# > 90th %	Mode	Geo. Mean	Har. Mean	Kurtosis	Skewness
66	•	•	•	9.2	–.3

(Figure 7.9, summary Table 7.11 a-b). The Delta-1 price changes do appear to be stationary (Figure 7.10 and summary Table 7.12 a-b) and the time series does appear to contain serial dependencies as shown

Figure 7.3

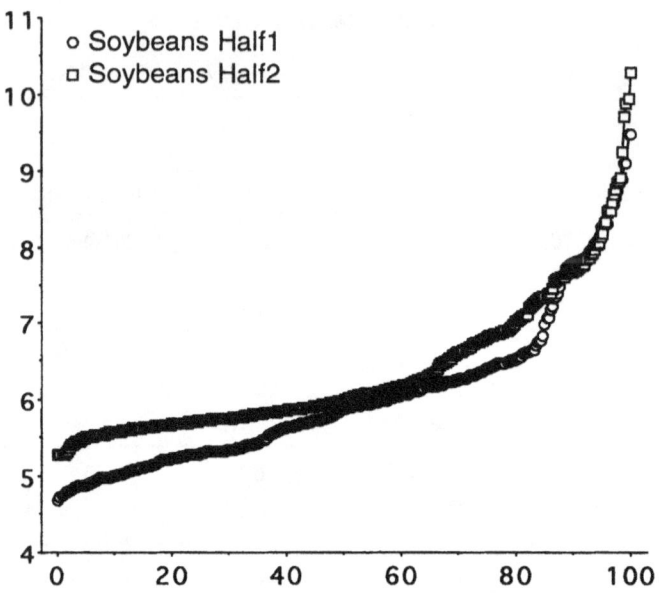

Table 7.3a

X_1: Soybeans Half1

Mean	Std. Dev.	Std. Error	Variance	Coef. Var.	Count
6	1	.1	1	16.7	332

Minimum	Maximum	Range	Sum	Sum of Sqr.	# Missing
4.7	9.5	4.8	2006.7	12465.9	0

# < 10th %	10th %	25th %	50th %	75th %	90th %
32	5	5.3	5.9	6.4	7.8

# > 90th %	Mode	Geo. Mean	Har. Mean	Kurtosis	Skewness
33	5.3	6	5.9	1.1	1.2

Table 7.3b

X₂: *Soybeans Half2*

Mean	Std. Dev.	Std. Error	Variance	Coef. Var.	Count
6.4	.9	.1	.8	14.4	332

Minimum	Maximum	Range	Sum	Sum of Sqr.	# Missing
5.3	10.3	5	2111.3	13703.8	0

# < 10th %	10th %	25th %	50th %	75th %	90th %
32	5.6	5.7	6	6.8	7.7

# > 90th %	Mode	Geo. Mean	Har. Mean	Kurtosis	Skewness
33	•	6.3	6.2	2.4	1.5

Figure 7.4

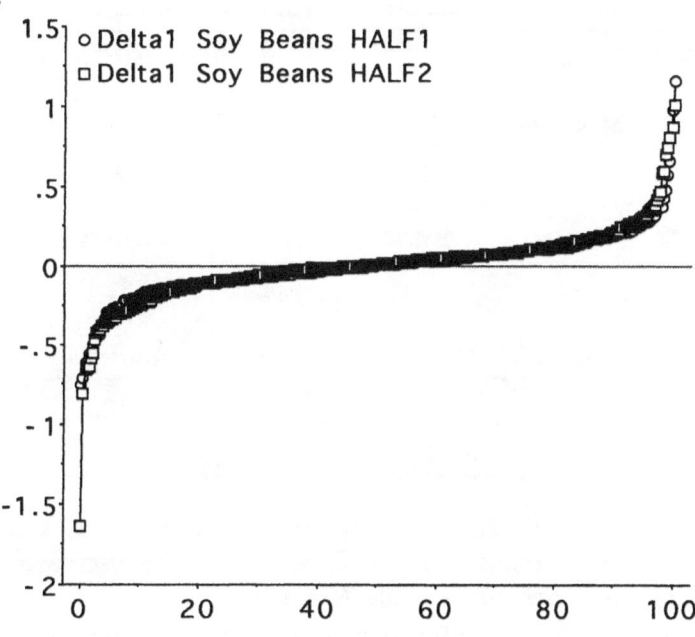

Table 7.4a-b

X_1: *Delta-1 Soybeans Half1*

Mean	Std. Dev.	Std. Error	Variance	Coef. Var.	Count
−1.6E−3	.2	1.1E−2	3.8E−2	−11993.5	331

Minimum	Maximum	Range	Sum	Sum of Sqr.	# Missing
−.8	1.2	1.9	−.5	12.5	1

# < 10th %	10th %	25th %	50th %	75th %	90th %
32	−.2	−.1	2.0E−3	.1	.2

# > 90th %	Mode	Geo. Mean	Har. Mean	Kurtosis	Skewness
33	•	•	•	6.8	.5

X_2: *Delta-1 Soybeans Half2*

Mean	Std. Dev.	Std. Error	Variance	Coef. Var.	Count
1.4E−3	.2	1.3E−2	.1	16602.5	332

Minimum	Maximum	Range	Sum	Sum of Sqr.	# Missing
−1.6	1	2.6	.5	17.4	0

# < 10th %	10th %	25th %	50th %	75th %	90th %
33	−.2	−.1	9.0E−3	.1	.2

# > 90th %	Mode	Geo. Mean	Har. Mean	Kurtosis	Skewness
33	•	•	•	10	−.7

by the asymmetrical differential spectrum (Figure 7.11) and the RPC digram transition matrix (Table 7.13). The price increases and decreases are summarized in Table 7.14 a-b.

Corn (Figure 7.12) and wheat (Figure 7.13) also appear to be stationary and contain serial dependencies (corn—Figure 7.14 and Table 7.15; wheat—Figure 7.15, Table 7.16).

Figure 7.5

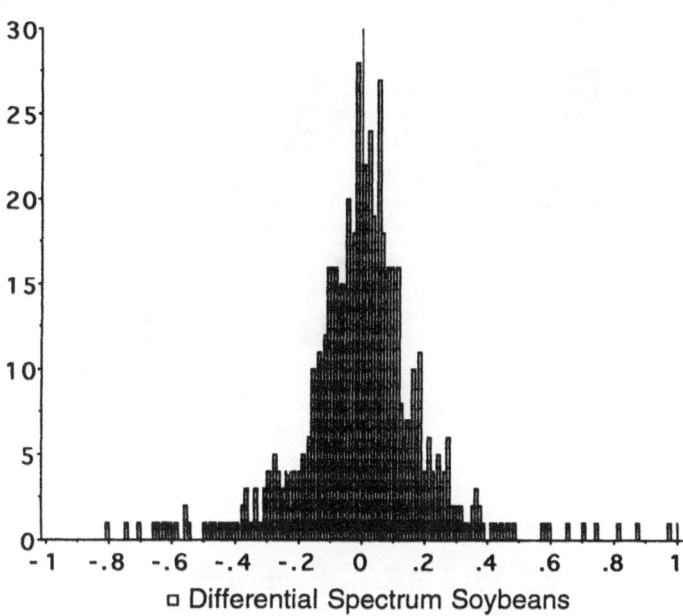

□ Differential Spectrum Soybeans

Table 7.5

	A	B	C	D	E	F	G	H	I
1	n	Digram	Expec Prob	(E)xpec Freq	(O)bser Freq	(E-O)	(E-O)^2	((E-O)^2)/E	SUM
2	662	11	0.1667	110.3554	153	-42.6446	1818.56191	16.4791384	
3		12	0.3333	220.6446	170	50.6446	2564.87551	11.6244654	28.1036038
4		21	0.3333	220.6446	169	51.6446	2667.16471	12.0880579	40.1916617
5		22	0.1667	110.3554	170	-59.6446	3557.47831	32.2365585	52.2797196
6									
7	n=			662	662				
8								Chi-Square	84.5162781

Currencies

The cumulative probability density functions for the closing prices of Deutschenmark, Japanese Yen, Swiss Franc, and British Pound are shown in Figures 7.16 to 7.19, respectively. The cumulative probability density functions for Half1 and Half2 for prices (Figures 7.20 to 7.22) and Delta-1 price changes (Figures 7.23 to 7.25) are stationary and the differential spectrums are asymmetrical (see Figures 7.26 to 7.28), which suggests that the time series contain serial dependencies. This is confirmed by the significant RPC digram Chi-Square values (see Tables 7.17 to 7.19).

Table 7.6a-b

X_1: 1 – RPC Soybeans

Mean	Std. Dev.	Std. Error	Variance	Coef. Var.	Count
–.1	.2	9.3E–3	2.8E–2	–116.5	323

Minimum	Maximum	Range	Sum	Sum of Sqr.	# Missing
–1.6	0	1.6	–46.1	15.5	17

# < 10th %	10th %	25th %	50th %	75th %	90th %
32	–.3	–.2	–.1	–4.3E–2	–1.6E–2

# > 90th %	Mode	Geo. Mean	Har. Mean	Kurtosis	Skewness
30	–1.8E–2	•	•	21.4	–3.5

X_2: 2 – RPC Soybeans

Mean	Std. Dev.	Std. Error	Variance	Coef. Var.	Count
.1	.2	8.5E–3	2.4E–2	115.2	340

Minimum	Maximum	Range	Sum	Sum of Sqr.	# Missing
2.0E–3	1.2	1.2	46	14.5	0

# < 10th %	10th %	25th %	50th %		75th %
33	2.0E–2	4.4E–2	.1	.2	.3

# > 90th %	Mode	Geo. Mean	Har. Mean	Kurtosis	Skewness
34	3.0E–2	.1	3.6E–2	12.9	3.1

Gold and Black Gold

The cumulative probability density functions for Comex Gold and Crude Oil Light are shown in Figures 7.29 and 7.30. The cumulative probability density functions for Half1 and Half2 for prices (Figures 7.31 and 7.32) and Delta-1 price changes (Figures 7.33 and 7.34) sug-

Figure 7.6

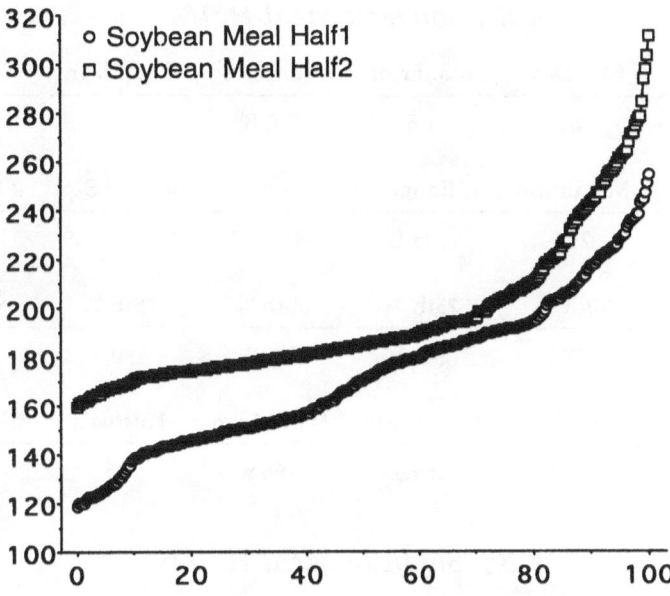

Figure 7.7

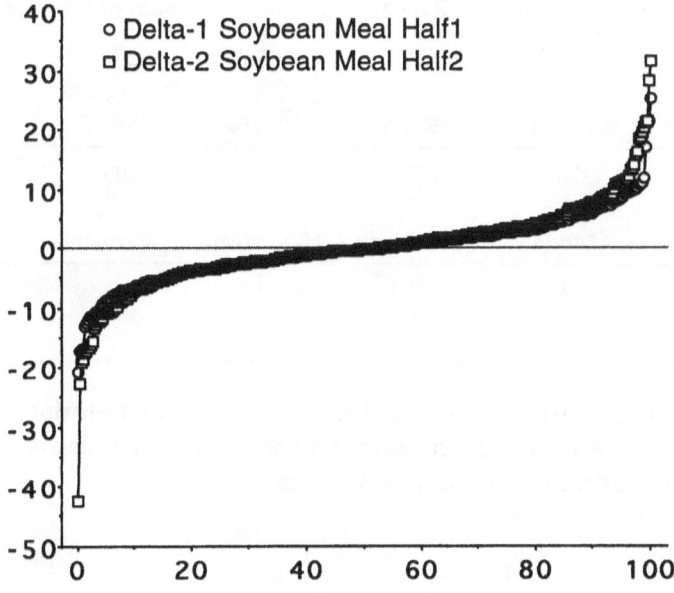

Table 7.7a-b

X_1: *Soybean Meal Half1*

Mean	Std. Dev.	Std. Error	Variance	Coef. Var.	Count
172	30.2	1.6	909.8	17.5	349

Minimum	Maximum	Range	Sum	Sum of Sqr.	# Missing
118.9	254	135.1	60013.7	10636524.9	0

# < 10th %	10th %	25th %	50th %	75th %	90th %
35	138.2	148.1	170.2	190.9	216.9

# > 90th %	Mode	Geo. Mean	Har. Mean	Kurtosis	Skewness
35	•	169.4	166.9	–.5	.4

X_2: *Soybean Meal Half2*

Mean	Std. Dev.	Std. Error	Variance	Coef. Var.	Count
195	29.2	1.6	852.8	15	349

Minimum	Maximum	Range	Sum	Sum of Sqr.	# Missing
159.4	310.7	151.3	68065.1	13571439	0

# < 10th %	10th %	25th %	50th %	75th %	90th %
35	170.3	175.7	184.8	203.7	243.4

# > 90th %	Mode	Geo. Mean	Har. Mean	Kurtosis	Skewness
35	•	193.1	191.4	2	1.6

gest that these time series are stationary. The differential spectrums (Figures 7.35 and 7.36) are asymmetrical, which indicates that the time series contain serial dependencies.

Table 7.8a-b

X_1: *Delta-1 Soybean Meal Half1*

Mean	Std. Dev.	Std. Error	Variance	Coef. Var.	Count
–.2	5.5	.3	30.2	–2378.7	348

Minimum	Maximum	Range	Sum	Sum of Sqr.	# Missing
–20.7	25.2	45.9	–80.4	10498.7	1

# < 10th %	10th %	25th %	50th %	75th %	90th %
35	–6.4	–3.3	–.2	2.5	5.9

# > 90th %	Mode	Geo. Mean	Har. Mean	Kurtosis	Skewness
35	•	•	•	2.9	.3

X_2: *Delta-1 Soybean Meal Half2*

Mean	Std. Dev.	Std. Error	Variance	Coef. Var.	Count
.1	7.1	.4	51	12649.6	349

Minimum	Maximum	Range	Sum	Sum of Sqr.	# Missing
–42.3	31.5	73.8	19.7	17743.7	0

# < 10th %	10th %	25th %	50th %	75th %	90th %
35	–7	–3	–.3	3.3	7.7

# > 90th %	Mode	Geo. Mean	Har. Mean	Kurtosis	Skewness
35	–2	•	•	5.4	–.1

Figure 7.8

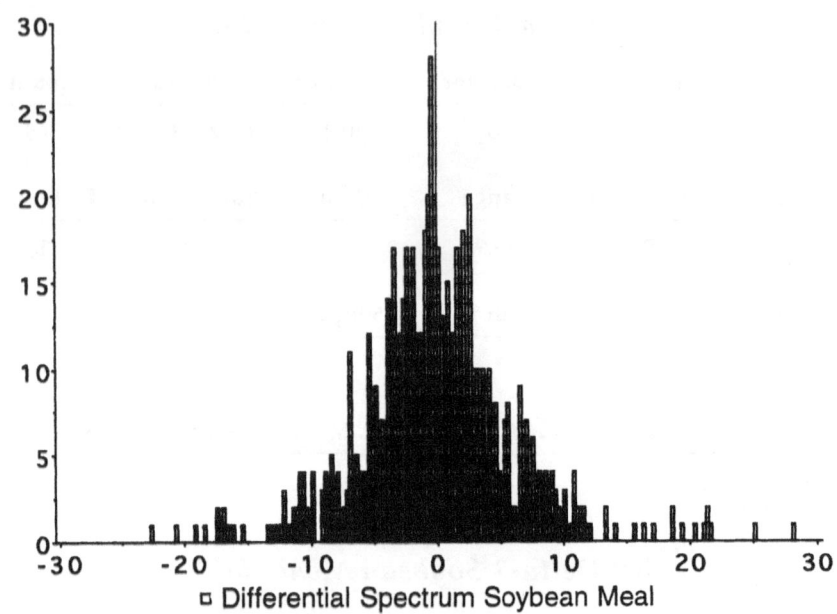

□ Differential Spectrum Soybean Meal

Table 7.9

	A	B	C	D	E	F	G	H	I
1	n	Digram	Expec Prob	(E)xpec Freq	(O)bser Freq	(E-O)	(E-O)^2	((E-O)^2)/E	SUM
2	696	11	0.1667	116.0232	202	-85.9768	7392.01014	63.711483	
3		12	0.3333	231.9768	173	58.9768	3478.26294	14.9940121	78.7054951
4		21	0.3333	231.9768	173	58.9768	3478.26294	14.9940121	93.6995072
5		22	0.1667	116.0232	148	-31.9768	1022.51574	8.81302824	108.693519
6									
7	n=			696	696				
8								Chi-Square	117.506547

Table 7.10a-b

X_1: 1 RPC Soybean Meal

Mean	Std. Dev.	Std. Error	Variance	Coef. Var.	Count
–4.1	4.5	.2	20.3	–108.8	376

Minimum	Maximum	Range	Sum	Sum of Sqr.	# Missing
–42.3	0	42.3	–1556.4	14049.2	0

# < 10th %	10th %	25th %	50th %	75th %	90th %
38	–9.9	–5.4	–2.8	–1.1	–.4

# > 90th %	Mode	Geo. Mean	Har. Mean	Kurtosis	Skewness
36	•	•	•	14.8	–2.8

X_2: 2 RPC Soybean Meal

Mean	Std. Dev.	Std. Error	Variance	Coef. Var.	Count
4.7	4.8	.3	22.6	102	321

Minimum	Maximum	Range	Sum	Sum of Sqr.	# Missing
.1	31.5	31.4	1495.7	14193.1	55

# < 10th %	10th %	25th %	50th %	75th %	90th %
31	.6	1.6	3.1	6.6	10

# > 90th %	Mode	Geo. Mean	Har. Mean	Kurtosis	Skewness
31	2.7	2.8	1.3	7	2.3

Figure 7.9

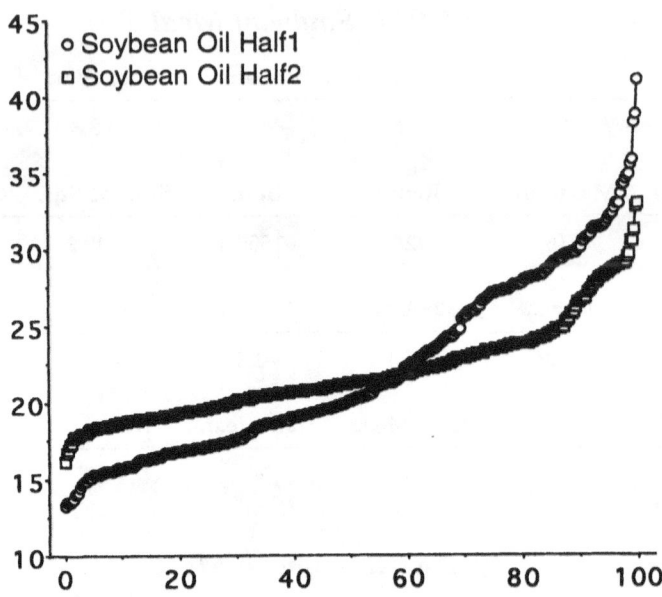

Table 7.11a

X₁: Soybean Oil Half1

Mean	Std. Dev.	Std. Error	Variance	Coef. Var.	Count
22	5.8	.3	33.1	26.2	348

Minimum	Maximum	Range	Sum	Sum of Sqr.	# Missing
13.3	41.1	27.7	7648.1	179558.8	0

# < 10th %	10th %	25th %	50th %	75th %	90th %
35	15.8	17.1	20.1	27.1	30.1

# > 90th %	Mode	Geo. Mean	Har. Mean	Kurtosis	Skewness
35	•	21.3	20.6	−.4	.7

Table 7.11b

X_2: *Soybean Oil Half2*

Mean	Std. Dev.	Std. Error	Variance	Coef. Var.	Count
21.9	3	.2	9.2	13.8	348

Minimum	Maximum	Range	Sum	Sum of Sqr.	# Missing
16.1	33	16.8	7614.2	169775.3	0

# < 10th %	10th %	25th %	50th %	75th %	90th %
35	18.7	19.6	21.3	23.4	26.5

# > 90th %	Mode	Geo. Mean	Har. Mean	Kurtosis	Skewness
35	18.9	21.7	21.5	1	1.1

Figure 7.10

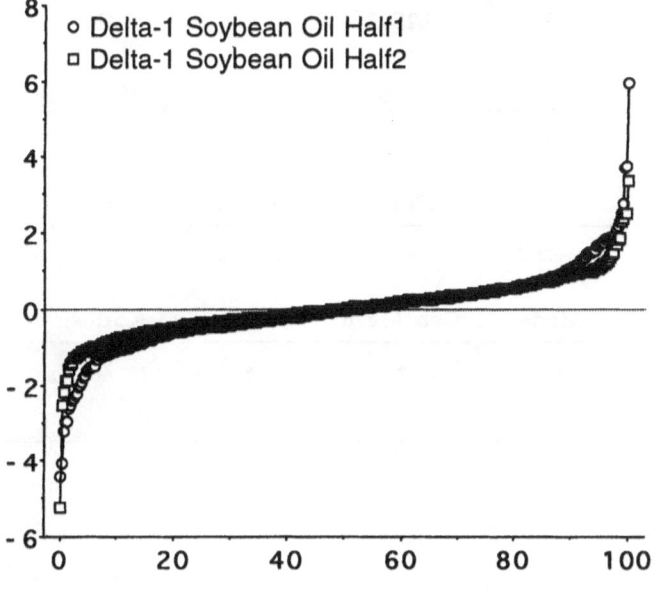

Table 7.12a-b

X_1: Delta-1 Soybean Oil Half1

Mean	Std. Dev.	Std. Error	Variance	Coef. Var.	Count
–2.9E–2	1	.1	1.1	–3569	347

Minimum	Maximum	Range	Sum	Sum of Sqr.	# Missing
–4.4	5.9	10.3	–10.2	379.6	1

# < 10th %	10th %	25th %	50th %	75th %	90th %
35	–1.1	–.5	–.1	.5	1.1

# > 90th %	Mode	Geo. Mean	Har. Mean	Kurtosis	Skewness
35	–.1	•	•	4.9	.3

X_2: Delta-1 Soybean Oil Half2

Mean	Std. Dev.	Std. Error	Variance	Coef. Var.	Count
3.8E–2	.8	4.2E–2	.6	2026.3	348

Minimum	Maximum	Range	Sum	Sum of Sqr.	# Missing
–5.2	3.4	8.6	13.3	210.2	0

# < 10th %	10th %	25th %	50th %	75th %	90th %
35	–.8	–.4	1.0E–2	.5	.9

# > 90th %	Mode	Geo. Mean	Har. Mean	Kurtosis	Skewness
35	–.4	•	•	7.1	–.5

Figure 7.11

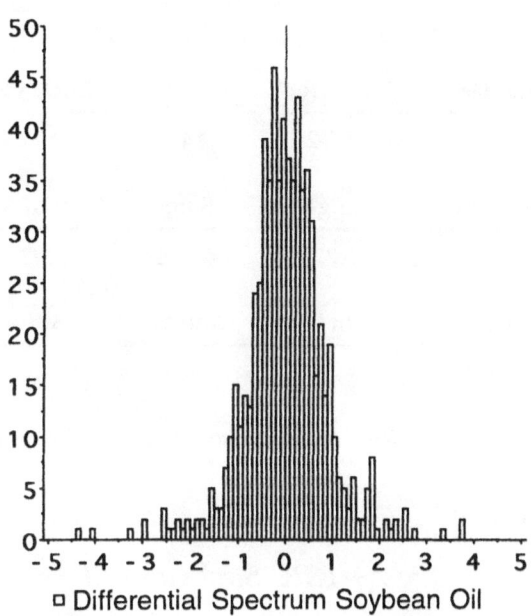

□ Differential Spectrum Soybean Oil

Table 7.13

	A	B	C	D	E	F	G	H	I
1	n	Digram	Expec Prob	(E)xpec Freq	(O)bser Freq	(E-O)	(E-O)^2	((E-O)^2)/E	SUM
2	694	11	0.1667	115.6898	184	-68.3102	4666.28342	40.3344411	
3		12	0.3333	231.3102	170	61.3102	3758.94062	16.2506479	56.585089
4		21	0.3333	231.3102	169	62.3102	3882.56102	16.7850835	73.3701725
5		22	0.1667	115.6898	171	-55.3102	3059.21822	26.4432839	90.1552561
6									
7	n=			694	694				
8								Chi-Square	116.59854

Table 7.14a-b

X_1: 1 – RPC Soybean Oil

Mean	Std. Dev.	Std. Error	Variance	Coef. Var.	Count
–.6	.7	3.5E–2	.4	–106.3	354

Minimum	Maximum	Range	Sum	Sum of Sqr.	# Missing
–5.2	0	5.2	–220.4	292	0

# < 10th %	10th %	25th %	50th %	75th %	90th %
35	–1.2	–.8	–.5	–.2	–.1

# > 90th %	Mode	Geo. Mean	Har. Mean	Kurtosis	Skewness
29	•	•	•	12.1	–2.9

X_2: 2 – RPC Soybean Oil

Mean	Std. Dev.	Std. Error	Variance	Coef. Var.	Count
.7	.7	3.6E–2	.4	101.7	341

Minimum	Maximum	Range	Sum	Sum of Sqr.	# Missing
1.0E–2	5.9	5.9	223.6	297.8	13

# < 10th %	10th %	25th %	50th %	75th %	90th %
32	.1	.2	.5	.9	1.5

# > 90th %	Mode	Geo. Mean	Har. Mean	Kurtosis	Skewness
33	•	.4	.2	14.2	2.9

Figure 7.12

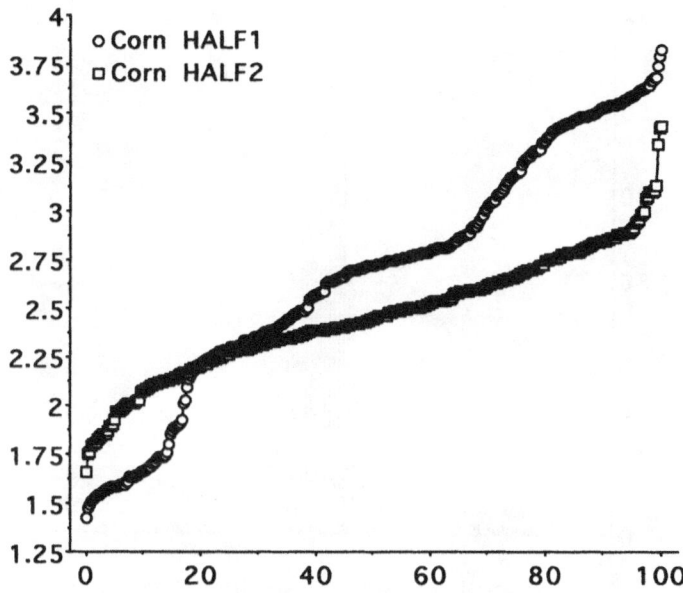

Figure 7.13

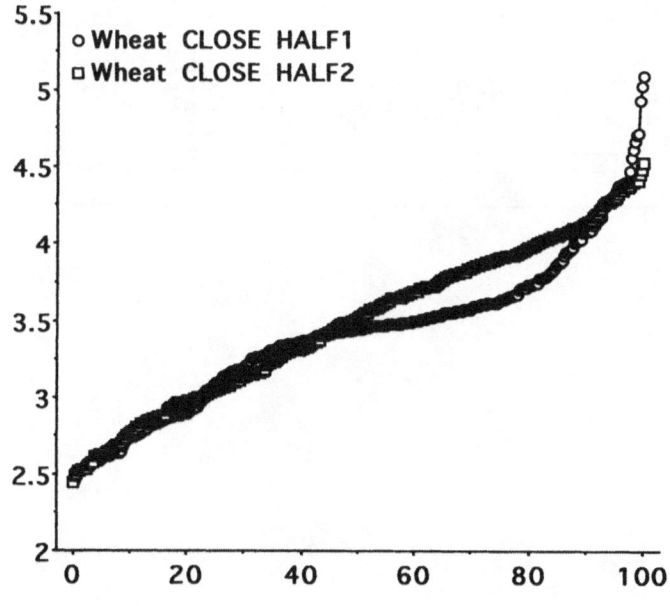

Figure 7.14

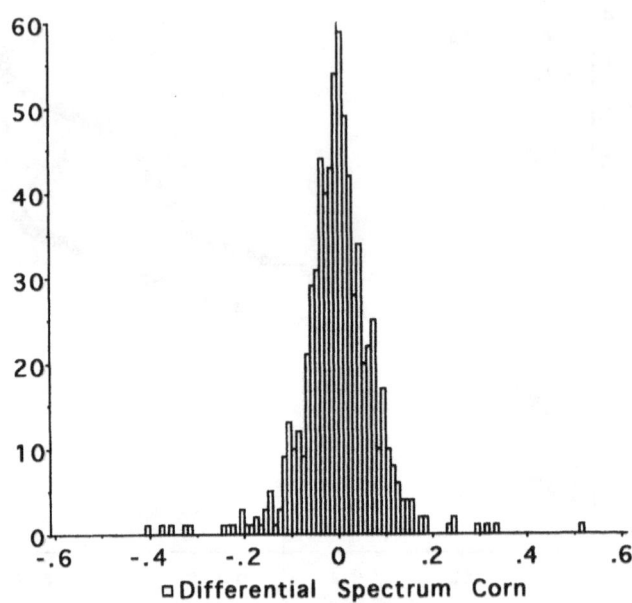

Table 7.15

	A	B	C	D	E	F	G	H	I
1	n	Digram	Expec Prob	(E)xpec Freq	(O)bser Freq	(E-O)	(E-O)^2	((E-O)^2)/E	SUM
2	696	11	0.1667	116.0232	191	-74.9768	5621 52054	48 4516936	
3		12	0.3333	231.9768	168	63.9768	4093.03094	17.6441391	66.0958328
4		21	0.3333	231.9768	167	64.9768	4221 98454	18.2000292	84 295862
5		22	0.1667	116.0232	170	-53.9768	2913 49494	25.1113134	102 495891
6									
7	n=			696	696				
8								Chi-Square	127 607205

Figure 7.15

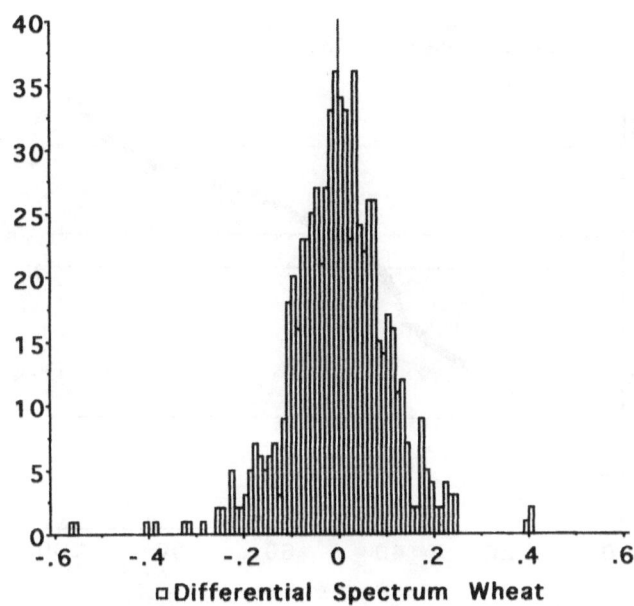

□ Differential Spectrum Wheat

Table 7.16

	A	B	C	D	E	F	G	H	I
1	n	Digram	Expec Prob	(E)xpec Freq	(O)bser Freq	(E-O)	(E-O)^2	((E-O)^2)/E	SUM
2	696	11	0.1667	116.0232	171	-54.9768	3022.44854	26.0503808	
3		12	0.3333	231.9768	183	48.9768	2398.72694	10.3403743	36.3907551
4		21	0.3333	231.9768	183	48.9768	2398.72694	10.3403743	46.7311293
5		22	0.1667	116.0232	159	-42.9768	1847.00534	15.919276	57.0715036
6									
7	n=			696	696				
8								Chi-Square	72.9907796

Figure 7.16

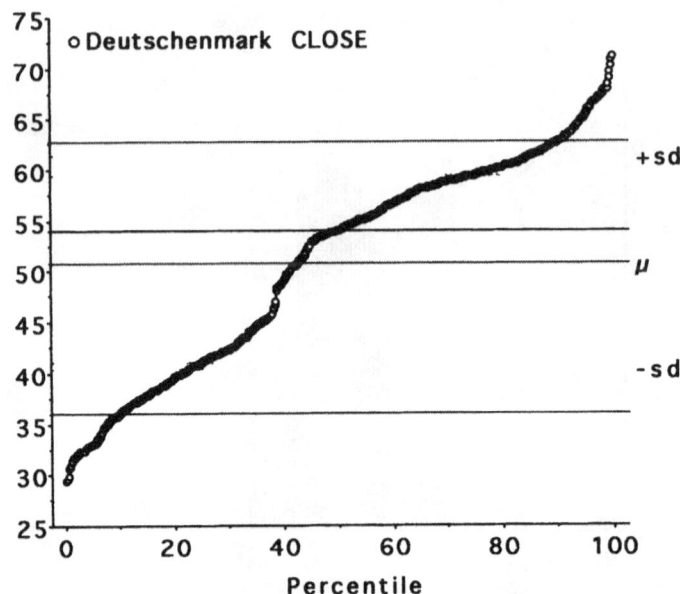

Figure 7.17

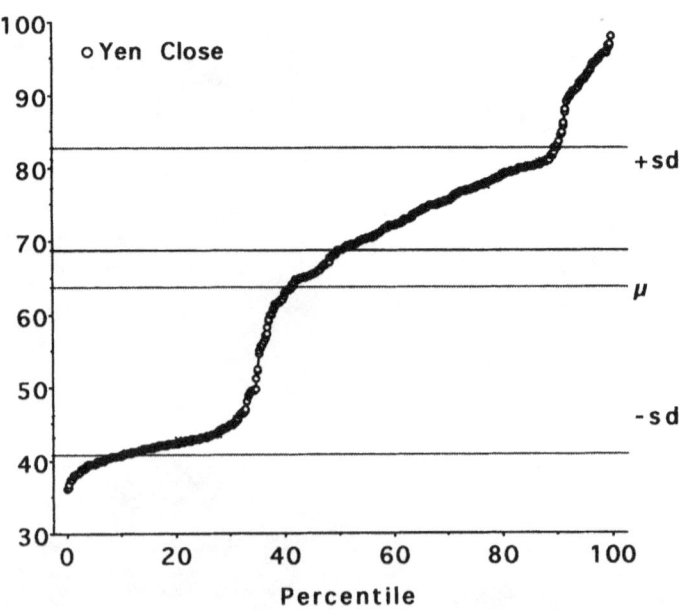

Figure 7.18

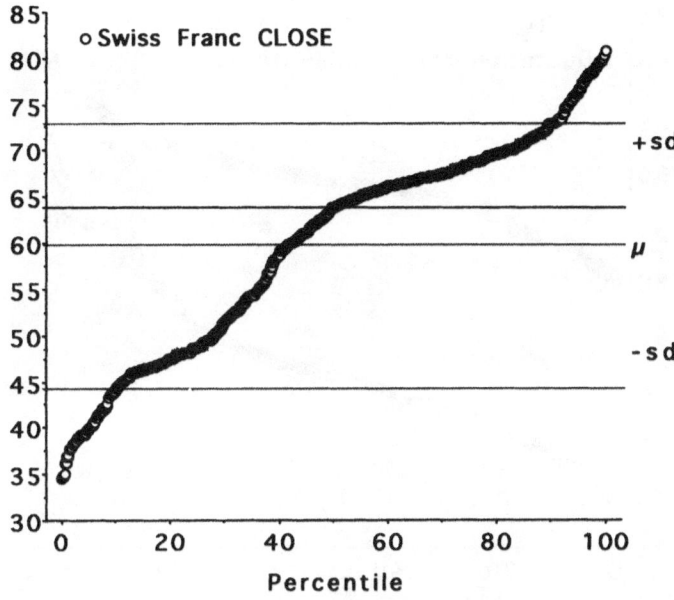

Figure 7.19

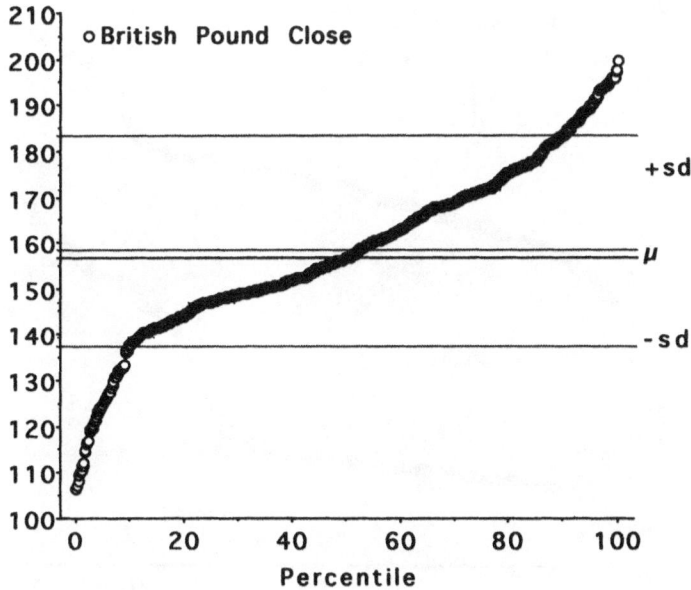

Figure 7.20

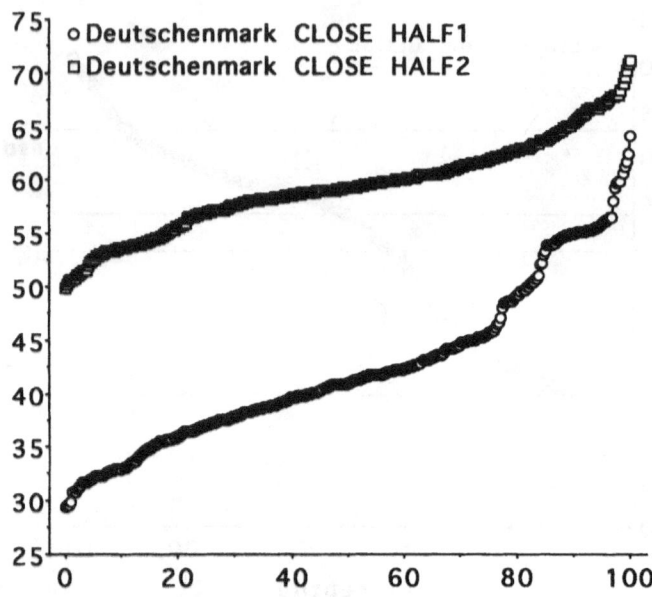

Figure 7.21

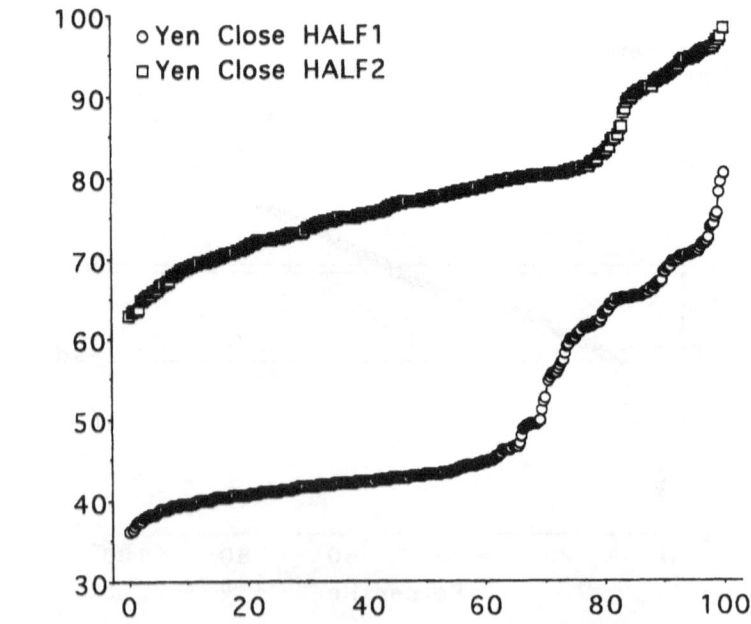

Figure 7.22

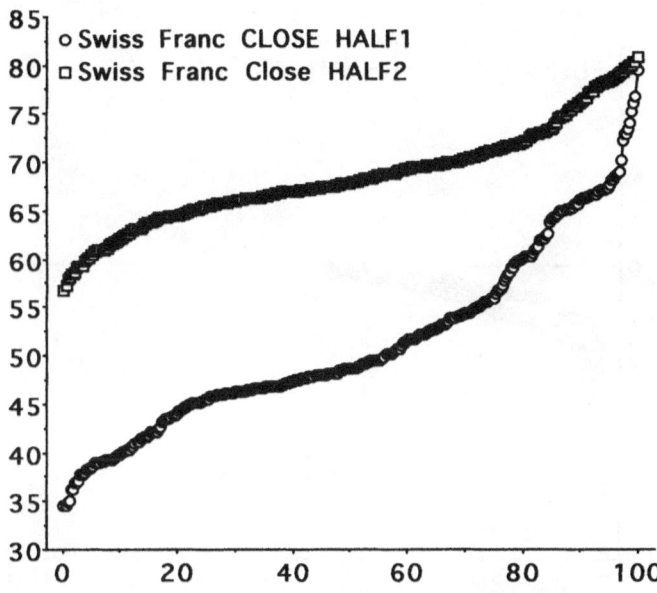

Figure 7.23

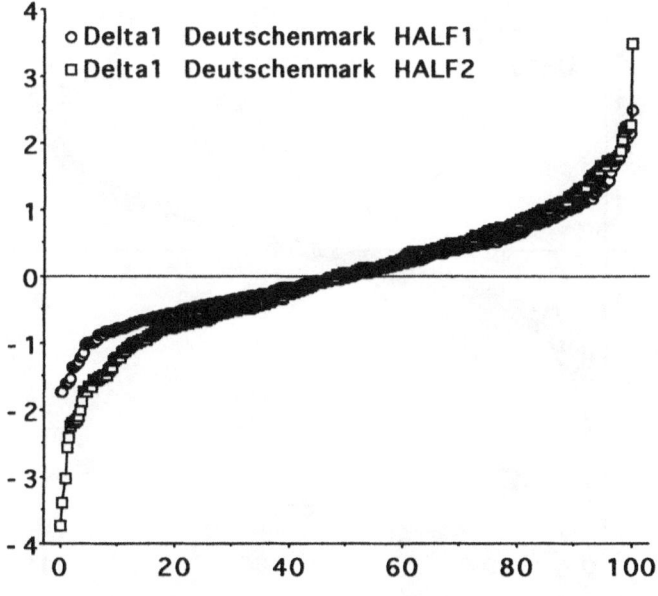

Figure 7.24

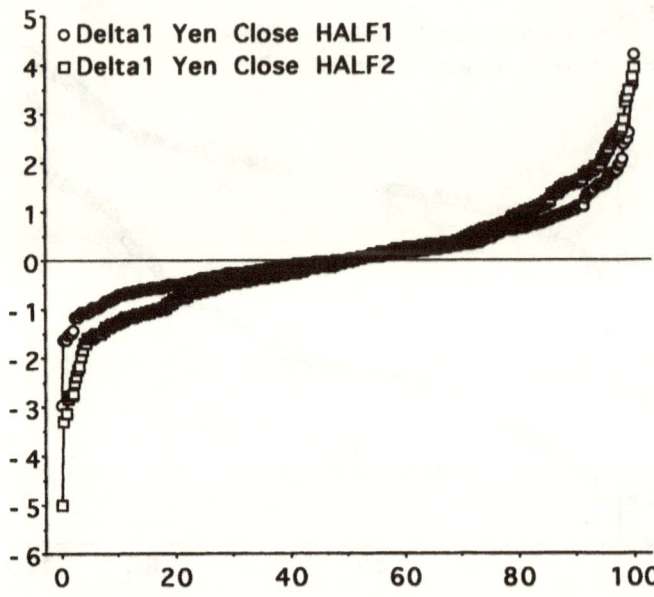

Figure 7.25

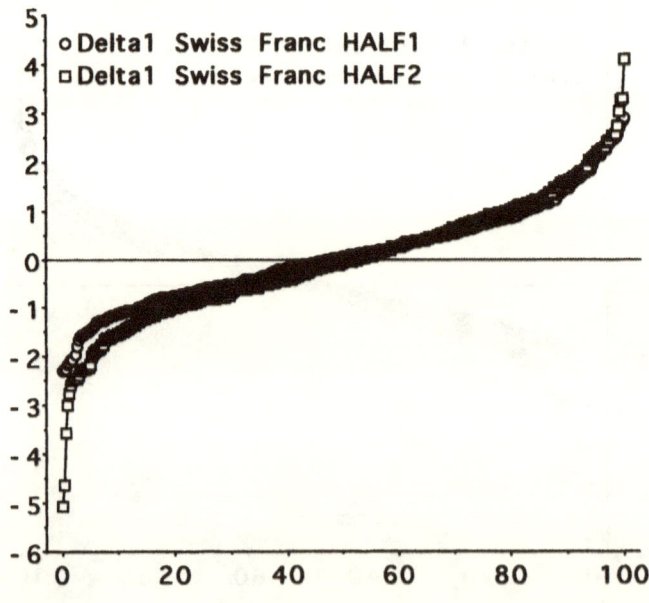

Figure 7.26

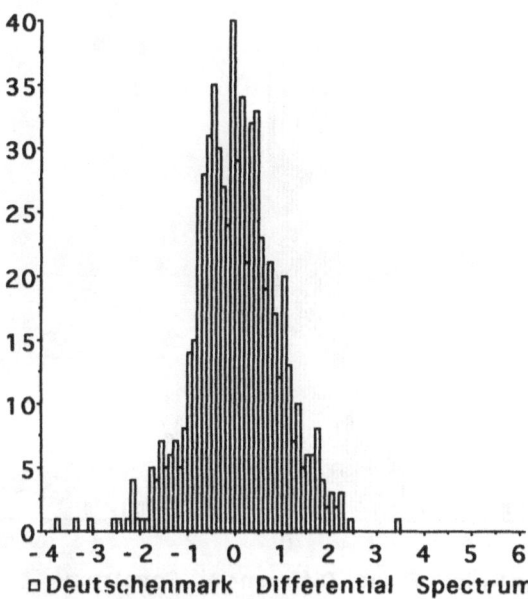

□Deutschenmark Differential Spectrum

Figure 7.27

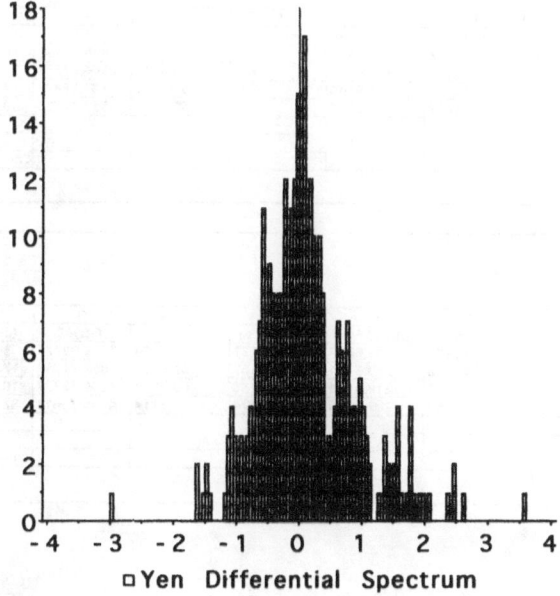

□Yen Differential Spectrum

Figure 7.28

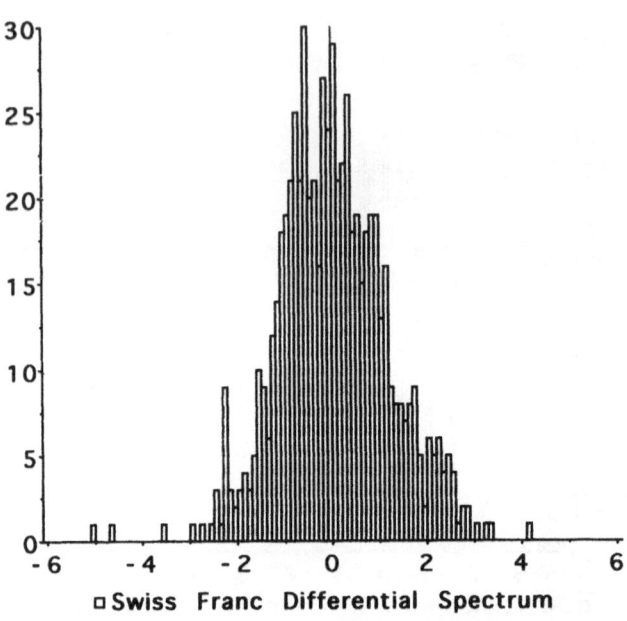

□ Swiss Franc Differential Spectrum

Table 7.17

	A	B	C	D	E	F	G	H	I
1	n	Digram	Expec Prob	(E)xpec Freq	(O)bser Freq	(E-O)	(E-O)^2	((E-O)^2)/E	SUM
2	662	11	0.1667	110.3554	153	-42.6446	1818.56191	16.4791384	
3		12	0.3333	220.6446	170	50.6446	2564.87551	11.6244654	28.1036038
4		21	0.3333	220.6446	169	51.6446	2667.16471	12.0880579	40.1916617
5		22	0.1667	110.3554	170	-59.6446	3557.47831	32.2365585	52.2797196
6									
7	n=			662	662				
8								Chi-Square	84.5162781

Table 7.18

	A	B	C	D	E	F	G	H	I
1	n	Digram	Expec Prob	(E)xpec Freq	(O)bser Freq	(E-O)	(E-O)^2	((E-O)^2)/E	SUM
2	658	11	0.1667	109.6886	171	-61.3114	3759.08777	34.270542	
3		12	0.3333	219.3114	157	62.3114	3882.71057	17.7040982	51.9746402
4		21	0.3333	219.3114	156	63.3114	4008.33337	18.2769038	70.2515441
5		22	0.1667	109.6886	174	-64.3114	4135.95617	37.7063448	88.5284479
6									
7	n=			658	658				
8								Chi-Square	126.234793

Table 7.19

	A π	B Digram	C Expec Prob	D (E)xpec Freq	E (O)bser Freq	F (E-O)	G (E-O)^2	H ((E-O)^2)/E	I SUM
2	666	11	0.1667	111.0222	160	-48.9778	2398.82489	21.6067137	
3		12	0.3333	221.9778	172	49.9778	2497.78049	11.2523887	32.8591024
4		21	0.3333	221.9778	178	43.9778	1934.04689	8.71279422	41.5718967
5		22	0.1667	111.0222	156	-44.9778	2023.00249	18.2216034	50.2846909
6									
7	n=			666	666				
8								Chi-Square	68.5062943

Figure 7.29

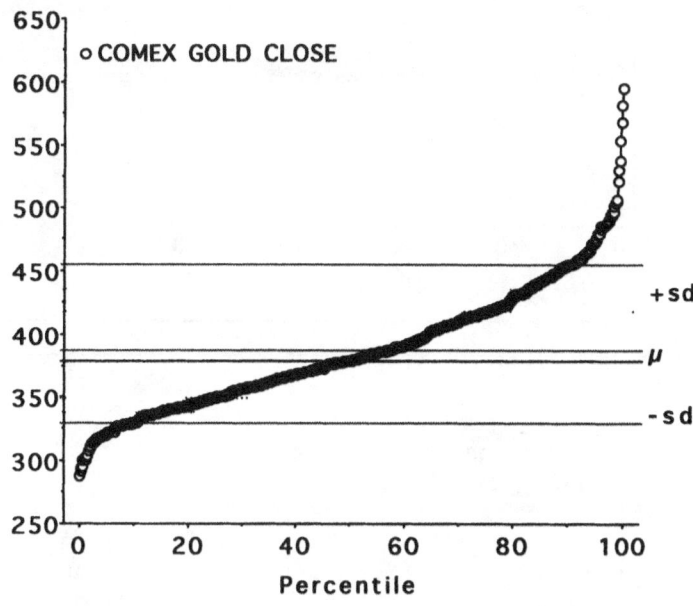

Figure 7.30

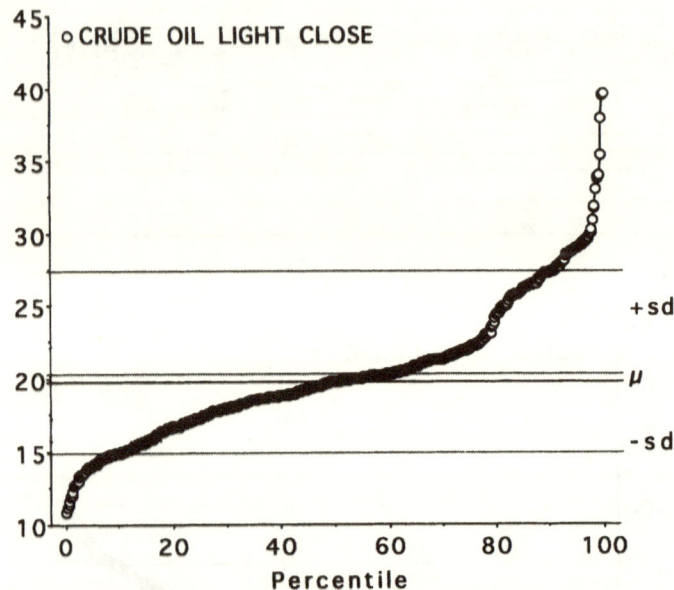

Figure 7.31

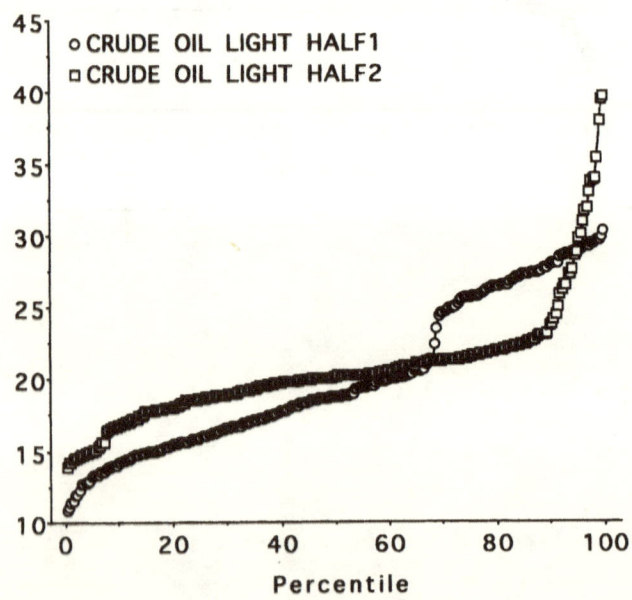

Figure 7.32

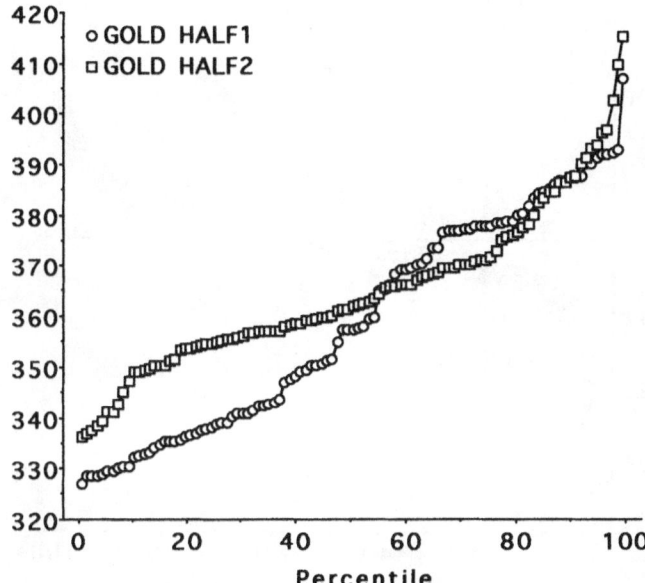

Figure 7.33

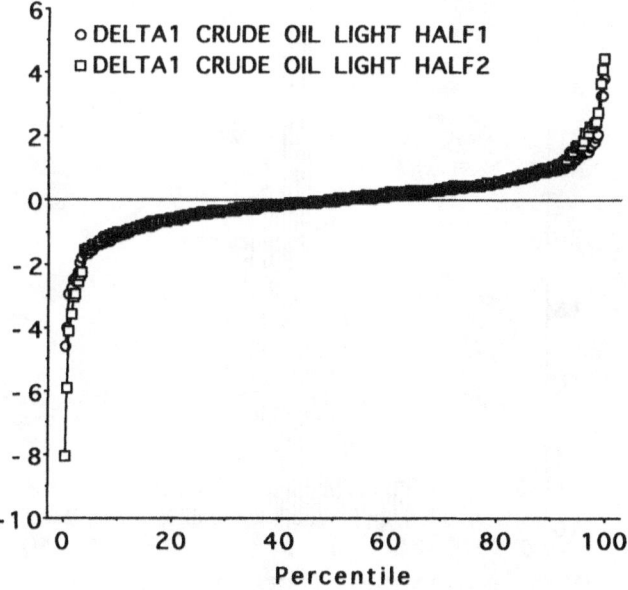

Figure 7.34

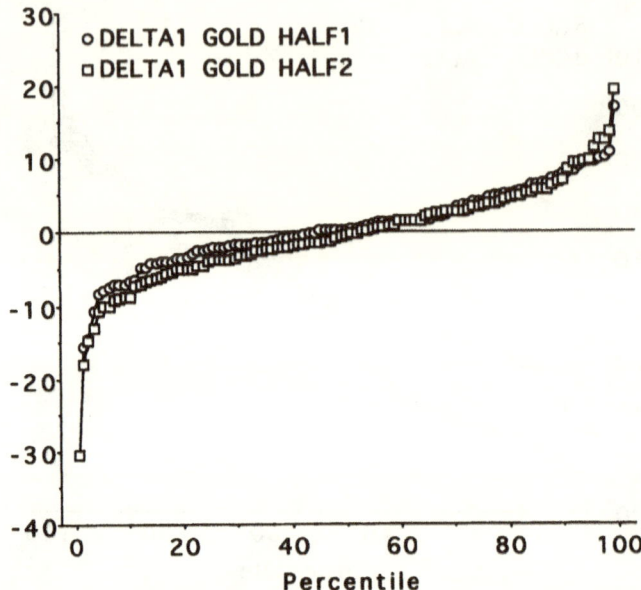

Figure 7.35

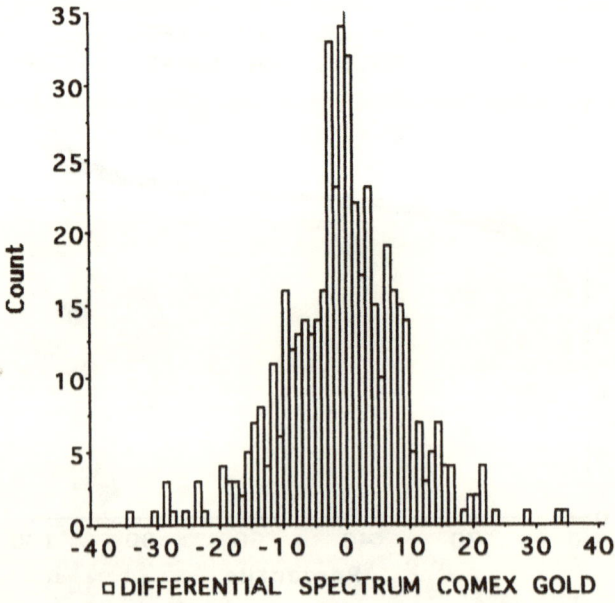

Figure 7.36

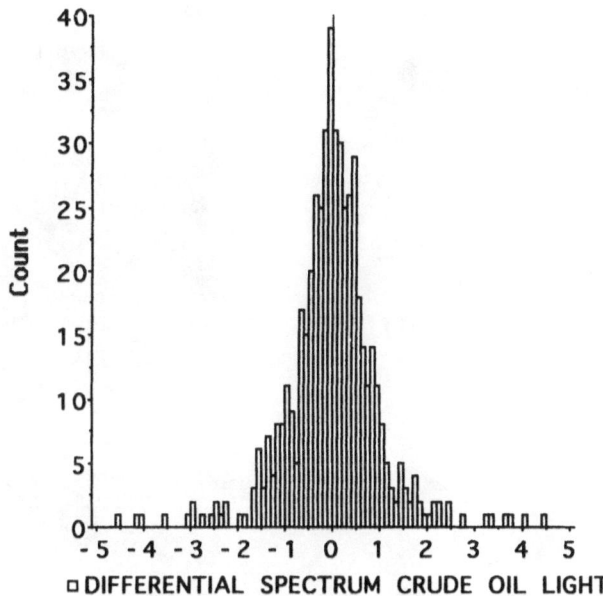

Indicators 8

It is important for traders to know something about the general state of the business cycle. A system of leading, coincident, and lagging indicators, which cover a wide variety of economic processes, was initially developed by the National Bureau of Economic Research in the 1930s. The monthly figures for these indicators and their components, as well as a variety of other economically important time series, are published by the U. S. Department of Commerce in *Business Conditions Digest*.[1] Selecting and classifying indicators is a place where empirical observation and economic theory come together to provide insights into the business climate and its fluctuations. The United States tends to exhibit more cycles than other market-oriented countries.[2] The Composite Index of Leading Economic Indicators (CI) is (or at least should be) important to traders and investors because of its 'look ahead' quality and because it is apparently sensitive to changes in the economy, especially as detected in the (aggregate) marketplace. The CI samples six major economic processes: (1) employment and unemployment; (2) production, income, consumption, and trade; (3) fixed capital investment; (4) inventories and inventory

---

1. The data used in this chapter were obtained from the BCI Historical Diskette which is available from Business Cycle Indicators Branch, Business Outlook Division (BE-52), Bureau of Economic Analysis, U.S. Department of Commerce, Washington, DC 20230. The telephone number is 202-523-0800.

2. An excellent introduction to business cycles and indicators can be found in Burns, Arthur F. and Mitchell, W.C. *Measuring Business Cycles*, New York: National Bureau of Economic Research, 1946.

investment; (5) prices, costs, and profits; and (6) money, credit, and interest rates. Average weekly hours in manufacturing and average weekly initial claims for unemployment insurance (inverted) are used to sample economic process 1. New orders for consumer goods and materials* are used to sample process 2; while vendor perform-ance, contracts and orders for new plants and equipment*, and build-ing permits for new private housing are used to sample process 3. Process 4 is sampled by change in inventories on hand and on order for durable goods, smoothed*, while process 5 is sampled by change in sensitive materials prices, smoothed* and the stock market index (S&P 500) and process 6 by money supply (M2)* and index of con-sumer expectations. The starred (*) items are in constant prices.

The monthly closing 'prices' of the Composite Index of Leading Economic Indicators for 1948 to 1992 are shown in Figure 8.1. The Delta-1 'price changes' are shown in Figure 8.2. They are presented in their raw form as this is the way that the values of the Composite Index are generally reported (i.e., the change from the previous month). The summary statistics are shown in Tables 8.1 and 8.2, re-spectively. The cumulative probability density function for the Com-posite Index for this same time frame is shown in Figure 8.3. The cumulative probability density functions for the individual compo-nent time series are shown in the following figures: average weekly hours in manufacturing (Figure 8.4), average weekly initial claims for unemployment insurance—inverted (Figure 8.5), new orders for con-sumer goods and materials (Figure 8.6), vendor performance (Figure 8.7), contracts and orders for new plants and equipment (Figure 8.8), building permits for new private housing (Figure 8.9), change in in-ventories on hand and on order for durable goods—smoothed (Fig-ure 8.10), change in sensitive materials prices—smoothed (Figure 8.11), the stock market index (Figure 8.12), money supply—M2 (Fig-ure 8.13), and index of consumer expectations (Figure 8.14). This last time series has monthly data for January 1978 to December 1992.

Several of the component time series cumulative probability density functions have very similar shapes. For example, although they differ in scale, the Composite Index (Figure 8.3) and money supply (Figure 8.13) have very similar shapes. If we convert these two time series to their respective standard scores and then plot the standard scores, the plots look very similar, as shown in Figure 8.15. This is confirmed by the high correlation coefficient (i.e., r = 0.99) as shown in Figure 8.16.

Figure 8.1

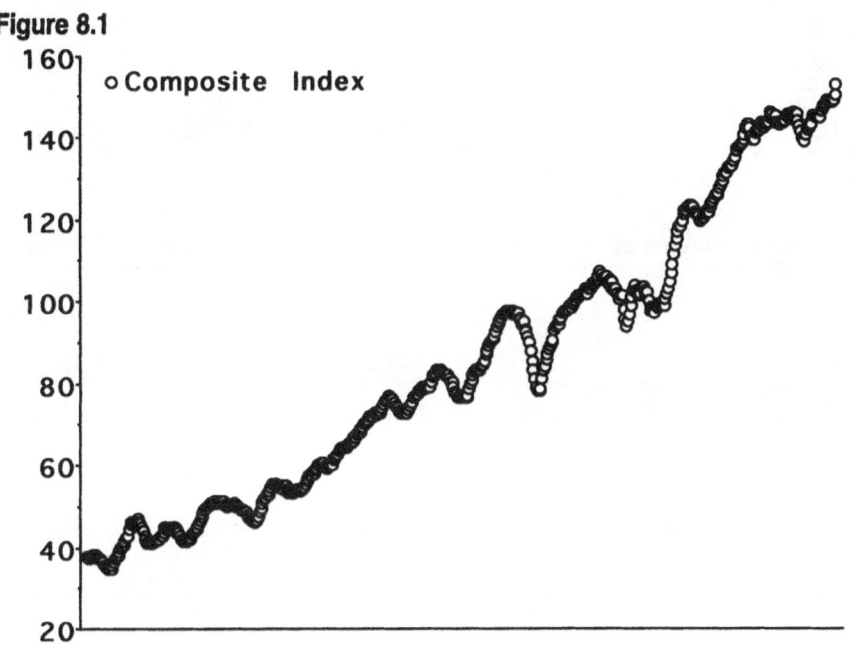

Figure 8.2

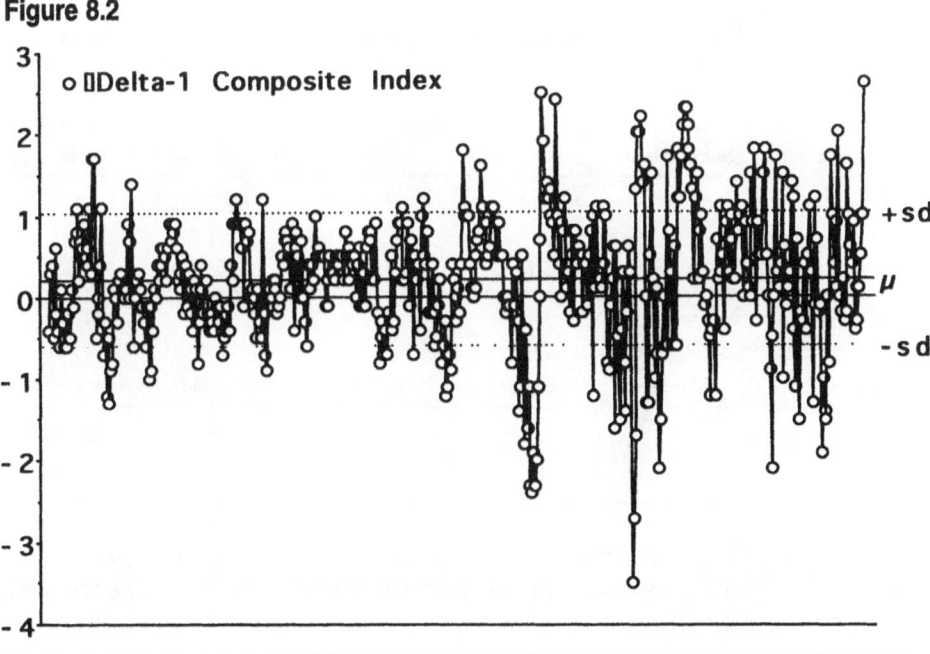

Table 8.1

X_1: Composite Index

Mean	Std. Dev.	Std. Error	Variance	Coef. Var.	Count
84.8778	33.9608	1.4614	1153.3345	40.0114	540

Minimum	Maximum	Range	Sum	Sum of Sqr.	# Missing
34.8	152.8	118	45834	4511935.36	3

# < 10th %	10th %	25th %	50th %	75th %	90th %
54	43.25	53.5	81.1	104.2	141.9

# > 90th %	Mode	Geo. Mean	Har. Mean	Kurtosis	Skewness
54	145.2	78.0257	71.4799	−.9927	.374

Table 8.2

X_1: Delta-1 Composite Index

Mean	Std. Dev.	Std. Error	Variance	Coef. Var.	Count
.2132	.8149	.0351	.664	382.2677	539

Minimum	Maximum	Range	Sum	Sum of Sqr.	# Missing
−3.5	2.6	6.1	114.9	381.75	4

# < 10th %	10th %	25th %	50th %	75th %	90th %
54	−.7	−.2	.2	.7	1.1

# > 90th %	Mode	Geo. Mean	Har. Mean	Kurtosis	Skewness
54	.5	•	•	1.467	−.3973

Unemployment (Figure 8.5), which theoretically samples process 1, and contracts and new orders plants/equipment (Figure 8.8), which theoretically samples process 3, have different scales, but very

Figure 8.3

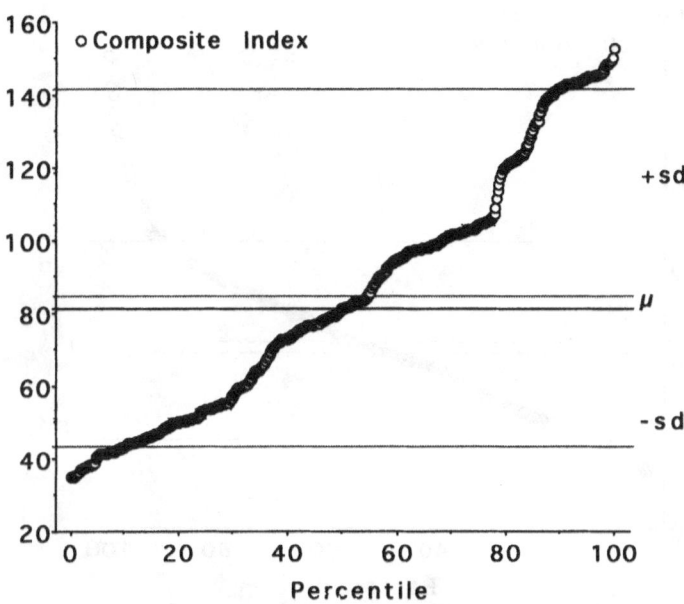

Figure 8.4

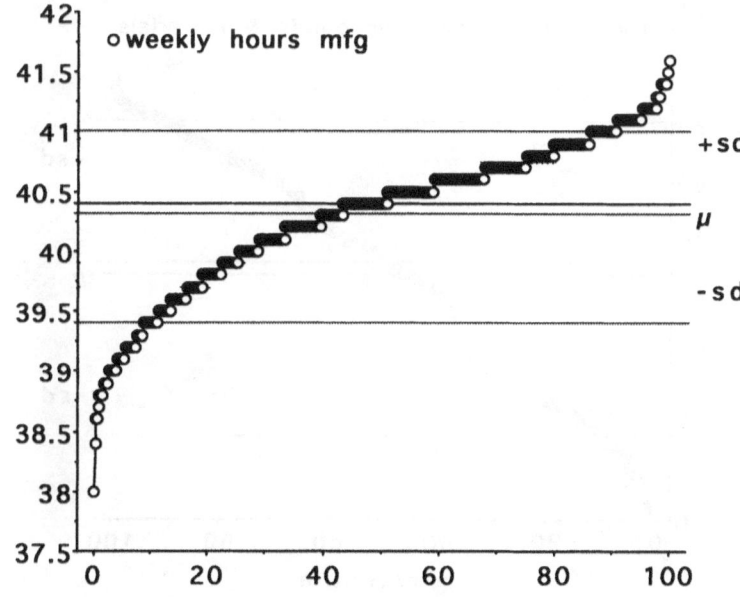

Figure 8.5

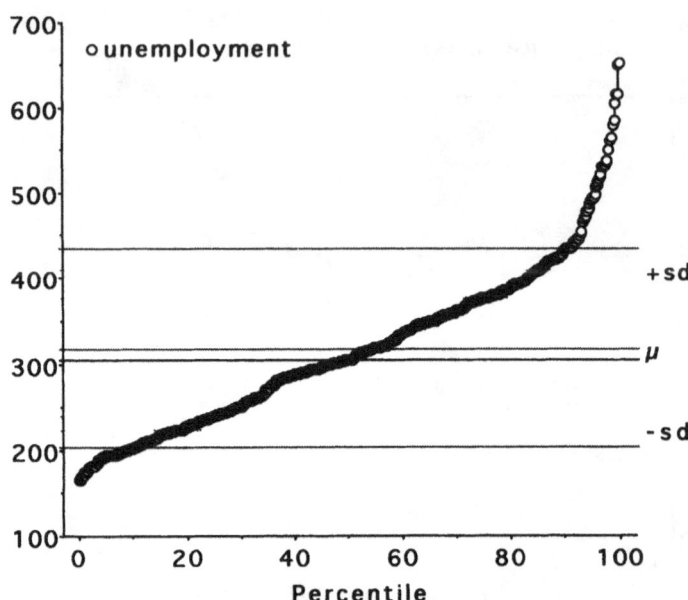

Figure 8.6

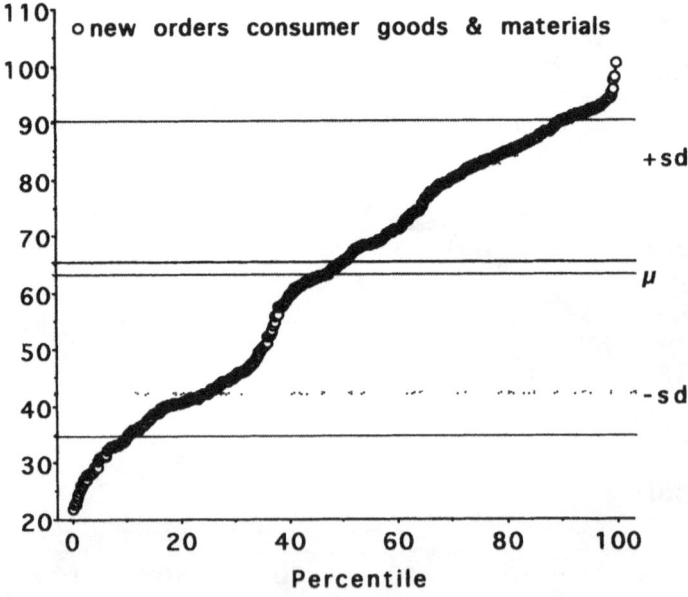

Figure 8.7

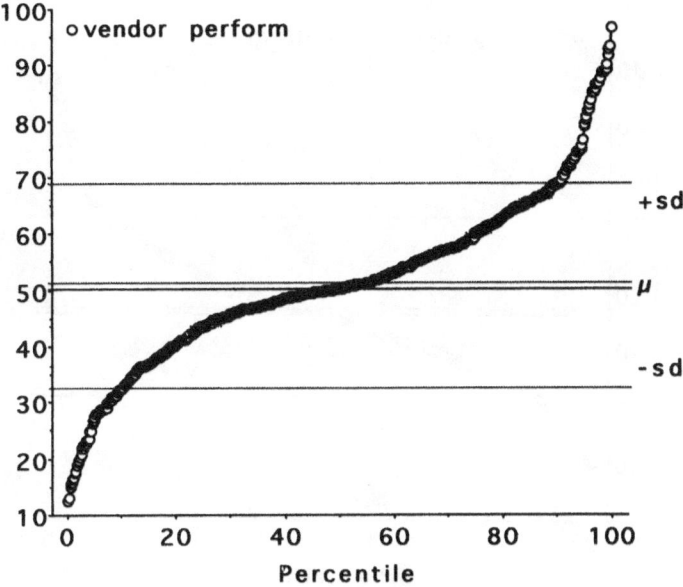

Figure 8.8

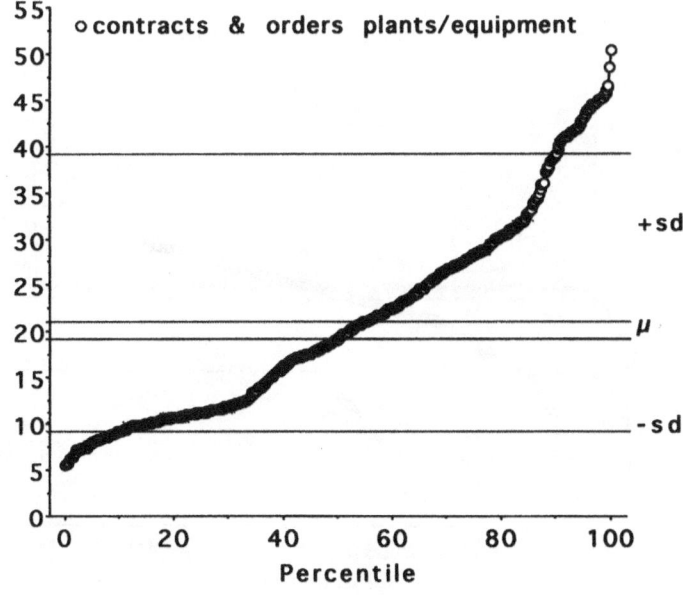

Figure 8.9

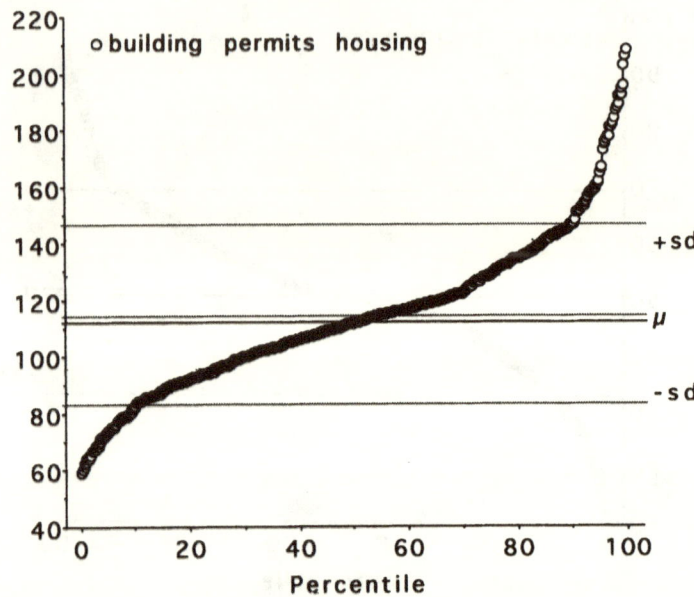

Figure 8.10

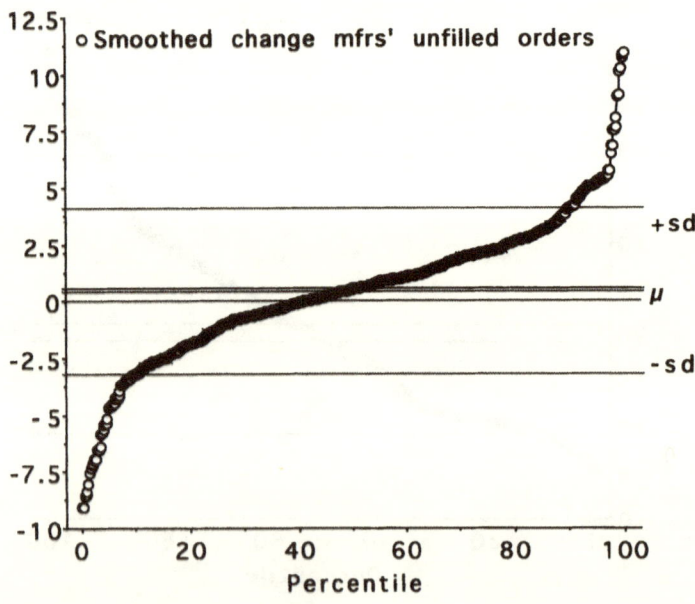

Figure 8.11

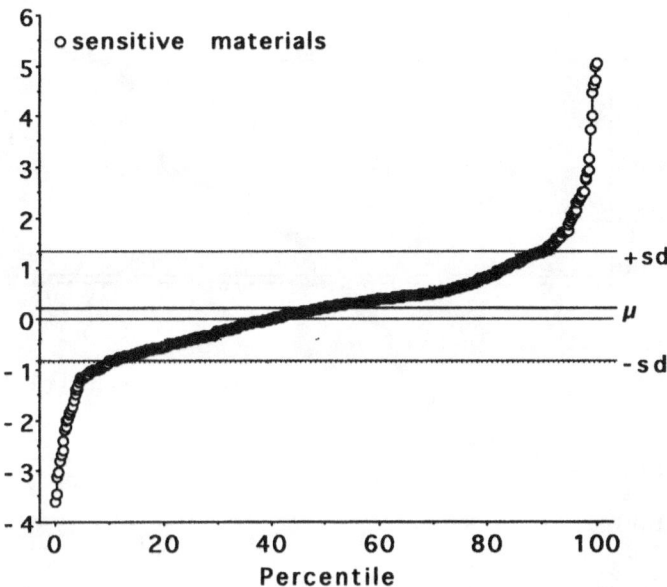

Figure 8.12

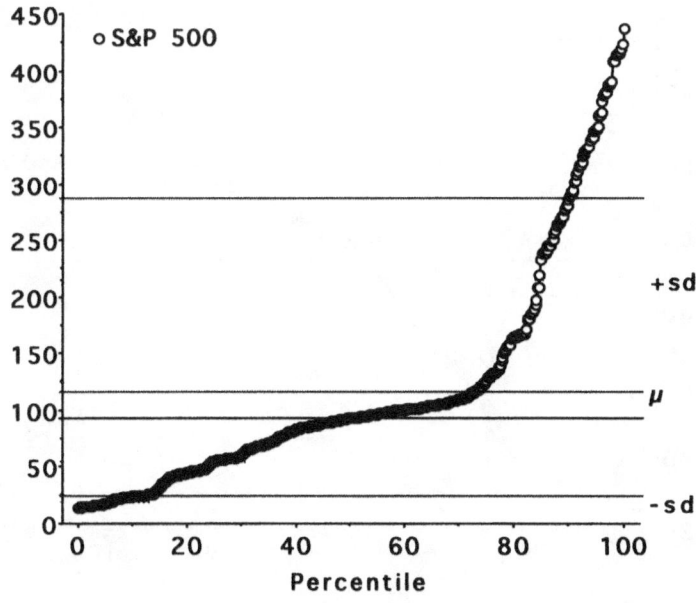

Figure 8.13

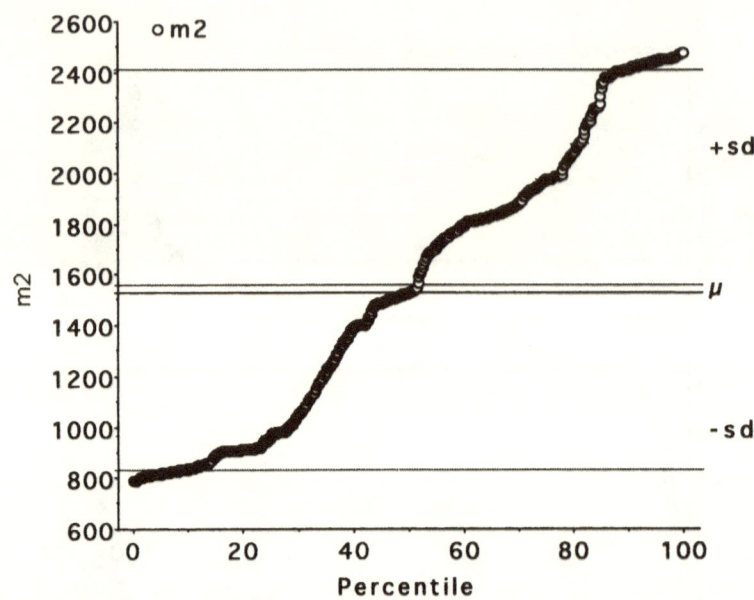

Figure 8.14

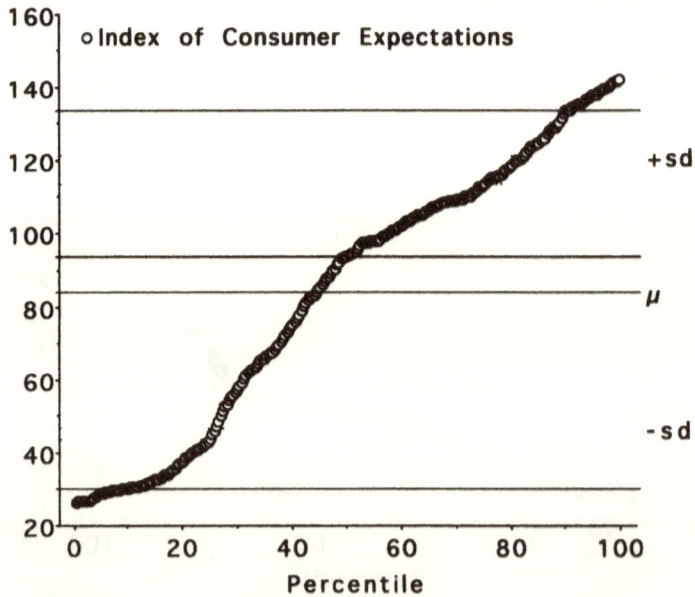

Figure 8.15

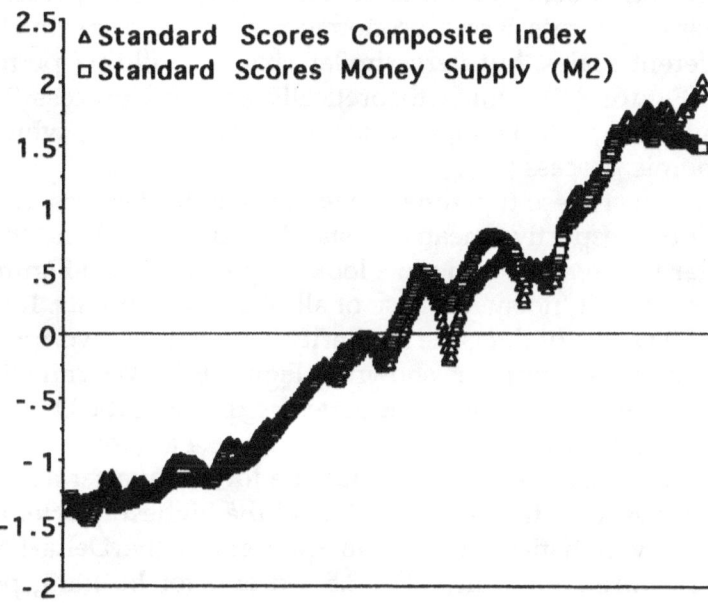

Figure 8.16

$$y = .9897x + 6.9399E-7, r^2 = .9795$$

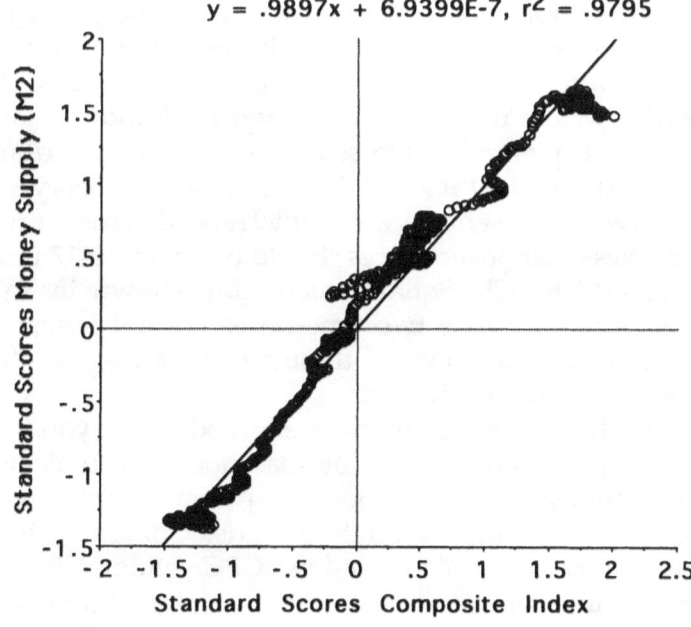

similar shapes. Similarly, sensitive materials (Figure 8.11) and changes in manufacturers' unfilled orders (Figure 8.10), which sample economic process 5 and economic process 4, respectively, also have different scales, but very similar shapes. Building permits for housing (Figure 8.9), which theoretically samples process 3, has a very similar shape to vendor performance (Figure 8.7), which samples economic process 5.

If we recast these two time series into standard scores (see Table 8.3 a-b for the respective mean and standard deviation) and then plot the standard scores, the two plots look very similar (See Figure 8.17). This figure is a bit 'noisy' because of all of the data points. If we plot the first 150 weeks of data, the similarity between the two time series can be seen more clearly, as shown in Figure 8.18. We can eliminate some of the 'jitter' in the two time series by dividing the Delta-1s into four approximately equal quarters (see Table 8.4 a-b for the summary statistics for the Delta-1s) and encode the lowest quarter as a '1', the next quarter as a '2', the next as a '3', and the highest quarter as a '4'. The summary statistics for the four quarters of the Delta-1 vendor performance are shown in Table 8.5 a-d and for building permits-housing in Table 8.6 a-d.

If we subtract the CPC category for building permits from the CPC category for vendors, we will obtain a unique score that represents the interrelationships of these two time series, as shown in Table 8.7. For example, in the first week, both time series were in their highest quarter, so the score was a '0'. The next week, the vendor time series was in the third quarter and the building starts time series in the first quarter, so the score was '2'. The distribution of these scores is shown in Table 8.8. Basic probability theory suggests that if these two time series are not interrelated (cross-correlated), then each of these composite scores should occur about 77 times (i.e., $540/7 = 77.14$). The Chi-Square calculation shown in Table 8.9 strongly suggests that these two time series are not independently distributed. For example, both of the time series were in the same CPC quarter about 31% of the time.

Based on these findings, if you are a modeler or you are using neural nets for pattern detection, you may not want to include both of these time series into your model. You might want to eliminate one of the time series or you might consider combining them into one composite time series. You could use CPC vendor–CPC building (see Table 8.7) time series to replace the two original time series. Or

Table 8.3a-b

X_1: Vendor Performance

Mean	Std. Dev.	Std. Error	Variance	Coef. Var.	Count
51.372	14.8234	.6379	219.7319	28.8549	540

Minimum	Maximum	Range	Sum	Sum of Sqr.	# Missing
12.4	96.8	84.4	27740.9	1543542.05	4

# < 10th %	10th %	25th %	50th %	75th %	90th %
54	32.5	43.55	50.2	60.05	68.85

# > 90th %	Mode	Geo. Mean	Har. Mean	Kurtosis	Skewness
54	49.9	49.0078	46.1896	.5129	.2195

X_2: Building Permits Housing

Mean	Std. Dev.	Std. Error	Variance	Coef. Var.	Count
114.3122	26.9819	1.1611	728.0252	23.6037	540

Minimum	Maximum	Range	Sum	Sum of Sqr.	# Missing
59.1	208.5	149.4	61728.6	7448739.04	4

# < 10th %	10th %	25th %	50th %	75th %	90th %
54	83.25	95.95	112	128.65	146.9

# > 90th %	Mode	Geo. Mean	Har. Mean	Kurtosis	Skewness
53	114.5	111.2571	108.2713	.8156	.7214

you could create a new time series using this data. For example, you could encode a (–1, 0, +1) as a '0', a (+2, +3) as a '+1', and a (–2, –3) as a '–1'. The new time series, recast in this manner, is shown in Figure 8.19 top. The first 25 data points of this plot are shown in Figure 8.19 bottom.

Figure 8.17

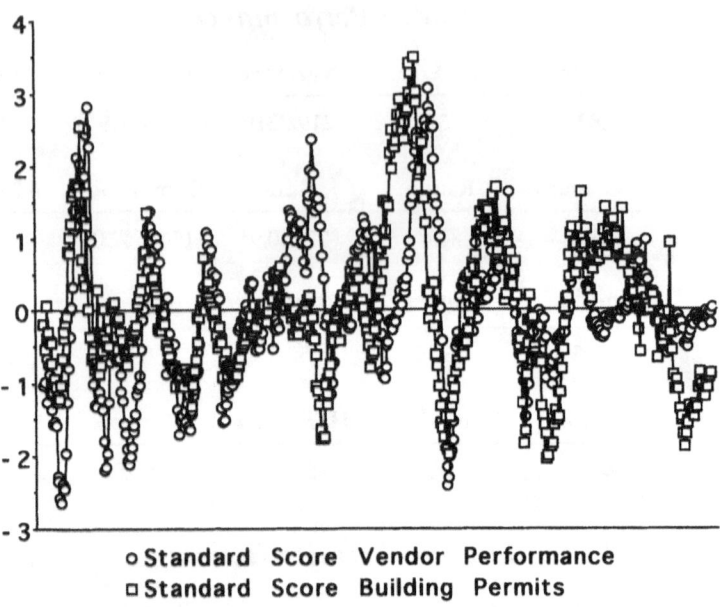

○ Standard Score Vendor Performance
□ Standard Score Building Permits

Figure 8.18

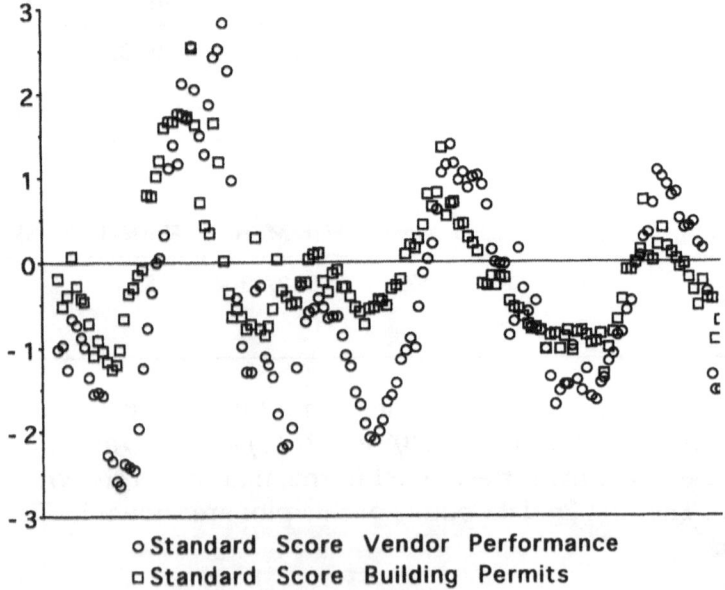

○ Standard Score Vendor Performance
□ Standard Score Building Permits

Table 8.4a-b

X_1: Delta-1 Vendor Performance

Mean	Std. Dev.	Std. Error	Variance	Coef. Var.	Count
.0957	4.7153	.2029	22.2341	4925.0808	540

Minimum	Maximum	Range	Sum	Sum of Sqr.	# Missing
−20.7	36.3	57	51.7	11989.15	4

# < 10th %	10th %	25th %	50th %	75th %	90th %
54	−4.65	−2.1	.3	2.15	4.7

# > 90th %	Mode	Geo. Mean	Har. Mean	Kurtosis	Skewness
54	.7	•	•	8.3789	.4964

X_2: Delta-1 Building Permits Housing

Mean	Std. Dev.	Std. Error	Variance	Coef. Var.	Count
.1774	9.2346	.3974	85.2785	5205.3267	540

Minimum	Maximum	Range	Sum	Sum of Sqr.	# Missing
−33.4	109.4	142.8	95.8	45982.1	4

# < 10th %	10th %	25th %	50th %	75th %	90th %
54	−9.05	−3.85	.15	4.2	8.65

# > 90th %	Mode	Geo. Mean	Har. Mean	Kurtosis	Skewness
54	1.2	•	•	36.4945	2.9131

The cumulative probability density functions for Half1 (January 1948–May 1970) and Half2 (June 1970–December 1992) for 'prices' and 'Delta-1 price changes' are shown in Figures 8.20 and 8.21, respectively. We can look at the behavior of the composite Index for the last 20 years by dividing Half2 into halves, so that Half1 contains data from June 1970 to September 1981 and Half2 contains data from

Table 8.5a-b

X_1: 1–Delta-1 Vendor Performance

Mean	Std. Dev.	Std. Error	Variance	Coef. Var.	Count
–5.3833	3.6158	.3147	13.0742	–67.1671	132

Minimum	Maximum	Range	Sum	Sum of Sqr.	# Missing
–20.7	–2.1	18.6	–710.6	5538.12	3

# < 10th %	10th %	25th %	50th %	75th %	90th %
13	–10.45	–6.85	–4	–2.9	–2.3

# > 90th %	Mode	Geo. Mean	Har. Mean	Kurtosis	Skewness
10	–3.5	•	•	3.7692	–1.8831

X_2: 2–Delta-1 Vendor Performance

Mean	Std. Dev.	Std. Error	Variance	Coef. Var.	Count
–.8737	.7121	.0617	.5071	–81.5069	133

Minimum	Maximum	Range	Sum	Sum of Sqr.	# Missing
–2.1	.3	2.4	–116.2	165.58	2

# < 10th %	10th %	25th %	50th %	75th %	90th %
12	–1.8	–1.4	–.9	–.3	.1

# > 90th %	Mode	Geo. Mean	Har. Mean	Kurtosis	Skewness
13	•	•	•	–1.1236	.0107

October 1981 to December 1992. The cumulative probability density functions for Half1 and Half2 for 'prices' and 'Delta-1 price changes' are shown in Figures 8.22 and 8.23, respectively (summary statistics, Tables 8.10 a-b and 8.11 a-b, respectively). The frequency distribution of 'prices' is shown in Figure 8.24 (June 1970 to December 1992). The differential spectrum as shown in Figure 8.25 is clearly not symmet-

Table 8.5c-d

X₃: 3–Delta-1 Vendor Performance

Mean	Std. Dev.	Std. Error	Variance	Coef. Var.	Count
1.1351	.5152	.0445	.2655	45.3909	134

Minimum	Maximum	Range	Sum	Sum of Sqr.	# Missing
.3	2.1	1.8	152.1	207.31	1

# < 10th %	10th %	25th %	50th %	75th %	90th %
12	.5	.7	1.05	1.5	1.9

# > 90th %	Mode	Geo. Mean	Har. Mean	Kurtosis	Skewness
12	.7	1.0001	.865	−1.0662	.2565

X₄: 4–Delta-1 Vendor Performance

Mean	Std. Dev.	Std. Error	Variance	Coef. Var.	Count
5.38	4.0235	.3463	16.1885	74.7861	135

Minimum	Maximum	Range	Sum	Sum of Sqr.	# Missing
2.2	36.3	34.1	726.3	6076.75	0

# < 10th %	10th %	25th %	50th %	75th %	90th %
13	2.5	3	4.1	6.2	8.9

# > 90th %	Mode	Geo. Mean	Har. Mean	Kurtosis	Skewness
13	•	4.5823	4.0766	25.0944	4.0529

rical and thus indicates that the Composite Index of Leading Economic Indicators contains serial dependencies. This is confirmed by the RPC digrams and trigrams transition matrices that diverge from independence (see Tables 8.12 and 8.13, respectively). The summary statistics for the 'price' decreases and increases are shown in Table 8.14 a-b. If we parse the Delta-1 time series into four unequal quarters

Table 8.6a-b

X_1: 1–Delta-1 Building Permits Housing

Mean	Std. Dev.	Std. Error	Variance	Coef. Var.	Count
–9.6474	5.8822	.5063	34.6003	–60.9718	135

Minimum	Maximum	Range	Sum	Sum of Sqr.	# Missing
–33.4	–3.9	29.5	–1302.4	17201.22	0

# < 10th %	10th %	25th %	50th %	75th %	90th %
13	–16.7	–11.325	–8.1	–5.6	–4.5

# > 90th %	Mode	Geo. Mean	Har. Mean	Kurtosis	Skewness
13	•	•	•	3.7459	–1.8795

X_2: 2–Delta-1 Buidling Permits Housing

Mean	Std. Dev.	Std. Error	Variance	Coef. Var.	Count
–1.7415	1.1751	.1011	1.3808	–67.4757	135

Minimum	Maximum	Range	Sum	Sum of Sqr.	# Missing
–3.8	.1	3.9	–235.1	594.45	0

# < 10th %	10th %	25th %	50th %	75th %	90th %
13	–3.4	–2.775	–1.7	–.6	–.3

# > 90th %	Mode	Geo. Mean	Har. Mean	Kurtosis	Skewness
12	•	•	•	–1.2539	–.1669

as shown in Figure 8.26, the distribution of the digrams using these CPC classes diverges from independence as shown in Table 8.15. The summary statistics for the four classes are shown in Table 8.16a-d.

Consumer Price Index represents the 'price' of a 'marketbasket' of goods and services. The goods and services are a cross-section of the goods and services that a typical urban family uses in its daily

Table 8.6c-d

X_3: 3–Delta-1 Building Permits Housing

Mean	Std. Dev.	Std. Error	Variance	Coef. Var.	Count
2.1037	1.2441	.1075	1.5477	59.1367	134

Minimum	Maximum	Range	Sum	Sum of Sqr.	# Missing
.2	4.2	4	281.9	785.93	1

# < 10th %	10th %	25th %	50th %	75th %	90th %
12	.5	1.1	2.05	3.1	3.8

# > 90th %	Mode	Geo. Mean	Har. Mean	Kurtosis	Skewness
11	1.2	1.6439	1.1165	−1.2441	.1073

X_4: 4–Delta-1 Building Permits Housing

Mean	Std. Dev.	Std. Error	Variance	Coef. Var.	Count
10.0769	10.1835	.8797	103.703	101.0579	134

Minimum	Maximum	Range	Sum	Sum of Sqr.	# Missing
4.3	109.4	105.1	1350.3	27399.29	1

# < 10th %	10th %	25th %	50th %	75th %	90th %
12	4.7	5.8	7.5	10.7	16.06

# > 90th %	Mode	Geo. Mean	Har. Mean	Kurtosis	Skewness
13	•	8.3916	7.5051	66.0577	7.2285

life. The selection of goods and services has remained constant. The Consumer Price Index is expressed in constant dollars, where 1982–1984 represent 100. Changes in the Consumer Price Index is the government's main gauge of inflation.

The cumulative probability density functions for the monthly 'prices' of the Consumer Price Index for 1948 to 1992 are shown in

Table 8.7

	CPC Vendor Performance	CPC Building Permits	CPD Vendor–CPC Building
1	4.000	4.000	0
2	3.000	1.000	2.000
3	1.000	3.000	–2.000
4	4.000	4.000	0
5	2.000	1.000	1.000
6	2.000	2.000	0
7	2.000	2.000	0
8	1.000	1.000	0
9	1.000	1.000	0
10	3.000	4.000	–1.000

Table 8.8

CPC Vendor–CPC Building	Count	Percent
–5	0	0%
–4	0	0%
–3	26	4.8148%
–2	52	9.6296%
–1	108	20%
0	169	31.2963%
1	101	18.7037%
2	58	10.7407%
3	26	4.8148%
4	0	0%

Figure 8.27 and the summary statistics are shown in Table 8.17. The cumulative probability density function for monthly 'prices' for the Consumer Price Index for 1972 to 1992 is shown in Figure 8.28. The Delta-1 'price changes' for this same time period are shown in Figure 8.29. This is the normal method that the government reports data about the Consumer Price Index; i.e., as changes from month to month. The cumulative probability density functions for Half1 and Half2 for this time period are shown in Figure 8.30 and the Delta-1

Table 8.9

	A	B	C	D	E
1					
2	Difference	Observed	Expected	O-E	((O-E)^2/E)
3	-3	26	77.14	-51.1429	33.906
4	-2	52	77.14	-25.1429	8.195
5	-1	108	77.14	30.8571	12.343
6	0	169	77.14	91.8571	109.378
7	1	101	77.14	23.8571	7.378
8	2	58	77.14	-19.1429	4.750
9	3	26	77.14	-51.1429	33.906
10					
11	n=	540	540.00		
12					
13				Chi-Square=	209.855439
14		28.9727761			
15		3.46410162			
16	z=	25.5086745			

'price changes' for Half1 and Half2 are shown in Figure 8.31. These data in these figures strongly suggest that the Consumer Price Index is stationary for this time period. The summary statistics are shown in Tables 8.18 a-b and 8.19 a-b, respectively. The differential spectrum for the Consumer Price Index is shown in Figure 8.32. The differential spectrum is not symmetrical and this indicates that the Consumer Price Index contains serial dependencies. If we divide the Consumer Price Index into four approximately equal parts and encode the 'price changes' in the lowest quarter as a '1', the next quarter as a '2', etc., and then collect the coded values into a digram transition matrix, it is clear that the observed matrix diverges from independence, as shown in Table 8.20.

Money supply (M2) and the Composite Index of Leading Economic Indicators are highly intercorrelated (see Figure 8.16) and each is also highly correlated with the Consumer Price Index, as shown in Figures 8.33 and 8.34, respectively. The correlation coefficients are 0.94 and 0.91, respectively. Remember that the correlation coefficient for Money Supply and the Composite Index was even higher.

Figure 8.19

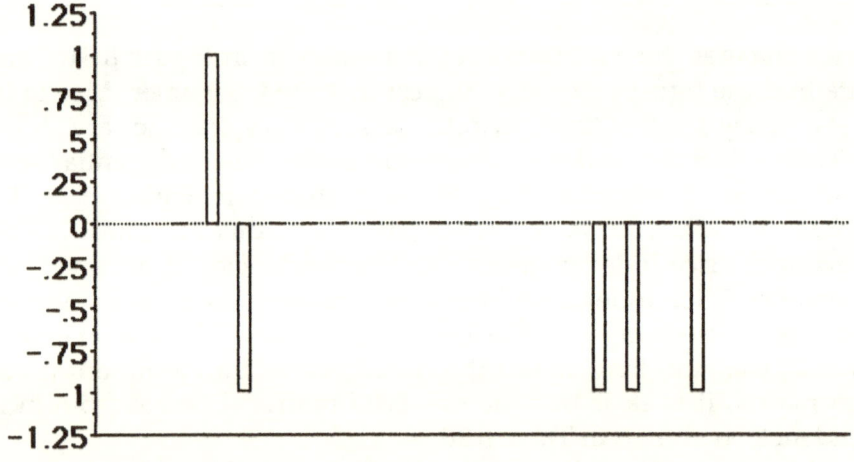

Figure 8.20

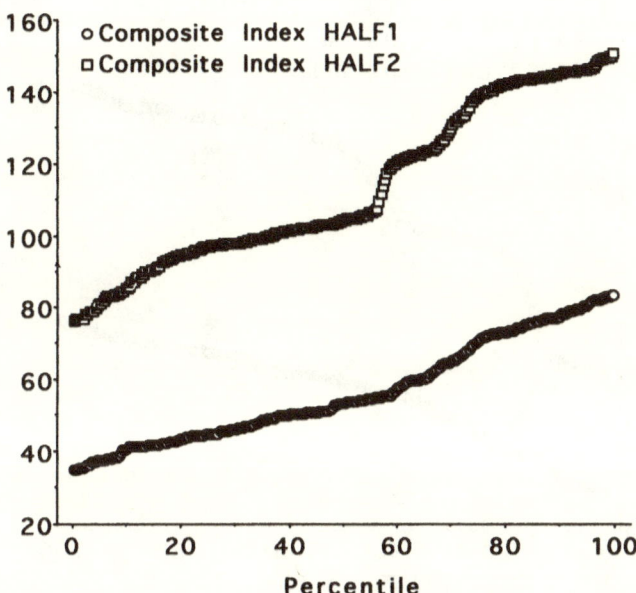

Figure 8.21

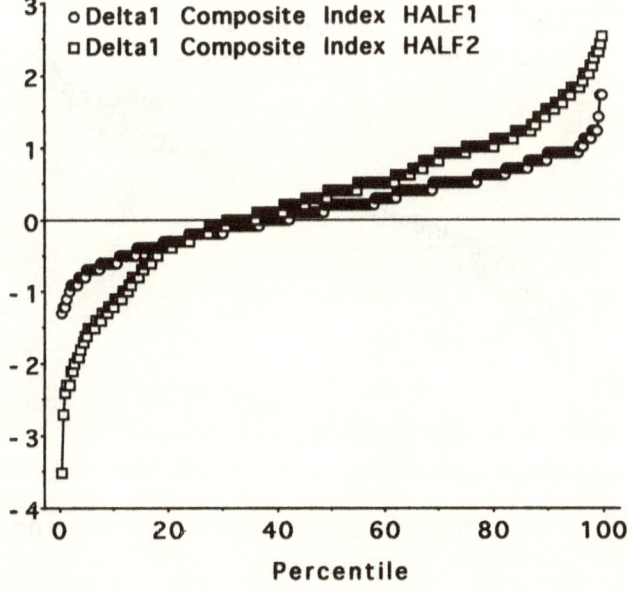

Figure 8.22

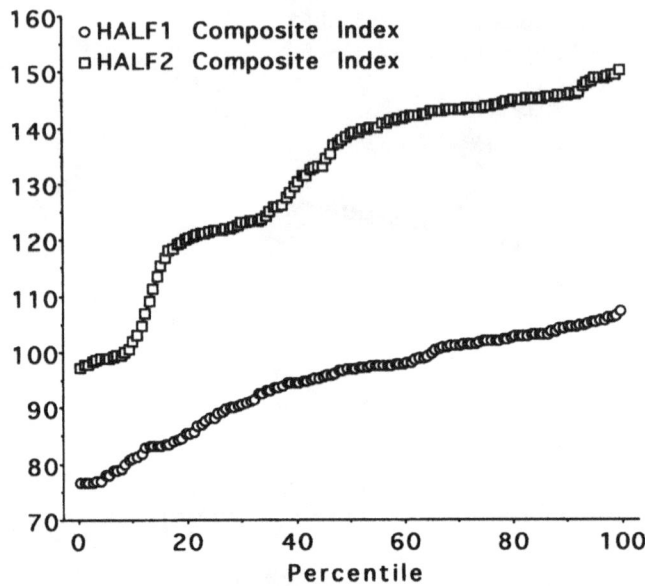

Figure 8.23

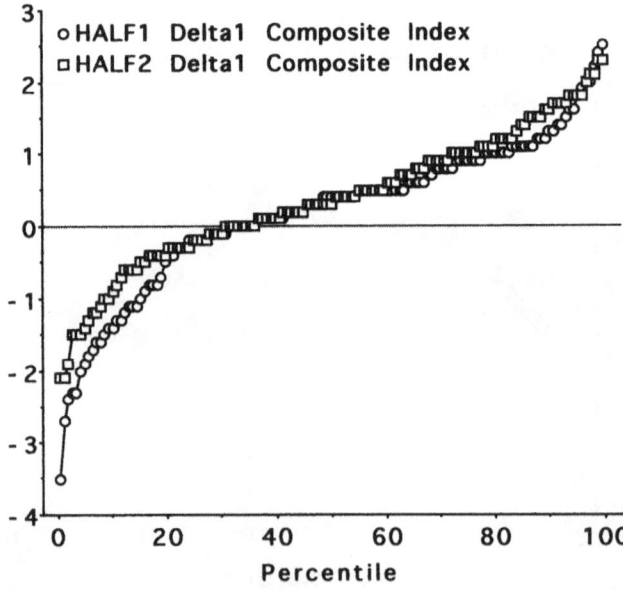

Table 8.10a-b

X_1: *Half1 Composite Index*

Mean	Std. Dev.	Std. Error	Variance	Coef. Var.	Count
94.5889	8.5884	.7392	73.7613	9.0798	135

Minimum	Maximum	Range	Sum	Sum of Sqr.	# Missing
76.5	107.1	30.6	12769.5	1217736.83	0

# < 10th %	10th %	25th %	50th %	75th %	90th %
13	81.1	88.4	96.9	101.875	104.3

# > 90th %	Mode	Geo. Mean	Har. Mean	Kurtosis	Skewness
13	•	94.1841	93.7619	–.7459	–.6274

X_2: *Half2 Composite Index*

Mean	Std. Dev.	Std. Error	Variance	Coef. Var.	Count
131.323	15.5862	1.3414	242.9289	11.8686	135

Minimum	Maximum	Range	Sum	Sum of Sqr.	# Missing
97.2	150.2	53	17728.6	2360724.76	0

# < 10th %	10th %	25th %	50th %	75th %	90th %
13	101.8	121.725	138.7	143.675	145.8

# > 90th %	Mode	Geo. Mean	Har. Mean	Kurtosis	Skewness
13	145.2	130.3276	129.2529	–.454	–.8489

Table 8.11a-b

X_1: *Half1 Delta-1 Composite Index*

Mean	Std. Dev.	Std. Error	Variance	Coef. Var.	Count
.1881	1.0635	.0915	1.1311	565.2511	135

Minimum	Maximum	Range	Sum	Sum of Sqr.	# Missing
−3.5	2.5	6	25.4	156.34	0

# < 10th %	10th %	25th %	50th %	75th %	90th %
13	−1.4	−.2	.4	.9	1.3

# > 90th %	Mode	Geo. Mean	Har. Mean	Kurtosis	Skewness
13	.5	•	•	.8462	−.7588

X_2: *Half2 Delta-1 Composite Index*

Mean	Std. Dev.	Std. Error	Variance	Coef. Var.	Count
.3563	.9358	.0805	.8758	262.6527	135

Minimum	Maximum	Range	Sum	Sum of Sqr.	# Missing
−2.1	2.3	4.4	48.1	134.49	0

# < 10th %	10th %	25th %	50th %	75th %	90th %
13	−.9	−.2	.3	1	1.6

# > 90th %	Mode	Geo. Mean	Har. Mean	Kurtosis	Skewness
13	0	•	•	−.1277	−.2594

Figure 8.24

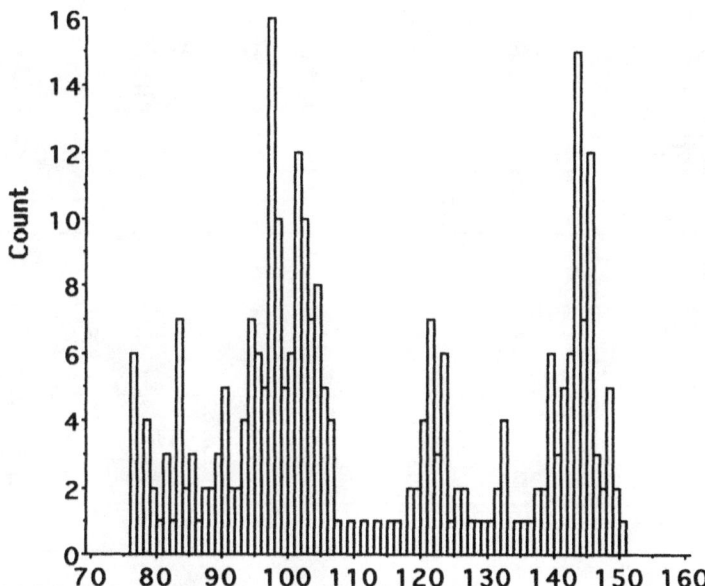

Figure 8.25

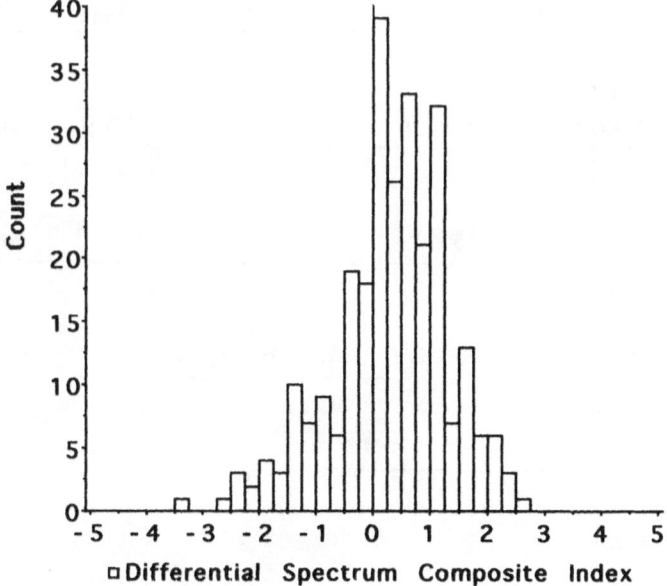

□Differential Spectrum Composite Index

Table 8.12

	A	B	C	D	E	F	G	H	I
1	n	Digram	Expec Prob	(E)xpec Freq	(O)bser Freq	(E-O)	(E-O)^2	((E-O)^2)/E	SUM
2	269	11	0.1667	44.8423	56	-11.1577	124.494269	2.7762686	
3		12	0.3333	89.6577	42	47.6577	2271.25637	25.3325299	28.1087985
4		21	0.3333	89.6577	41	48.6577	2367.57177	26.4067868	54.5155853
5		22	0.1667	44.8423	130	-85.1577	7251.83387	161.718598	80.9223721
6									
7		n=			269	269			
8								Chi-Square	242.64097

Table 8.13

	A	B	C	D	E	F	G	H	I
1	n	Trigram	Expec Prob	(E)xpec Freq	(O)bser Freq	(E-O)	(E-O)^2	((E-O)^2)/E	SUM
2	268	111	0.042	11.168	38.000	-26.832	719.980	64.471	
3		112	0.125	33.500	18.000	15.500	240.250	7.172	71.642
4		121	0.208	55.832	15.000	40.832	1667.288	29.862	101.505
5		122	0.125	33.500	27.000	6.500	42.250	1.261	102.766
6		211	0.125	33.500	17.000	16.500	272.250	8.127	110.893
7		212	0.208	55.832	24.000	31.832	1013.304	18.149	129.042
8		221	0.125	33.500	26.000	7.500	56.250	1.679	130.721
9		222	0.042	11.168	103.000	-91.832	8433.197	755.151	
10									
11								Chi-Square	885.872

Table 8.14a

X_1: 1 – RPC Price Decrease

Mean	Std. Dev.	Std. Error	Variance	Coef. Var.	Count
–.75	.7453	.0753	.5555	–99.3772	98

Minimum	Maximum	Range	Sum	Sum of Sqr.	# Missing
–3.5	0	3.5	–73.5	109.01	74

# < 10th %	10th %	25th %	50th %	75th %	90th %
10	–1.87	–1.2	–.45	–.2	0

# > 90th %	Mode	Geo. Mean	Har. Mean	Kurtosis	Skewness
0	0	•	•	.9218	–1.1348

Table 8.14b

X_2: 2 – RPC Price Increase

Mean	Std. Dev.	Std. Error	Variance	Coef. Var.	Count
.8686	.5862	.0447	.3437	67.4931	172

Minimum	Maximum	Range	Sum	Sum of Sqr.	# Missing
.1	2.6	2.5	149.4	188.54	0

# < 10th %	10th %	25th %	50th %	75th %	90th %
13	.2	.4	.8	1.2	1.73

# > 90th %	Mode	Geo. Mean	Har. Mean	Kurtosis	Skewness
17	.5	.6555	.4411	.024	.797

Figure 8.26

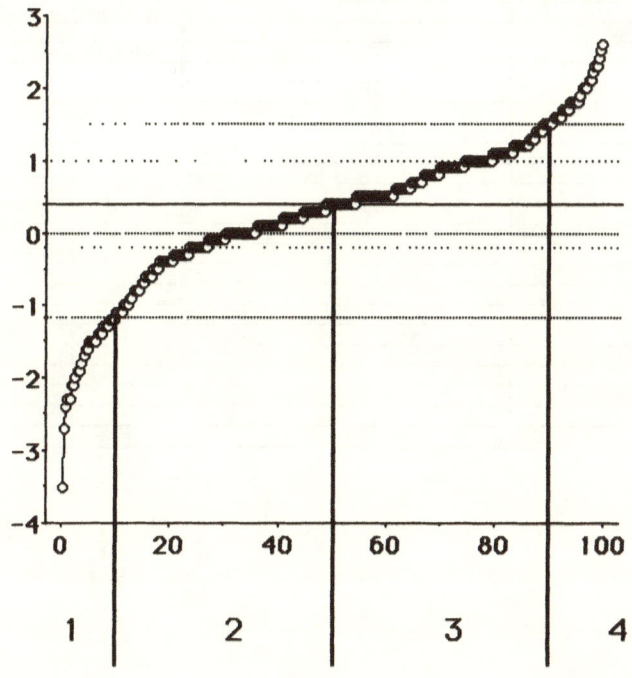

Table 8.15

	A	B	C	D	E	F	G	H
1	Symbols	Frequency	Probability					
2	1	27	0.1007					
3	2	119	0.4440					
4	3	68	0.2537					
5	4	54	0.2015					
6	Sum	268						
7								
8								
9								
10	Digrams	(E)xpected	(O)bserved	(E-O)	(E-O)^2	((E-O)^2)/E	Sum	
11	11	2.720	8.000	-5.280	27.877	10.248		
12	12	11.989	13.000	-1.011	1.023	0.085	10.334	
13	13	6.851	4.000	2.851	8.127	1.186	11.520	
14	14	5.440	2.000	3.440	11.836	2.176	13.695	
15	21	11.989	14.000	-2.011	4.045	0.337	11.857	
16	22	52.840	64.000	-11.160	124.556	2.357	14.214	
17	23	30.194	26.000	4.194	17.590	0.583	14.797	
18	24	23.978	15.000	8.978	80.598	3.361	18.158	
19	31	6.851	4.000	2.851	8.127	1.186	15.983	
20	32	30.194	26.000	4.194	17.590	0.583	16.566	
21	33	17.254	26.000	-8.746	76.497	4.434	20.999	
22	34	13.701	12.000	1.701	2.895	0.211	21.211	
23	41	5.440	1.000	4.440	19.716	3.624	24.835	
24	42	23.978	15.000	8.978	80.598	3.361	28.196	
25	43	13.701	12.000	1.701	2.895	0.211	28.408	
26	44	10.881	26.000	-15.119	228.596	21.010	49.417	
27								
28	n=	268	268			Chi-Square	49.4171037	
29								
30								
31						9.94153949	5.47722558	
32								
33					z=	4.46431391		
34								
35								
36								
37								
38								
39								
40								
41								
42								
43								
44								
45								
46								
47								
48								
49								
50								
51								4.46431391

Table 8.16a-b

X_1: 1–CPC < 10%

Mean	Std. Dev.	Std. Error	Variance	Coef. Var.	Count
−1.763	.5379	.1035	.2893	−30.5116	27

Minimum	Maximum	Range	Sum	Sum of Sqr.	# Missing
−3.5	−1.2	2.3	−47.6	91.44	92

# < 10th %	10th %	25th %	50th %	75th %	90th %
3	−2.38	−2.075	−1.6	−1.4	−1.22

# > 90th %	Mode	Geo. Mean	Har. Mean	Kurtosis	Skewness
3	−1.5	•	•	2.1274	−1.4277

X_2: 2–CPC > 10% < 50%

Mean	Std. Dev.	Std. Error	Variance	Coef. Var.	Count
−.1092	.4115	.0377	.1693	−376.6678	119

Minimum	Maximum	Range	Sum	Sum of Sqr.	# Missing
−1.1	.4	1.5	−13	21.4	0

# < 10th %	10th %	25th %	50th %	75th %	90th %
12	−.8	−.3	0	.2	.4

# > 90th %	Mode	Geo. Mean	Har. Mean	Kurtosis	Skewness
10	0	•	•	−.1746	−.7875

Table 8.16c-d

X_3: 3–CPC > 50% < 90%

Mean	Std. Dev.	Std. Error	Variance	Coef. Var.	Count
.7441	.1958	.0237	.0383	26.308	68

Minimum	Maximum	Range	Sum	Sum of Sqr.	# Missing
.5	1	.5	50.6	40.22	51

# < 10th %	10th %	25th %	50th %	75th %	90th %
0	.5	.5	.8	.9	1

# > 90th %	Mode	Geo. Mean	Har. Mean	Kurtosis	Skewness
0	.5	.7177	.6914	−1.5692	−.0314

X_4: 4–CPC 9 > 0%

Mean	Std. Dev.	Std. Error	Variance	Coef. Var.	Count
1.5673	.415	.056	.1722	26.4805	55

Minimum	Maximum	Range	Sum	Sum of Sqr.	# Missing
1.1	2.6	1.5	86.2	144.4	64

# < 10th %	10th %	25th %	50th %	75th %	90th %
1	1.1	1.2	1.5	1.8	2.2

# > 90th %	Mode	Geo. Mean	Har. Mean	Kurtosis	Skewness
5	1.1	1.517	1.471	−.4269	.7036

Figure 8.27

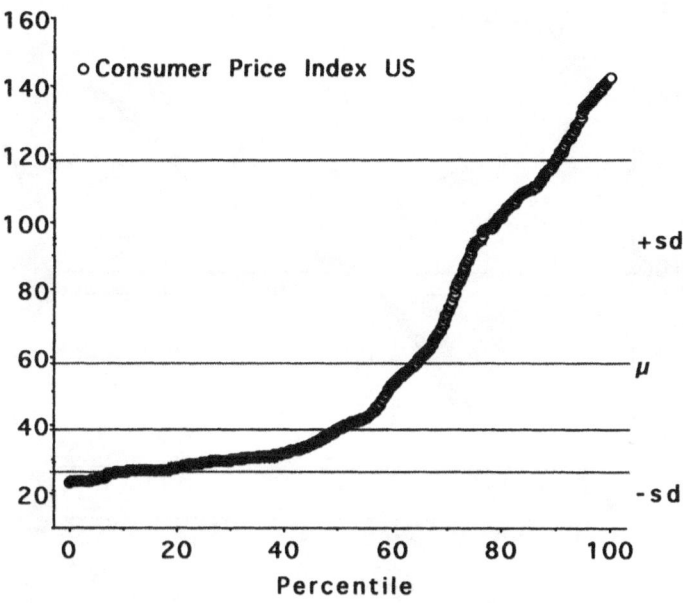

Table 8.17

X₁: Consumer Price Index US

Mean	Std. Dev.	Std. Error	Variance	Coef. Var.	Count
58.1276	37.0449	1.5942	1372.3262	63.7304	540

Minimum	Maximum	Range	Sum	Sum of Sqr.	# Missing
23.4	142	118.6	31388.9	2564245.01	6

# < 10th %	10th %	25th %	50th %	75th %	90th %
51	26.5	29	38.9	93.3	118.25

# > 90th %	Mode	Geo. Mean	Har. Mean	Kurtosis	Skewness
54	26.9	48.2995	41.4158	−.7043	.8943

Figure 8.28

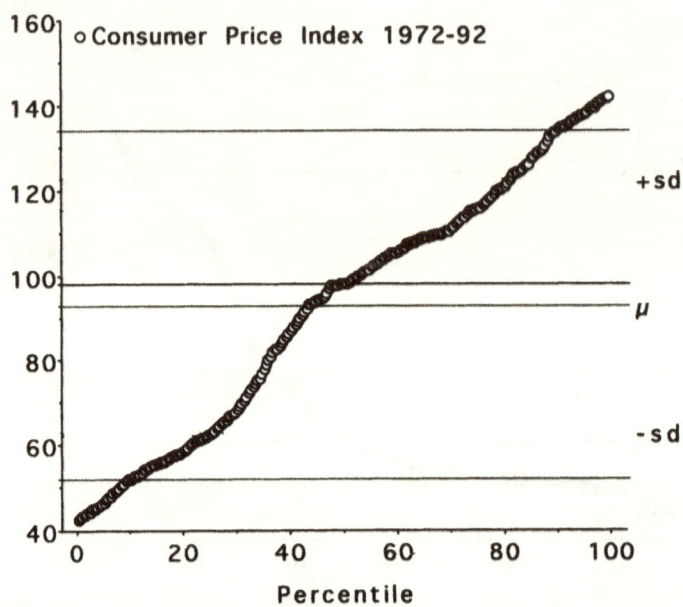

Figure 8.29

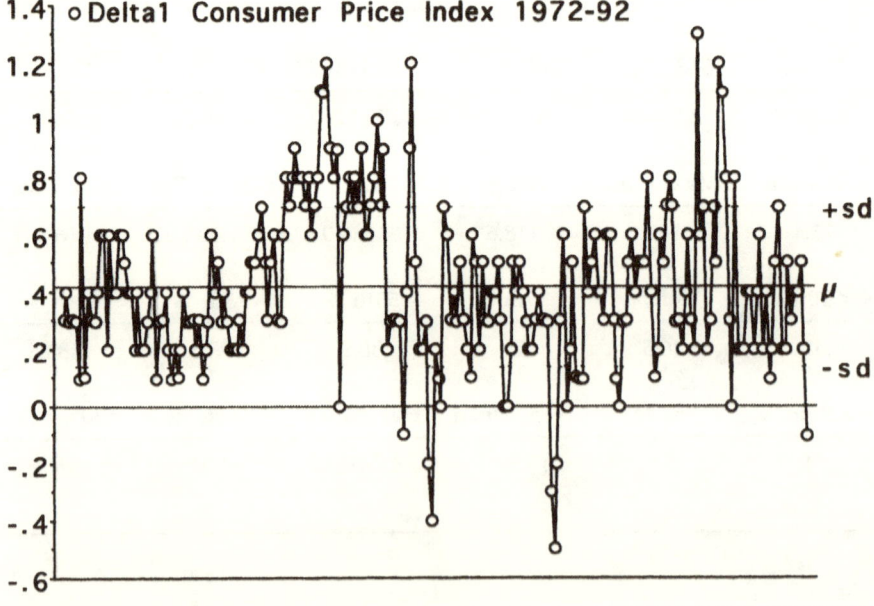

Figure 8.30

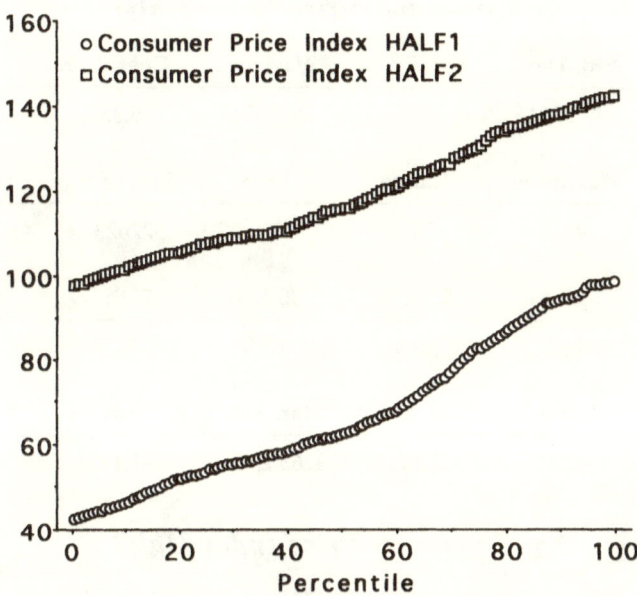

Figure 8.31

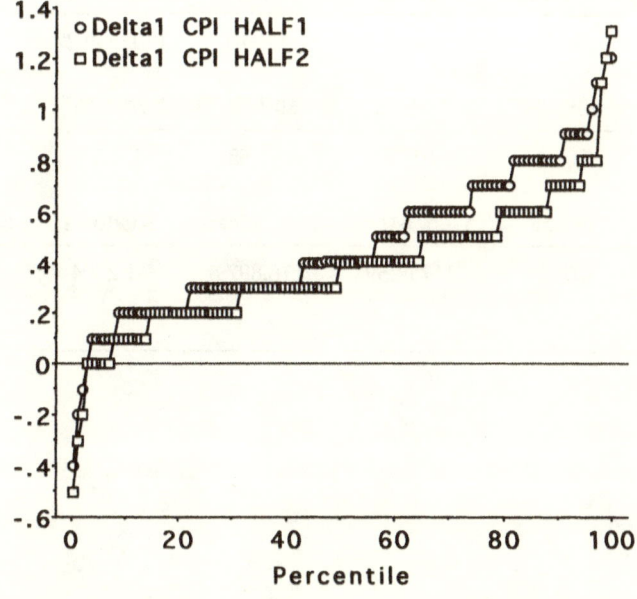

Table 8.18a-b

X_1: *Consumer Price Index Half1*

Mean	Std. Dev.	Std. Error	Variance	Coef. Var.	Count
67.2683	17.0564	1.557	290.9218	25.3558	120

Minimum	Maximum	Range	Sum	Sum of Sqr.	# Missing
42.6	98.2	55.6	8072.2	577623.14	0

# < 10th %	10th %	25th %	50th %	75th %	90th %
12	46.4	53.9	62.3	82.7	94.15

# > 90th %	Mode	Geo. Mean	Har. Mean	Kurtosis	Skewness
12	82.7	65.1917	63.2248	−1.1121	.4255

X_2: *Consumer Price Index Half2*

Mean	Std. Dev.	Std. Error	Variance	Coef. Var.	Count
118.3675	13.3962	1.2229	179.4577	11.3174	120

Minimum	Maximum	Range	Sum	Sum of Sqr.	# Missing
97.8	142	44.2	14204.1	1702659.27	0

# < 10th %	10th %	25th %	50th %	75th %	90th %
12	101.6	107.7	115.55	130.15	138

# > 90th %	Mode	Geo. Mean	Har. Mean	Kurtosis	Skewness
12	105.3	117.6259	116.8979	−1.2204	.2842

Table 8.19a-b

X_1: Delta-1 CPI Half1

Mean	Std. Dev.	Std. Error	Variance	Coef. Var.	Count
.4622	.2867	.0263	.0822	62.0334	119

Minimum	Maximum	Range	Sum	Sum of Sqr.	# Missing
−.4	1.2	1.6	55	35.12	1

# < 10th %	10th %	25th %	50th %	75th %	90th %
10	.2	.3	.4	.7	.8

# > 90th %	Mode	Geo. Mean	Har. Mean	Kurtosis	Skewness
11	•	•	•	.1514	.2426

X_2: Delta-1 CPI Half 2

Mean	Std. Dev.	Std. Error	Variance	Coef. Var.	Count
.3731	.2619	.024	.0686	70.1945	119

Minimum	Maximum	Range	Sum	Sum of Sqr.	# Missing
−.5	1.3	1.8	44.4	24.66	1

# < 10th %	10th %	25th %	50th %	75th %	90th %
10	.1	.2	.4	.5	.7

# > 90th %	Mode	Geo. Mean	Har. Mean	Kurtosis	Skewness
9	•	•	•	2.0948	.3407

Figure 8.32

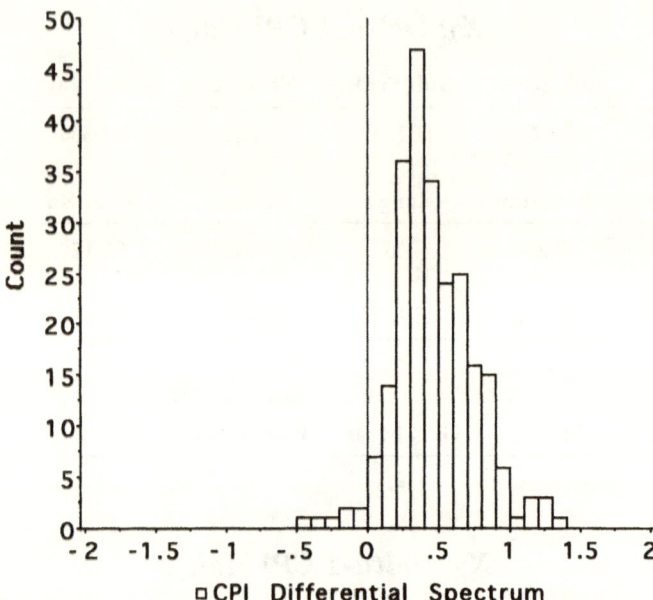

Table 8.20

	A	B	C	D	E	F	G	H
1	Symbols	Frequency	Probability					
2	1	27	0.1134					
3	2	86	0.3613					
4	3	54	0.2269					
5	4	71	0.2983					
6	Sum	238						
7								
8								
9								
10	Digrams	(E)xpected	(O)bserved	(E-O)	(E-O)^2	((E-O)^2)/E	Sum	
11	11	3.063	8.000	-4.937	24.374	7.957		
12	12	9.756	12.000	-2.244	5.034	0.516	8.473	
13	13	6.126	1.000	5.126	26.276	4.289	12.763	
14	14	8.055	6.000	2.055	4.221	0.524	13.287	
15	21	9.756	13.000	-3.244	10.522	1.078	13.841	
16	22	31.076	42.000	-10.924	119.342	3.840	17.681	
17	23	19.513	22.000	-2.487	6.187	0.317	17.999	
18	24	25.655	9.000	16.655	277.404	10.813	28.811	
19	31	6.126	3.000	3.126	9.772	1.595	19.594	
20	32	19.513	18.000	1.513	2.288	0.117	19.711	
21	33	12.252	19.000	-6.748	45.534	3.716	23.427	
22	34	16.109	14.000	2.109	4.449	0.276	23.704	
23	41	8.055	4.000	4.055	16.440	2.041	25.745	
24	42	25.655	13.000	12.655	160.161	6.243	31.987	
25	43	16.109	12.000	4.109	16.886	1.048	33.036	
26	44	21.181	42.000	-20.819	433.444	20.464	53.500	
27								
28	n=	238	238			Chi-Square	53.4997919	
29								
30								
31						10.3440603	5.47722558	
32								
33					z=	4.86683474		
34								
35								
36								
37								
38								
39								
40								
41								
42								
43								
44								
45								
46								
47								
48								
49								
50								
51								4.86683474

Figure 8.33

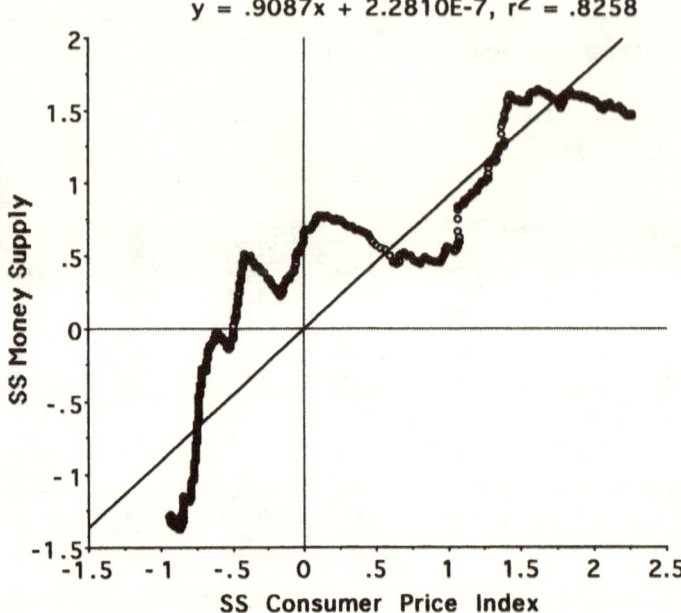

$$y = .9087x + 2.2810E\text{-}7, \ r^2 = .8258$$

SS Money Supply vs. SS Consumer Price Index

Figure 8.34

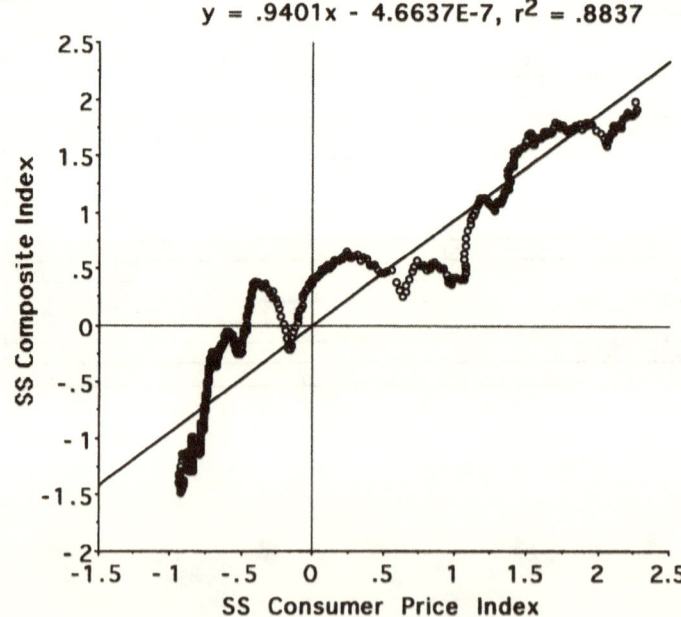

$$y = .9401x - 4.6637E\text{-}7, \ r^2 = .8837$$

SS Composite Index vs. SS Consumer Price Index

Comparisons   9

The techniques described in the previous chapters of this book and in my previous book, *Mathematics of Technical Analysis*, are designed to answer two crucial questions: (1) Is the time series stationary? and (2) Is the time series a random walk (i.e., are the transitions random and independent)? I believe that it is important to answer the first question because most (all that I am aware of) statistical techniques and technical tools assume that your data are stationary. If they are not stationary, then these tools will not work properly. It is important to note that this does not mean that if you apply a specific statistical or technical tool to a nonstationary time series that you will not get back a result; it means that the result will probably not be meaningful. Therefore, if you test your time series with the techniques described above and discover that it is not stationary, it is a good idea to stop and think carefully. This means that the underlying 'rules' that generate the time series change over time, often without warning or external sign. Consequently, if you are planning to use statistical techniques or technical tools to develop a trading strategy, then it is probably best to find another time series to analyze.

If, on the other hand, you test your time series and discover that it is stationary, then the next step is to test your time series to determine if it is independent or if it contains serial dependencies. If it is independent, then it means that there are no 'rules' to discover; that is, that the sequential relationships are determined by change alone.

Conversely, if your time series is stationary and thus contains serial dependencies, then these serial dependencies represent the underlying 'rules'. Logically, you can then attempt to use technical tools, such as moving averages, oscillators, Stochastics, RSI, etc., to attempt to discover what these 'rules' are, so that you can use this information to develop a workable and, hopefully, profitable trading strategy.

In parallel, you should test your time series to determine if it is random or nonrandom. It is noteworthy that all of the economically important time series that we tested in the previous chapters in this book appear to be random, but many contain significant serial dependencies. My colleagues and I found essentially the same story when we analyzed data from the nervous system. This is a bit difficult to reconcile. How can a time series be random and still contain significant serial dependencies? First, it is important to remember that independence-dependence refers to sequential relationships between a number of data points, while randomness refers to the selection of a single data point.

I think that an analogy will help here. Think of our time series as a series of 'words'. The 'word' length can vary, much like it does in the English language. If we use the relative or category price change methods described in Chapter 1 and discover that our serial dependencies are five days long, then this implies that our 'word' length is about five days. Now, if we carry this analogy one step further, if I tell you that the 'word' begins with an F, then you generate (based on experience) probability density functions for the second and successive letters in the 'word'. This probability density function will not contain all of the letters in the alphabet. For example, in English, my *Merriam-Webster OnLine Dictionary*, suggests that the next letter will be drawn from the following pool of letters: a, e, i, l, o, r, and u. The probability of occurrence of the members of this pool varies. The probability of occurrence of the other letters of the alphabet is essentially zero. Similarly, there is a pool of letters, with varying probabilities of occurrence, for the third place in the word, and for the fourth, and the fifth. It is important to note that when the second letter in the 'word' is chosen, then the probability density functions of the third, fourth, and fifth letter of the 'word' are altered.

Let us consider a hypothetical time series that we have analyzed and determined that it is random, but contains serial dependencies that are five days long. Figure 9.1 shows a 'word' in this time series.

Remember that a box and whiskers plot as shown in this figure represents probability density functions. This implies that on Day-1, the 'price' is set at 87.11. The fact that this time series is not independent means that, under normal circumstances, the 'price' on succeeding days is constrained. That is, it is very likely that that 'price' for the second day will be contained in the probability density function shown for Day-2 and that the 'price' for Day-3 will be contained in the cumulative probability density function shown.

The 'price' on any succeeding day is partially determined (i.e., dependent on) by previous prices and will be selected from a probability density function that varies in size. This latter represents the random aspect of the time series. It is important not to try to carry this analogy too far, but it does serve to highlight these important relationships.

The tests described in Chapter 1 for testing for stationarity and randomness cannot be easily used for the discovery of patterns. On the other hand, the relative and category price change methods can be used to identify patterns of prices that occur more often than would be predicted by chance alone. These patterns could potentially

Figure 9.1

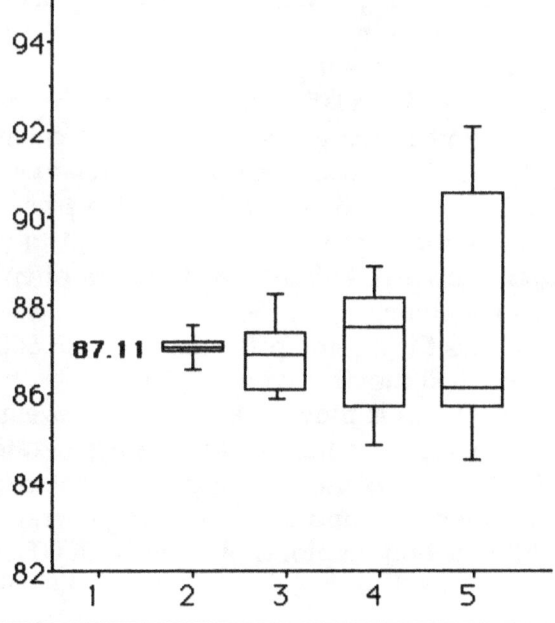

be valuable in developing a trading strategy. Therefore, I will show how these patterns compare to more traditional technical tools, such as moving averages, etc.

Figure 9.2 shows the closing prices for General Motors for 6/15/90 to 6/17/91 as well as a 20-day moving average based on these prices. An exploded version of the portion of this time series that is contained in the dashed square is shown in Figure 9.3 top. Using the category price method described in Chapter 1, I divided the GM closing Delta-1 price changes into four approximately equal quarters. Price changes from –$2.375 to –$0.625 are encoded as a '1', –$0.50 to –$0.125 as a '2', $0.0 to $0.375 as a '3', and $0.50 to $2.25 as a '4'. The summary statistics for the quarters are shown in Table 9.1a-d. The bottom of Figure 9.2 shows the relationship between two sequential price changes. The tens place encodes the quarter into which the first price change occurs. The units place encodes the quarter into which the second price change occurs. This might be easier to see in Figure 9.4, which shows the first 30 prices shown in Figure 9.3. In Figure 9.4, the open circles represent the price and the open squares the moving average. Note that the arrow points to a '33', which means that a price change of $0.0 to $0.375 was followed by a price change of the same magnitude and this occurred at the point indicated on the price time line. Another set of relationships is shown in Figure 9.5, which is an exploded version of the price line shown in Figure 9.2 solid square. In this case, I used the category price change method and the figure shows trigrams.

Figure 9.6 shows the relationship between the price time line and a 10- and 40-day moving average for GM closing prices for 9/3/91 to 9/3/92 and the occurrence of the relative price change tetragrams 1,1,1,2. The occurrence of the relative price change tetragram 2,2,2,2 is shown in Figure 9.7. It is important to note that both of these tetragrams occurred about two times as often as would be predicted by change alone.

Figure 9.8 shows the closing prices for the S&P 500 Stock Index, a 7-day and 23-day bell moving average, and the Stochastic's 14-day FastK and FastD. The bell moving average is a weighted moving average with the weights adjusted so as to generate a Gaussian curve. Figure 9.9 top shows the closing prices for 9/3/91 to 9/3/93, the middle plot shows the 7 and 23 bell moving average for the same time period, while the bottom plot shows the 14-day FastK and FastD Stochastic plot. Figures 9.10 through 9.12 show the relationship be-

Figure 9.2

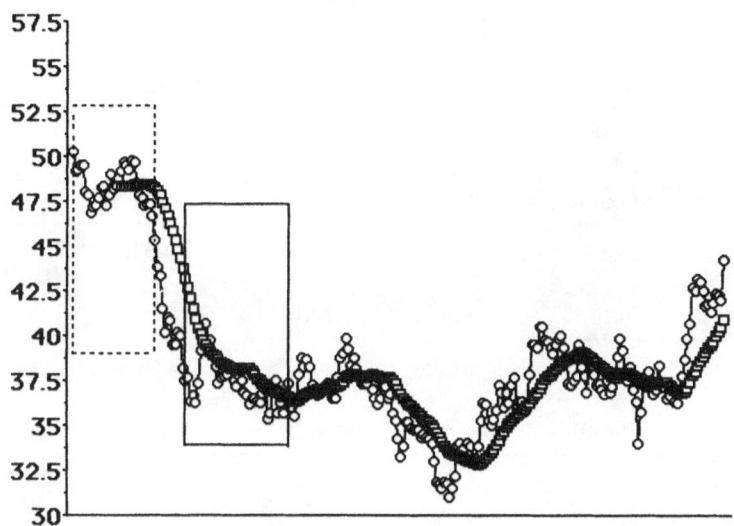

Figure 9.3

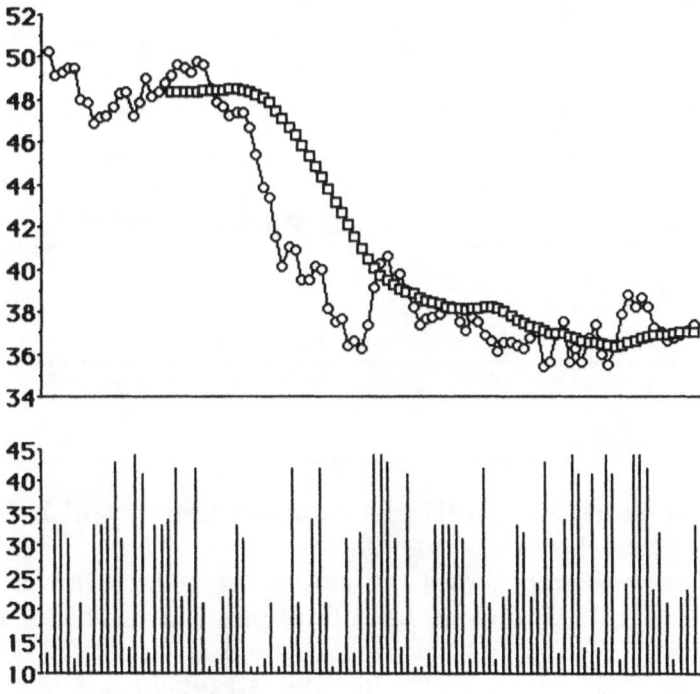

Table 9.1a-b

X_1: 1–Delta-1 GM

Mean	Std. Dev.	Std. Error	Variance	Coef. Var.	Count
−1.0059	.3943	.0493	.1555	−39.2001	64

Minimum	Maximum	Range	Sum	Sum of Sqr.	# Missing
−2.375	−.625	1.75	−64.375	74.5469	0

# < 10th %	10th %	25th %	50th %	75th %	90th %
6	−1.5125	−1.25	−.875	−.6875	−.625

# > 90th %	Mode	Geo. Mean	Har. Mean	Kurtosis	Skewness
0	−.625	•	•	1.1241	−1.1891

X_2: 2–Delta-1 GM

Mean	Std. Dev.	Std. Error	Variance	Coef. Var.	Count
−.2702	.1435	.0182	.0206	−53.1148	62

Minimum	Maximum	Range	Sum	Sum of Sqr.	# Missing
−.5	−.125	.375	−16.75	5.7812	2

# < 10th %	10th %	25th %	50th %	75th %	90th %
0	−.5	−.375	−.25	−.125	−.125

# > 90th %	Mode	Geo. Mean	Har. Mean	Kurtosis	Skewness
0	−.125	•	•	−1.2956	−.4025

tween the FastK and FastD plots and the trigrams 1,1,1; 2,1,2; and 2,2,2, respectively.

Figure 9.13 top shows the closing price for Comex Gold for 1/02/80 to 6/29/90, while the bottom plot shows RSI for the same time period. Figure 9.14 shows an exploded view of this data for 6/1/89 to 12/29/89 where the top plot is 14-day FastK (crosses) and

Table 9.1c-d

X₃: 3–Delta-1 GM

X_3: 3–Delta-1 GM

Mean	Std. Dev.	Std. Error	Variance	Coef. Var.	Count
.1885	.1418	.0179	.0201	75.21	63

Minimum	Maximum	Range	Sum	Sum of Sqr.	# Missing
0	.375	.375	11.875	3.4844	1

# < 10th %	10th %	25th %	50th %	75th %	90th %
0	0	.0312	.25	.3438	.375

# > 90th %	Mode	Geo. Mean	Har. Mean	Kurtosis	Skewness
0	•	•	•	−1.3757	−.0198

X₄: 4–Delta-1 GM

X_4: 4–Delta-1 GM

Mean	Std. Dev.	Std. Error	Variance	Coef. Var.	Count
1.004	.4484	.0565	.2011	44.6661	63

Minimum	Maximum	Range	Sum	Sum of Sqr.	# Missing
.5	2.25	1.75	63.25	75.9688	1

# < 10th %	10th %	25th %	50th %	75th %	90th %
0	.5	.625	.875	1.25	1.75

# > 90th %	Mode	Geo. Mean	Har. Mean	Kurtosis	Skewness
4	.625	.9155	.8399	−.1372	.8677

FastD (diamonds), the middle plot shows RSI, and the bottom plot shows the sequential pairs of symbols, where the symbols represent Delta-1 price changes divided into approximately equal quarters (see summary Table 9.2 a-d).

Figure 9.15 shows the relationship between the RSI for Comex Gold (top) and the category price digrams (bottom): 11 (circles), 22

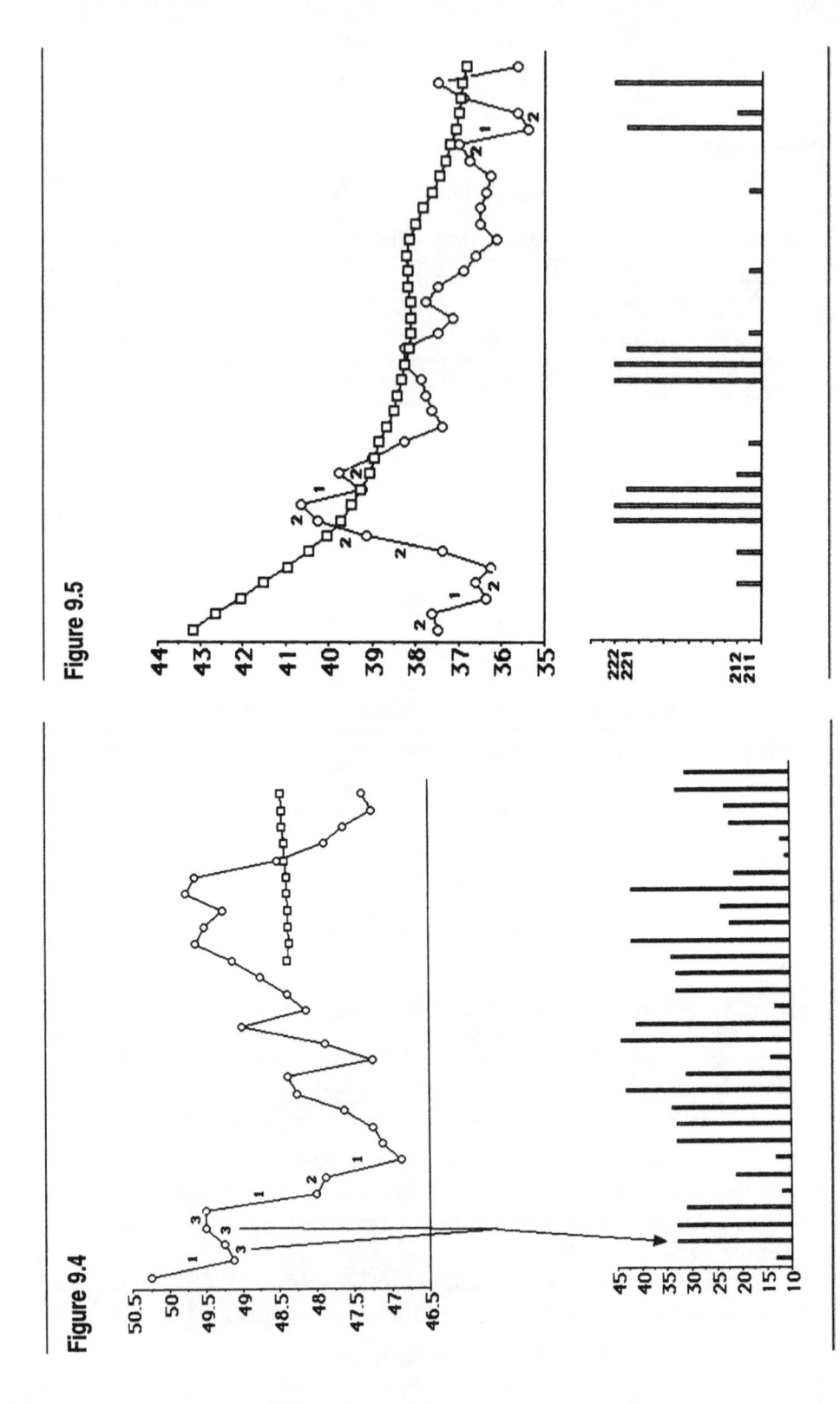

Figure 9.4

Figure 9.5

Figure 9.6

Figure 9.7

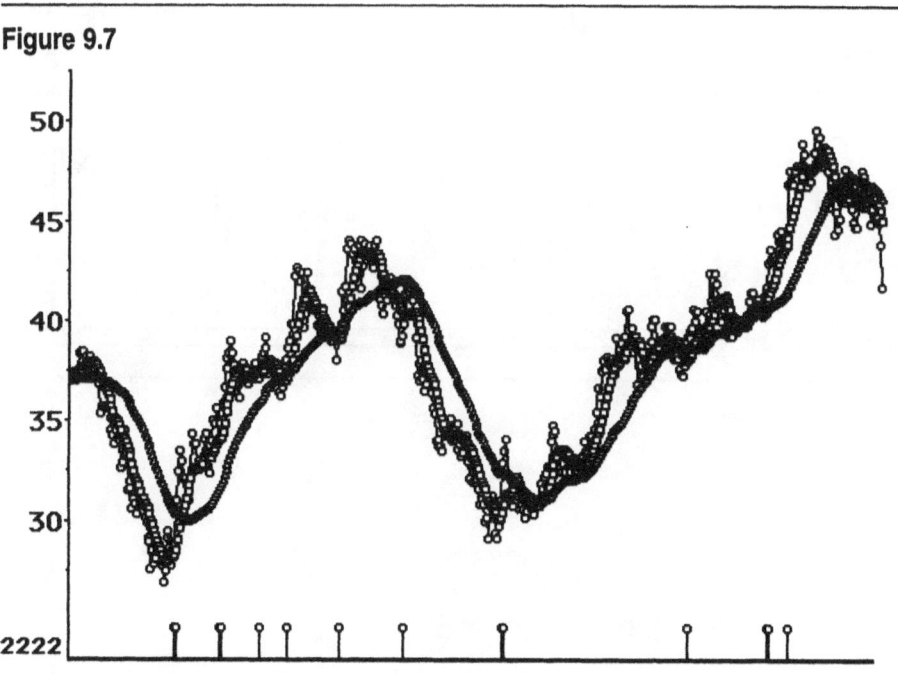

(squares), 33 (triangles), and 44 (diamonds). Figure 9.16 shows the same relationship for the category price digrams and Stochastic 14-day FastK (crosses) and FastD (diamonds).

The closing prices for T-Bonds for 10/18/90 to 3/20/92 are shown in Figure 9.17 middle, with a 7- and 20-day bell moving average. The top of Figure 9.17 shows the Stochastics FastD and FastK, while the bottom shows RSI. Figure 9.18 shows the relationship between RSI (top) for 6/3/91 and 3/13/92 and the relative price digram 21 (circles) and the 21 digrams that grow into the trigram 2,1,2 (squares) at the bottom of the figure.

Figure 9.8

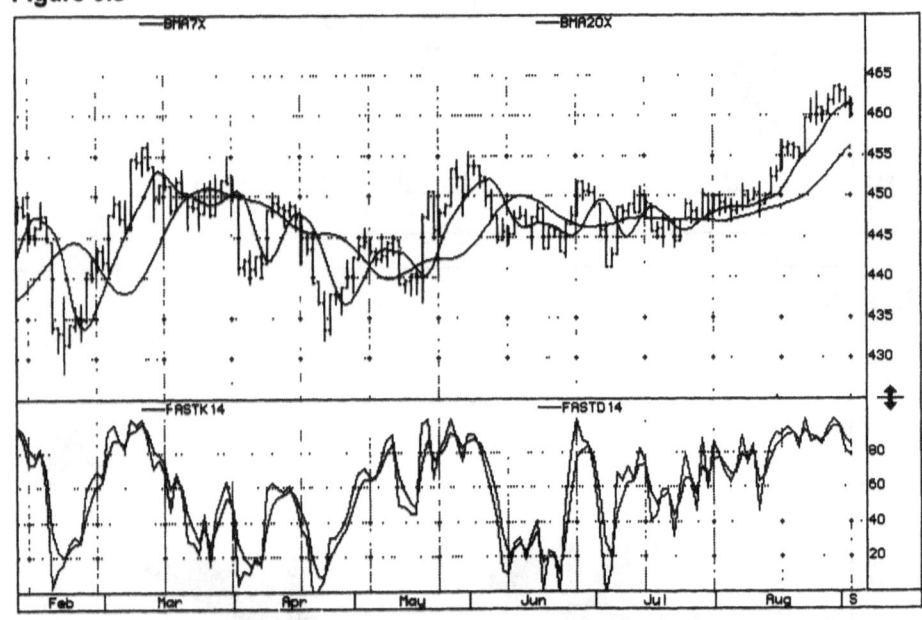

Figure 9.9

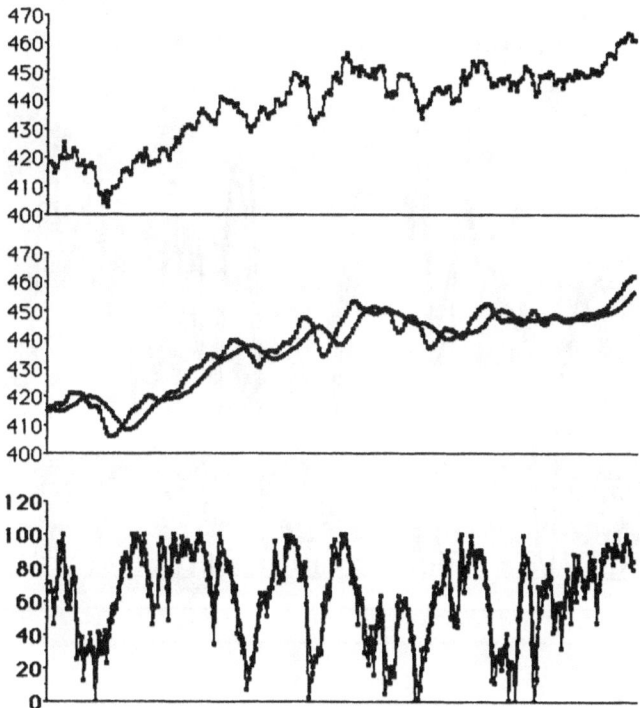

Figure 9.10

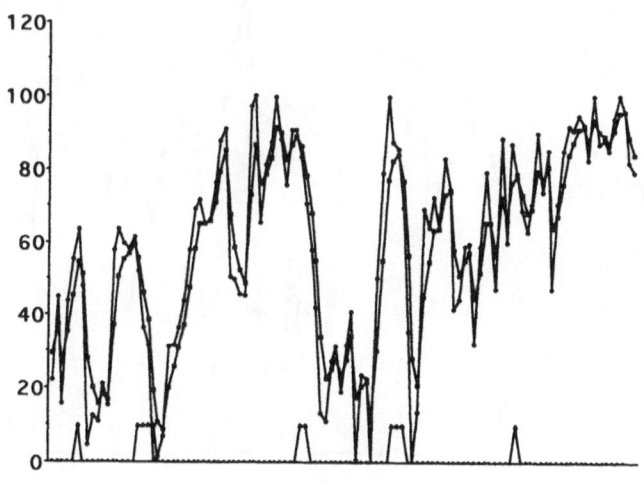

Figure 9.11

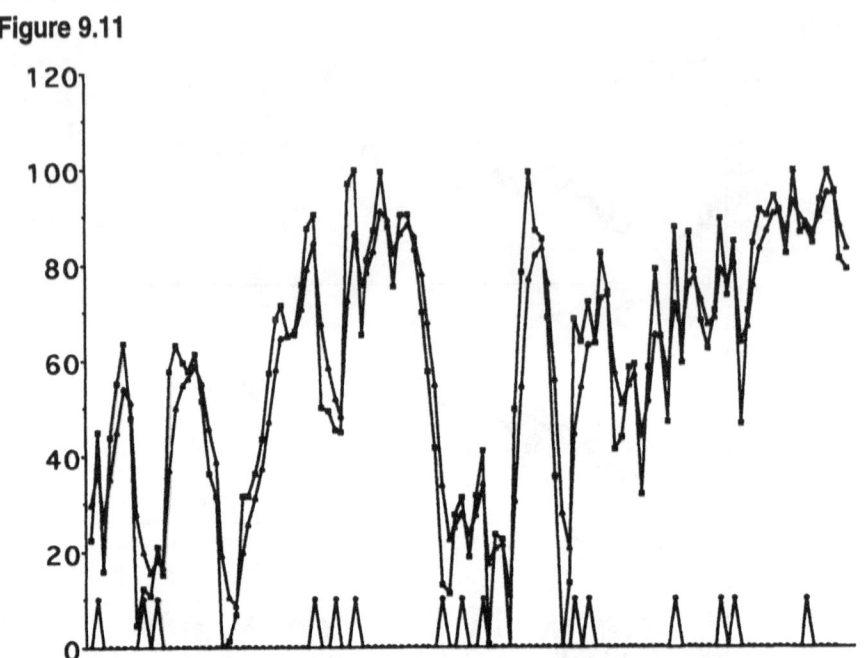

Figure 9.12

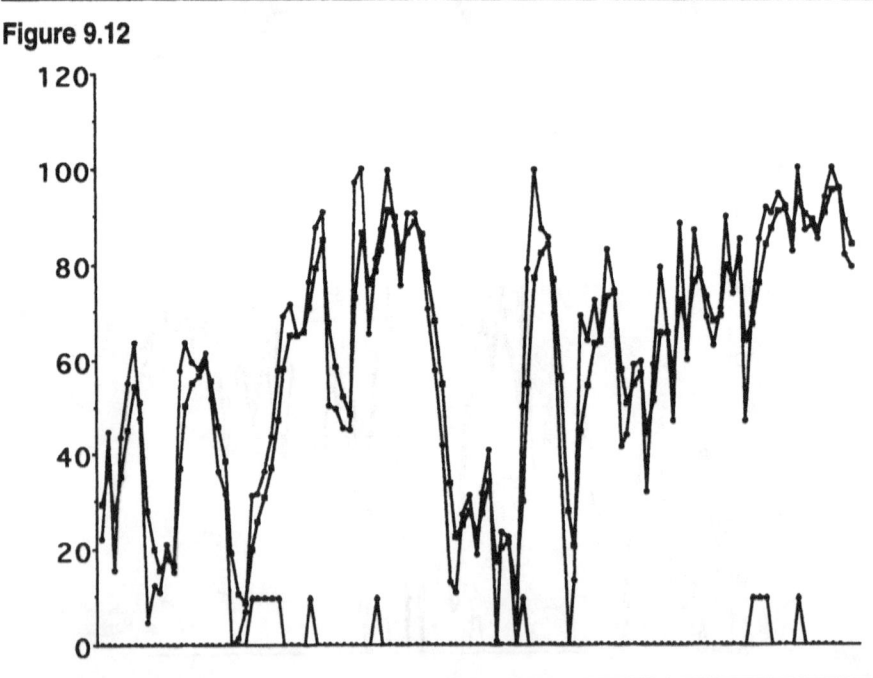

Figure 9.13

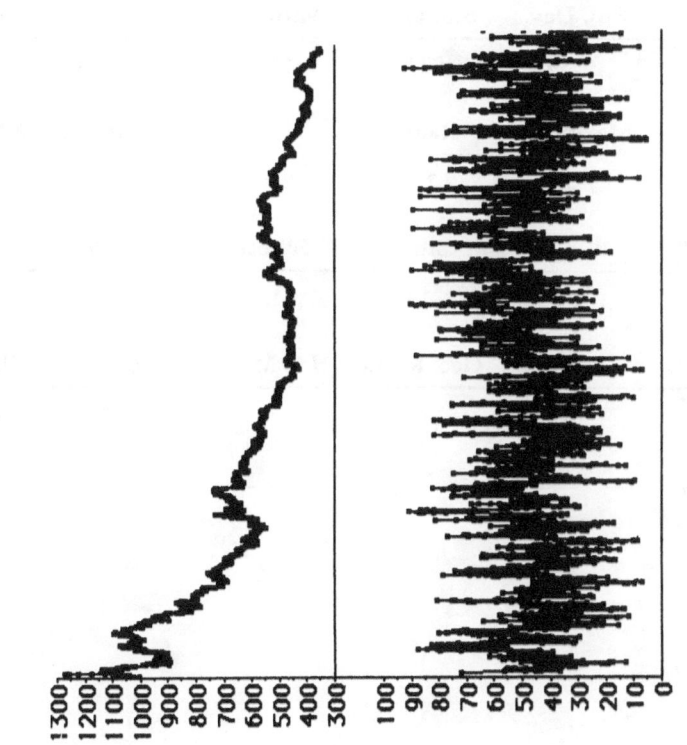

Figure 9.14

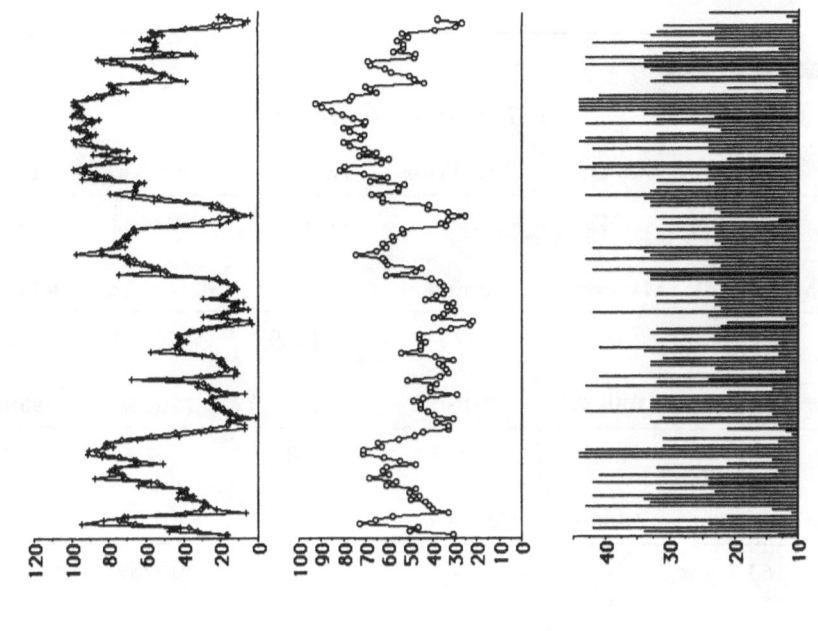

Table 9.2a-b

X_1: 1–Comex Gold Delta-1

Mean	Std. Dev.	Std. Error	Variance	Coef. Var.	Count
–8.2753	6.8426	.267	46.8207	–82.6862	657

Minimum	Maximum	Range	Sum	Sum of Sqr.	# Missing
–50	–2.7	47.3	–5436.9	75706.61	10

# < 10th %	10th %	25th %	50th %	75th %	90th %
66	–15.74	–9.75	–5.8	–4	–3.1

# > 90th %	Mode	Geo. Mean	Har. Mean	Kurtosis	Skewness
63	•	•	•	10.7759	–2.7795

X_2: 2–Comex Gold Delta-1

Mean	Std. Dev.	Std. Error	Variance	Coef. Var.	Count
–1.2715	.6989	.0271	.4885	–54.9696	667

Minimum	Maximum	Range	Sum	Sum of Sqr.	# Missing
–2.6	–.2	2.4	–848.1	1403.73	0

# < 10th %	10th %	25th %	50th %	75th %	90th %
67	–2.3	–1.9	–1.1	–.7	–.4

# > 90th %	Mode	Geo. Mean	Har. Mean	Kurtosis	Skewness
56	–1	•	•	–1.09	–.2828

Table 9.2c-d

X_3: 3–Comex Gold Delta-1

Mean	Std. Dev.	Std. Error	Variance	Coef. Var.	Count
1.0879	.7764	.0302	.6028	71.3621	663

Minimum	Maximum	Range	Sum	Sum of Sqr.	# Missing
–.1	2.5	2.6	721.3	1183.75	663

# < 10th %	10th %	25th %	50th %	75th %	90th %
65	.1	.4	1	1.8	2.2

# > 90th %	Mode	Geo. Mean	Har. Mean	Kurtosis	Skewness
66	0	•	•	–1.1592	.1821

X_4: 4–Comex Gold Delta-1

Mean	Std. Dev.	Std. Error	Variance	Coef. Var.	Count
7.6658	6.4448	.254	41.5352	84.0715	644

Minimum	Maximum	Range	Sum	Sum of Sqr.	# Missing
2.6	50	47.4	4936.8	64551.86	23

# < 10th %	10th %	25th %	50th %	75th %	90th %
60	3	3.7	5.5	8.7	14.64

# > 90th %	Mode	Geo. Mean	Har. Mean	Kurtosis	Skewness
64	3.5	6.1333	5.2174	12.4052	2.9605

Figure 9.15

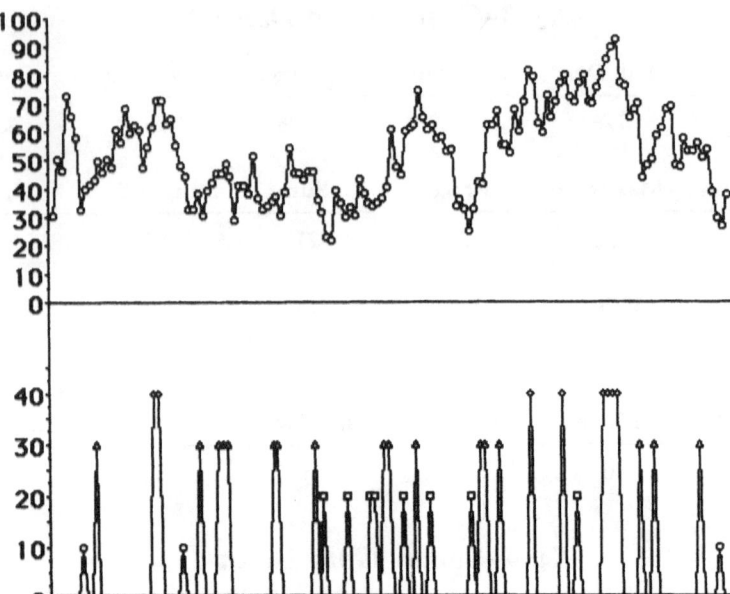

Figure 9.16

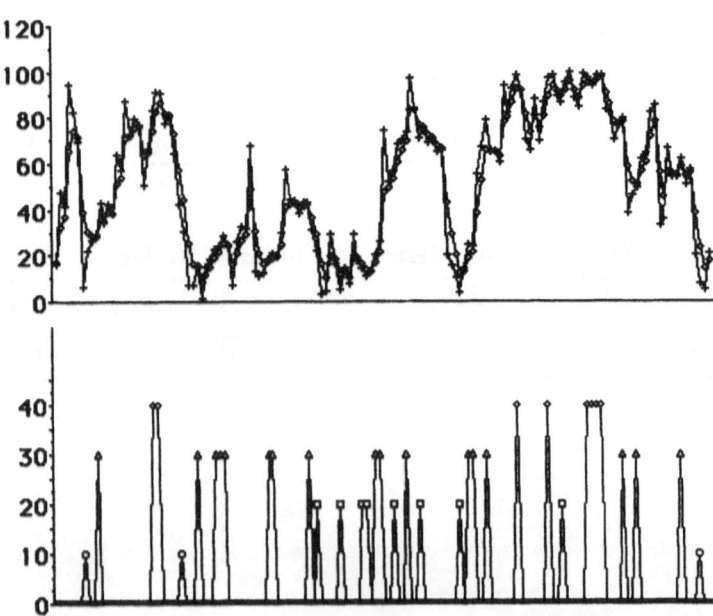

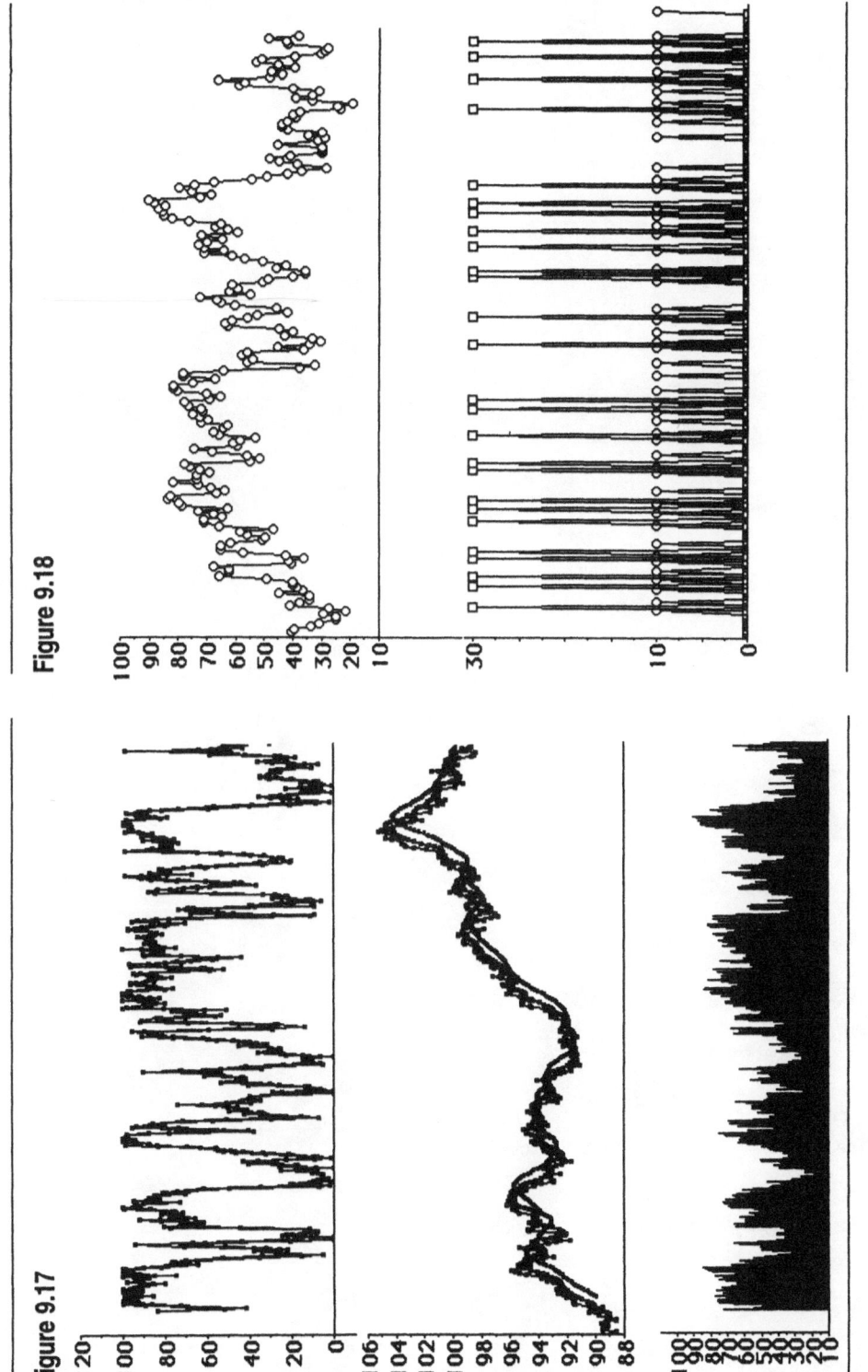

Figure 9.17

Figure 9.18

Index